Husrev Taha Sencar · Nasir Memon
Editors



# Digital Image Forensics

## There is More to a Picture than Meets the Eye

Springer

*Editors*
Husrev Taha Sencar
Computer Engineering Department
TOBB University of Economics
and Technology
Sogutozu Cad. 43
06560 Ankara
Turkey

Nasir Memon
Department of Computer
and Information Science
Polytechnic University
Brooklyn
NY 11201
USA

ISBN 978-1-4899-9232-1 ISBN 978-1-4614-0757-7 (eBook)
DOI 10.1007/978-1-4614-0757-7
Springer New York Heidelberg Dordrecht London


Softcover reprint of the hardcover 1st edition 2013


Printed on acid-free paper

Springer is part of Springer Science+Business Media (www.springer.com)

# Preface

It has now been over 100 years since photographs started to be used as a visual record of events, people, and places. Over the years, this humble beginning burgeoned into a technological revolution in photographic technology—digital imaging. Today, with an increasing volume of images being captured across an ever-expanding range of devices and innovative technologies that enable fast and easy dissemination, digital images are now ubiquitous in modern life.

In parallel to advances in technology, we have socially come to understand events in a far more visual way than ever before. As a result, digital media in general and digital images in particular are now relied upon as the primary source for news, entertainment, and information. They are used as evidence in a court of law, as part of medical records, or as financial documents. This dependence on digital media, however, has also brought with it a whole new set of issues and challenges which were either not as apparent before or were non-existent. Today, more than ever, people realize that they cannot simply accept photographs at face value. There is often more to an image than meets the eye.

Digital image forensics is a young and emerging field that concerns with obtaining quantitative evidence on the origin and veracity of digital images. In practice, digital image forensics can be defined simply as a process that consists of several steps. The first step starts with the finding of image evidence in a suspect device and organization of this extracted evidence for more efficient search. This is followed by analysis of the evidence for source attribution and authentication, and in the last step a forensic expert gives testimony in court regarding investigative findings. The goal of our book is to present a comprehensive overview and understanding of all aspects of digital image forensics by including the perspectives of researchers, forensics experts, and law enforcement personnel and legal professionals. To the best of our knowledge this is the first book to provide a holistic view of digital image forensics.

To address different aspects of digital image forensics, we organized our book into three parts. Part I starts by tackling the question of how digital images are created in a digital camera. This question is answered in two chapters by focusing on the hardware and processing elements of a digital camera. Next, we address the question of how images are stored by visiting different image formats and studying their characteristics. The last chapter in this part describes techniques for extracting and recovering image evidence from storage volumes.

Part II of the book provides a scientifically and scholarly sound treatment of state-of-the-art techniques proposed for forensic analysis of images. This part comprises six chapters that are focused on two main problems, namely, the image source attribution and image authenticity verification problem. The first of the three chapters related to source attribution considers class-level characteristics and the following two chapters examine individual characteristics that include image sensor noise and physical defects in the light path of a camera or scanner. The subsequent chapters in this part expand on specific research questions at the core of image authenticity and integrity verification. These chapters present characteristics of natural images and describe techniques for detecting doctored images and discrimination of synthesized or recaptured images from real images.

In Part III, practical aspects of image forensics are considered. The first chapter of this part explores legal issues by addressing questions regarding the validity of digital images in a courtroom. The second chapter focuses on counter-forensics and presents an attacker's perspective.

The availability of powerful media editing, analysis, and creation software, combined with the increase in computational power of modern computers, makes image modification and generation easy even for novice users. This trend is only expected to yield more automated and accurate procedures, making such capabilities available for everyone. Digital image forensics aims at strengthening the trust we place in digital images by providing the necessary tools and techniques to practitioners and experts in the field. The coverage of this book is intended to provide a greater understanding of the concepts, challenges, and opportunities related to this field of study. It is our sincere hope that this book will serve to enhance the knowledge of students and researchers in the field of engineering, forensic experts and law enforcement personnel, and photo enthusiasts who are interested or involved in the study, research, use, design, and development of techniques related to digital image forensics. Perhaps this publication will inspire its readers to contribute to the current discoveries in this emerging field.

Husrev Taha Sencar
Nasir Memon

# Digital Image Forensics

# Contents

# Part I
# Background on Digital Images

# Digital Camera Image Formation: Introduction and Hardware

**James E. Adams Jr. and Bruce Pillman**

**Abstract** A high-level overview of image formation in a digital camera is presented. The discussion includes optical and electronic hardware issues, highlighting the impact of hardware characteristics on the resulting images.

## 1 Introduction

Forensic analysis of digital images is supported by a deep understanding of the creation of those images. This chapter and the one following serve two purposes. One is to provide a foundation for the reader to more readily understand the later chapters in this book. The second purpose is to provide increased insight into digital camera image formation to encourage further research in image forensics.

This chapter deals primarily with digital still cameras, with Sect. 4 discussing how video sequence capture differs from still capture. Most of the discussion applies equally to still and video capture. In addition, the technology of digital still cameras and video cameras is converging, further reducing differences.

This chapter focuses on camera hardware and characteristics that relate to forensic uses, while processing algorithms will be discussed in the following chapter.

J. E. Adams Jr. (✉) · B. Pillman
Corporate Research and Engineering, Eastman Kodak Company,
Rochester, New York, USA
e-mail: james.e.adams@kodak.com

B. Pillman
e-mail: bruce.pillman@kodak.com

H. T. Sencar and N. Memon (eds.), *Digital Image Forensics*,
DOI: 10.1007/978-1-4614-0757-7_1,

Fig. 1 Optical image path of a digital camera

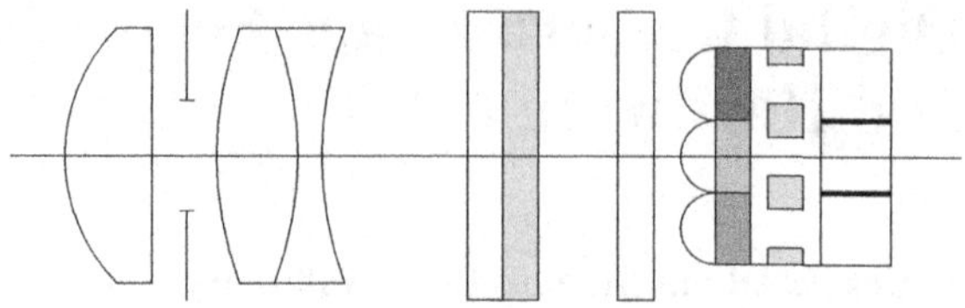


## 2 Optical Imaging Path

Figure 1 is an exploded diagram of a typical digital camera optical imaging path. The taking lens forms an image of the scene on the surface of the substrate in the sensor. Antialiasing (AA) and infrared (IR) cutoff filters prevent unwanted spatial and spectral scene components from being imaged. A cover glass protects the imaging surface of the sensor from dust and other environmental contaminants. The sensor converts the incident radiation into photocharges, which are subsequently digitized and stored as raw image data. Each of these components is discussed in detail below.

### *2.1 Taking Lens*

There are standard considerations when designing or selecting a taking lens that translate directly from film cameras to digital cameras, e.g., lens aberrations and optical material characteristics. Due to constraints introduced by the physical design of the individual pixels that compose the sensor, digital camera taking lenses must address additional concerns in order to achieve acceptable imaging results. It is these digital camera-specific considerations that will be discussed below.

#### 2.1.1 Image Space Telecentricity

Due to the optical characteristics of the sensor that will be discussed in Sect. 2.5, a key requirement of high-quality digital camera taking lenses is *telecentricity* [32]. More explicitly, image space telecentricity is a necessary or, at least, a highly desirable feature. Figure 2 illustrates this condition. An object in the scene to the left of a simple thin lens is imaged on the right. This thin lens is represented by its (coincident) *principal planes* passing through the middle of the lens. An aperture stop is placed at the front focal plane of the lens. By definition, any ray now passing through the center of the aperture stop (through the front focal point, F) will emerge from the lens parallel to the optical axis. Optically, this is known as having the *exit pupil* at infinity. (The exit pupil is the image of the aperture stop formed by the optics on the image side of the aperture.) This is the necessary condition of image space telecentricity. Three ray bundles are shown in Fig. 2. One ray bundle originates at the base of the

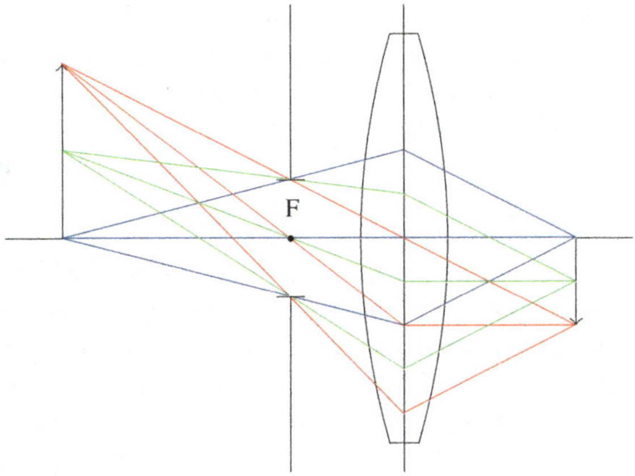


**Fig. 2** Image space telecentricity

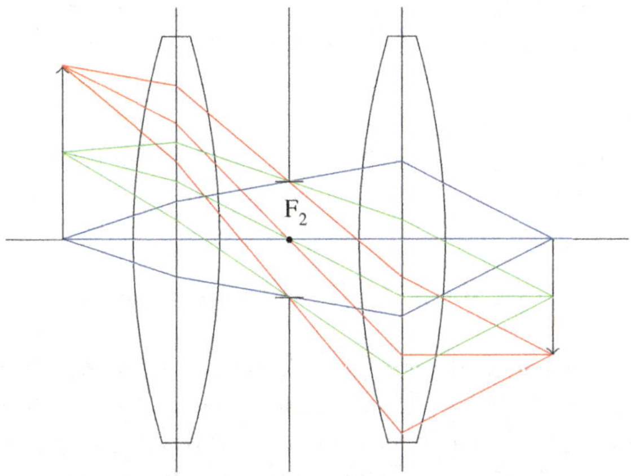


**Fig. 3** Image space telecentricity, two-lens system

object on the optical axis. The lens produces a corresponding ray bundle as the base of the image on the optical axis with the image ray bundle being, essentially, normal to the image plane. A second ray bundle originates from half-way up the object. This ray bundle is imaged by the lens at the half-way point of the (inverted) image with a ray bundle that is also normal to the image plane. Finally, a third ray bundle originates from the top of the object and is imaged at the top of the (inverted) image with a ray bundle also normal to the image plane. Inspection of the figure also shows that the ray bundles originating from the object are not normal to the object plane. Hence, in this situation telecentricity occurs in image space and not object space. Placing additional optics in front of the aperture as shown in Fig. 3 will change the system magnification and focal length, but not the telecentricity condition. The advantage of having an image space telecentric taking lens is that the cone of rays incident on each pixel is the same in size and orientation regardless of location within the image. As a result *lens falloff*, i.e., the darkening of image corners relative to the center of the image, is avoided. Even classic $\cos^4$ falloff is eliminated in the ideal case of perfect image space telecentricity.

With all its evident advantages, perfect image space telecentricity is frequently impractical to achieve. To maintain constant illumination over the entire image plane, the size of the rear element of the telecentric lens may become rather large. The design constraint of having the exit pupil at infinity means there is one fewer degree of freedom for addressing other lens issues, such as aberrations. To offset this, the number of optical elements in a telecentric lens tends to be greater than that of the equivalent conventional lens, making telecentric lenses generally more expensive. Sometimes, in the lens design process the location where the aperture stop needs to be placed is inaccessible (e.g., inside a lens) or impractical given other physical constraints of the system. In the low-cost imaging environment of consumer and mobile phone digital cameras, such cost and spatial footprint issues can become severe. As a consequence, perfect image space telecentricity in digital cameras is usually sacrificed, either partially or completely. The Four Thirds standard incorporates lenses that are "near telecentric" for DSLR cameras [5]. Although there are some minor image quality consequences of only being "nearly" telecentric, the engineering compromises are reasonable, as befitting a high-quality imaging system. In the case of low-end digital cameras, especially mobile phone cameras, the telecentric condition may be dismissed completely or addressed marginally. These latter cameras can freely exhibit lens falloff, which may be considered acceptable at the price point of the camera.

### 2.1.2 Point Spread Function Considerations

The basic imaging element of a digital camera sensor is a *pixel*, which for the present discussion can be considered to be a light bucket. The size of the photosensitive area of a pixel has a direct impact on the design goals of the taking lens. In Fig. 4 three pixels are shown with three different sizes of *point spread functions (PSF)*. The PSF is the image created by the taking lens of a point of light in the scene. Generally speaking, the higher the quality of the taking lens, the smaller the PSF. Degrees of freedom that affect the size of the PSF at the pixel are lens aberrations, including defocus; the size, shape, and placement of the aperture stop; the focal length of the lens; and the wavelength of light. These will be discussed in greater detail below. What follows is a greatly simplified discussion of incoherent (white) light imaging theory in order to discuss some concepts relevant to forensics. For a more detailed development of this topic the reader is directed to [6, 7].

In Fig. 4 the pixel is partitioned into a photosensitive region (gray) and non-photosensitive region (white). The latter refers to the region containing metal wires, light shields, and other non-imaging components. The quantity $b$ in Fig. 4a is related to the *fill factor* of the pixel. Assuming a square pixel and a square photosensitive region, the fill factor can be defined as $b^2$ with $b$ being relative to the full pixel width [9]. The range of the fill factor can be from zero (no photosensitive area) to unity (the entire pixel is photosensitive). The width of the PSF in Fig. 4a, $d$, is also with respect to the full pixel width. As can be seen in Fig. 4a, the PSF easily fits inside the photosensitive region of the pixel. In Fig. 4b, the PSF is as large as possible while

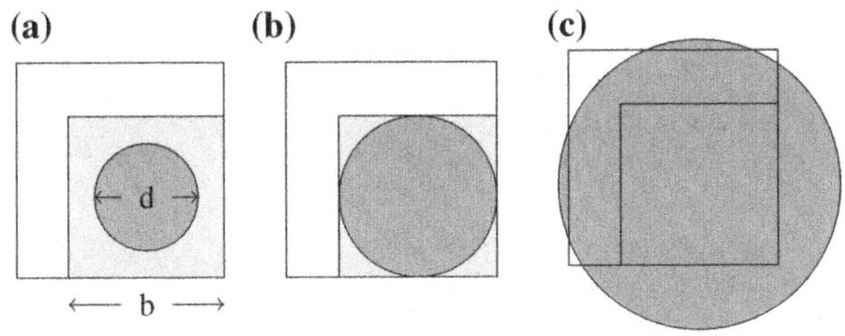


**Fig. 4** Pixels and point spread functions. **a** Small PSF. **b** Large PSF. **c** Oversized PSF

still fitting entirely into the photosensitive region. Recalling that the pixel is being treated as a light bucket, there is no difference in optical efficiency between Fig. 4a and b. Assuming each PSF is of the same input point of light, the same amount of energy is collected in both cases, resulting in the same number of photocharges being generated. In Fig. 4c the PSF is clearly larger than the photosensitive area of the pixel, resulting in a loss of optical efficiency. (Sect. 2.5.1 will discuss one solution to this problem.) For a high-quality taking lens, which is *diffraction limited*, i.e., having aberrations that are small enough to be ignored, the volume normalized PSF from a circular aperture is given in (1). For convenience, the notational shorthand $r = \sqrt{x^2 + y^2}$ is used.

$$p_{\text{somb}^2}(x, y) = \frac{\pi}{4d^2}\text{somb}^2\left(\frac{r}{d}\right) \tag{1}$$

The *sombrero function*, somb, is based on the first-order Bessel function of the first kind as shown in (2).

$$\text{somb}\left(\frac{r}{d}\right) = \frac{2J_1\left(\frac{\pi r}{d}\right)}{\left(\frac{\pi r}{d}\right)} \tag{2}$$

The portion of the PSF captured by the pixel can be expressed as the efficiency in (3). A numerical evaluation of this expression is shown in Fig. 5.

$$E_{\text{somb}^2} = \frac{\pi}{4d^2}\int_{-b/2}^{b/2}\int_{-b/2}^{b/2}\text{somb}^2\left(\frac{r}{d}\right)\mathrm{d}x\mathrm{d}y \tag{3}$$

Since the sombrero function is difficult to work with mathematically, the PSF is often modeled with simpler expressions. In (4) the PSF is modeled as a uniform circular disk of diameter $d$ via the cylinder (cyl) function given in (5). The PSF is again normalized to have a volume of unity.

$$p_{\text{cyl}}(x, y) = \frac{4}{\pi d^2}\text{cyl}\left(\frac{r}{d}\right) \tag{4}$$

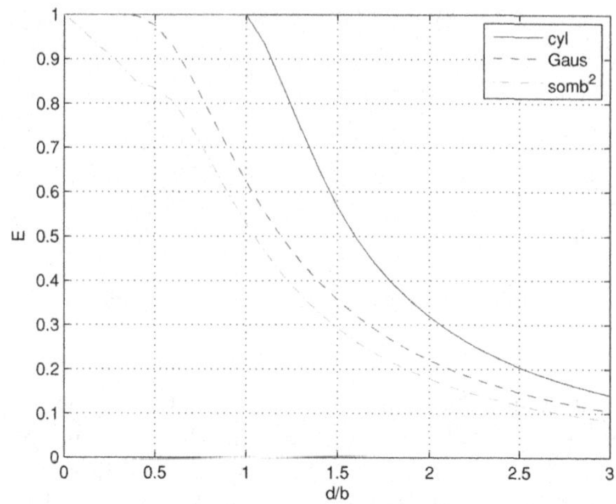


**Fig. 5** Relative light capturing efficiency for various PSF models

$$\operatorname{cyl}\left(\frac{r}{d}\right)=\begin{cases}1, & 0\le r<\dfrac{d}{2}\\[4pt] \dfrac{1}{2}, & r=\dfrac{d}{2}\\[4pt] 0, & r>\dfrac{d}{2}\end{cases} \tag{5}$$

In (6) the corresponding efficiency has three branches. When $d < b$, all of the energy is captured by the pixel and the efficiency is one. As the PSF grows in size the efficiency drops first due to the geometric mismatch between the circular PSF and the square photosensitive pixel region (second branch), and then finally due to simply exceeding the size of the pixel (third branch). A plot of the efficiency can be seen in Fig. 5. (For simplicity, the concept of *crosstalk*, which refers to signal meant for one pixel ending up in another [29], is ignored.)

$$E_{\mathrm{cyl}}=\begin{cases}1, & d\le b\\[4pt] 1+\dfrac{4}{\pi}\left[\dfrac{b}{d}\sqrt{1-\left(\dfrac{b}{d}\right)^2}-\cos^{-1}\left(\dfrac{b}{d}\right)\right], & b<d\le\sqrt{2}b\\[4pt] \dfrac{4}{\pi}\left(\dfrac{b}{d}\right)^2, & \sqrt{2}b<d\end{cases} \tag{6}$$

In (7) the PSF is modeled with a volume normalized Gaussian function using the Gaus function defined in (8). The corresponding efficiency is given in (9) and plotted in Fig. 5. In (9), erf is the standard error function. Example 1-D profiles of the PSF models discussed are drawn to scale in Fig. 6.

$$p_{\mathrm{Gaus}}(x,y)=\frac{1}{d^2}\operatorname{Gaus}\left(\frac{r}{d}\right) \tag{7}$$

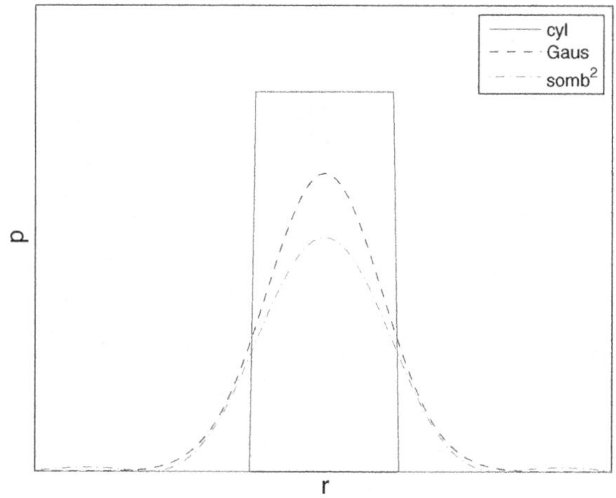


**Fig. 6** Profiles of various PSF models

$$\text{Gaus}\left(\frac{r}{d}\right) = \exp\left[-\pi\left(\frac{r}{d}\right)^2\right] \tag{8}$$

$$E_{\text{Gaus}} = \text{erf}^2\left(\frac{b\sqrt{\pi}}{2d}\right) \tag{9}$$

In addition to affecting the light capturing efficiency of the system, the nature of the PSF also affects the spatial imaging characteristics of the image capture system. What follows is a discussion from the perspective of an isolated pixel. In Sect. 2.2, imaging with an array of pixels will be examined. The spatial imaging effects of the PSF are usually modeled as a 2-D convolution as given in (10).

$$g(x, y) = f(x, y) * *p(x, y) \tag{10}$$

In (10), $f$ is the image formed at the pixel due to an ideal optical system, $p$ is the PSF, $g$ is the resulting image, and $**$ is a 2-D convolution operator. The frequency response of this system is computed by taking the Fourier transform of (10) and is given in (11).

$$G(\xi, \eta) = F(\xi, \eta)\, P(\xi, \eta) \tag{11}$$

Continuing with the three PSF models above, (12), (13), and (14) give the Fourier transforms as normalized frequency responses. In these expressions, the notational shorthand $\rho = \sqrt{\xi^2 + \eta^2}$ is used.

$$P_{\text{somb}^2}(\xi, \eta) = \frac{2}{\pi}\left[\cos^{-1}(\mathrm{d}\rho) - \mathrm{d}\rho\sqrt{1 - (\mathrm{d}\rho)^2}\right]\text{cyl}\left(\frac{\mathrm{d}\rho}{2}\right) \tag{12}$$

$$P_{\text{cyl}}(\xi, \eta) = \text{somb}(\mathrm{d}\rho) \tag{13}$$

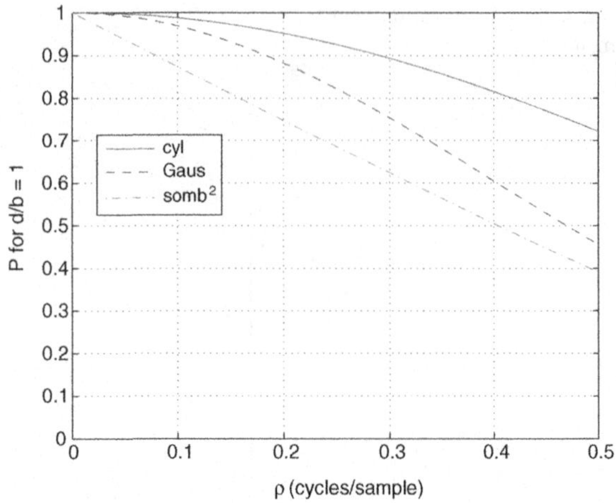


**Fig. 7** Spatial frequency responses of various PSF models

$$P_{\text{Gaus}}(\xi, \eta) = \text{Gaus}(d\rho) \tag{14}$$

An example plot of these responses for $d/b = 1$ is given in Fig. 7.

The PSF models of (4) and (7) are just that: models. Therefore, the parameter $d$ can be adjusted to suit the needs at hand without overdue concern about its physical meaning. In the case of (1), however, it is possible to relate $d$ to the physical parameters of the taking lens and the imaging system as a whole. Referring to Fig. 8, (15) shows that the width of the PSF is related to the ratio of the wavelength of light ($\lambda$) times the distance between the exit pupil and the image plane ($z_i$) to the width of the exit pupil ($l$). Implied by the size and location of the exit pupil are the size and location of the aperture stop and the focal length of the taking lens. Note that in the case of telecentric imaging, $z_i$ becomes infinite and this simplified development breaks down due to violation of underlying assumptions known collectively as the *Fresnel conditions*. The reader is again referred to [6, 7] for a more encompassing development of this topic.

$$d = \frac{\lambda z_i}{l} \tag{15}$$

Finally, the incorporation of the effects of taking lens aberrations into the foregoing analysis greatly complicates the mathematics and the reader is, once again, referred to [6, 7]. However, it is noted that the presence of aberrations will always reduce the magnitude of the spatial frequency response for all spatial frequencies, i.e.,

$$|P(\xi, \eta)|_{\text{with aberrations}} \le |P(\xi, \eta)|_{\text{without aberrations}}. \tag{16}$$

The consequences of (16) are a general broadening of the width of the PSF and a loss in fidelity of the spatial frequency response of the taking lens.

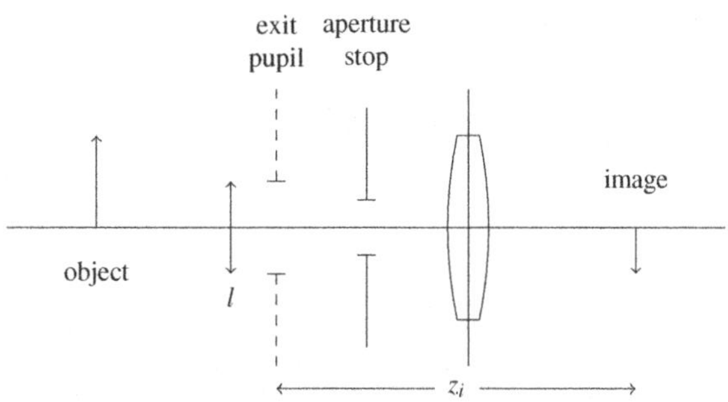


**Fig. 8** Taking lens image path for PSF computation

It will be clear from the foregoing analysis that the smaller the PSF, the better the quality of the image, both in terms of light sensitivity (signal-to-noise) and spatial fidelity (resolution). However, there are significant costs associated with maintaining a small PSF relative to the photosensitive area of the pixel. Today's digital cameras generally have rather small pixels with low fill factors resulting in target sizes of the PSF being challengingly small. The professional DSLR camera may have pixel sizes of the order of 5 μ while consumer mobile phone cameras may have pixel sizes of the order of 1.4 μ. Fill factors are generally around 0.5, although as will be discussed in Sect. 2.5.1, this liability can be significantly mitigated. While the professional DSLR market will support the cost of taking lenses with sufficient size and image quality to meet the larger pixel PSF requirements, seemingly everything is working against consumer digital cameras. Consumer devices are under constant pressure to be reduced in size and to cost less to manufacture. In the case of taking lenses, this almost invariably leads to a reduction in lens elements which, in turn, reduces the degrees of freedom available to the lens designer to control the size of the PSF. Making one or more of the lens surfaces aspheric restores some degrees of freedom, but at the price of more expensive manufacturing costs. As a consequence, it is not unusual for low-end consumer digital cameras to have PSFs that span two or three pixels. From the foregoing analysis this clearly results in significant losses of both signal-to-noise and spatial resolution. Some of these liabilities can be partially offset by some of the other components in the optical chain to be discussed below, but the size of the PSF relative to the pixel sets a significant limit on what the optical system can and cannot achieve.

## 2.2 Antialiasing Filter

The digital camera sensor consists of a rectilinear grid of pixels. As such it senses a *sampled* version of the image formed by the taking lens. Figure 9 shows a portion of the pixel array with the *pixel pitches*, $x_s$ and $y_s$, indicated. Continuing from (10), the PSF-modified image, $g$, is sampled as given in (17).

**Fig. 9** Array of pixels

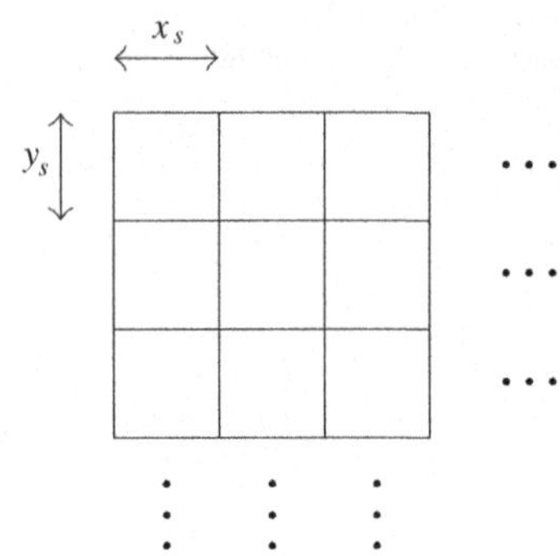


$$g_s(x, y) = [g(x, y) * *s(x, y)] \frac{1}{x_s y_s} \text{comb}\left(\frac{x}{x_s}, \frac{y}{y_s}\right) \tag{17}$$

The comb function is a shorthand notation for an array of delta functions. Sometimes, this function is casually referred to as a bed of nails.

$$\frac{1}{x_s y_s} \text{comb}\left(\frac{x}{x_s}, \frac{y}{y_s}\right) = \sum_{m=-\infty}^{\infty} \sum_{n=-\infty}^{\infty} \delta(x - m x_s)\, \delta(y - n y_s) \tag{18}$$

The function $s(x, y)$ in (17) describes the size and shape of the photosensitive region of the pixel. Recalling the use of $b$ to denote the relative fill factor of the pixel, $s(x, y)$ can be written in the following manner for a rectangular photosensitive region of dimensions $p_x \times p_y$. (For simplicity it is assumed that the photosensitive area is centered within the pixel.)

$$s(x, y) = \text{rect}\left(\frac{x}{b_x p_x}, \frac{y}{b_y p_y}\right) \tag{19}$$

The rect function used in (19) is defined below.

$$\text{rect}\left(\frac{x}{x_s}\right) = \begin{cases} 1, & \left|\frac{x}{x_s}\right| < \frac{1}{2} \\ \frac{1}{2}, & \left|\frac{x}{x_s}\right| = \frac{1}{2} \\ 0, & \left|\frac{x}{x_s}\right| > \frac{1}{2} \end{cases} \tag{20}$$

$$\text{rect}\left(\frac{x}{x_s}, \frac{y}{y_s}\right) = \text{rect}\left(\frac{x}{x_s}\right) \text{rect}\left(\frac{y}{y_s}\right) \tag{21}$$

Substituting (19) into (17) and taking the Fourier transform produces the frequency spectrum of the sampled image. A normalized version of this frequency spectrum is given in (22).

$$G_s(\xi,\eta) = \sum_{m=-\infty}^{\infty}\sum_{n=-\infty}^{\infty} \operatorname{sinc}\left[b_x p_x\left(\xi - \frac{m}{x_s}\right), b_y p_y\left(\eta - \frac{n}{y_s}\right)\right] G\left(\xi - \frac{m}{x_s}, \eta - \frac{n}{y_s}\right) \tag{22}$$

The sinc function in (22) is defined below.

$$\operatorname{sinc}\left(\frac{x - x_0}{x_s}\right) = \frac{\sin\left[\pi\left(\frac{x - x_0}{x_s}\right)\right]}{\left[\pi\left(\frac{x - x_0}{x_s}\right)\right]} \tag{23}$$

$$\operatorname{sinc}\left(\frac{x - x_0}{x_s}, \frac{y - y_0}{y_s}\right) = \operatorname{sinc}\left(\frac{x - x_0}{x_s}\right)\operatorname{sinc}\left(\frac{y - y_0}{y_s}\right) \tag{24}$$

In (22) it can be seen that replicas of the spectrum of $G$ are produced at regular intervals in frequency space. Each replica is modified by a sinc function, which acts as a kind of low-pass filter. Assuming a Gaussian model for the spectrum of $G$ and $x_s = 1$, a one-dimensional slice along the $\xi$-axis of frequency space is plotted in Fig. 10. The solid line represents just a series of Gaus functions without sinc scalars. The dashed line includes sinc scalars with $b_x = 1$, i.e., a fill factor of unity. It can be seen that the greatest effect that the fill factor can have on the Gaus functions is small. Smaller fill factors will reduce this effect even more. Returning to the Gaus functions, it can be seen that the energy from adjacent frequency replicas overlaps halfway between each spectrum peak. This (usually undesirable) signal overlap is called *aliasing*. Although it can take on a number of manifestations, aliasing in an image is usually associated with distortions within a higher frequency region, as shown in Fig. 11. In Fig. 11, the original image is of a chirped-frequency sinusoidal function of the form $A\cos\left(2\pi f r^2\right)$ where $A$ is an amplitude scalar, $f$ is a frequency scalar, and $r$ is the radial coordinate. As such, the only location that low-frequency circles should be occurring is near $r = 0$ at the center of the image. The repeated low-frequency circles throughout the rest of the image are the result of aliasing.

There are two fundamental ways to reduce aliasing. The first is to increase the sampling rate by decreasing the pixel pitch. In the present case this would amount to exchanging the sensor for one with smaller pixels. If the new sensor has pixels that are half as wide as the pixels on the previous sensor, this would give a pixel pitch of $x_s = 1/2$ relative to the original pixels. (A fill factor of unity is assumed.) If everything else is kept the same, the resulting 1-D frequency space response is given in Fig. 12. It can be seen in this figure that the signal mixing between adjacent frequency replicas has largely been eliminated. Unfortunately, this sensor substitution comes with a number of problems. First, as discussed in "Point Spread Function Considerations", shrinking the size of the pixel generally requires the PSF to also

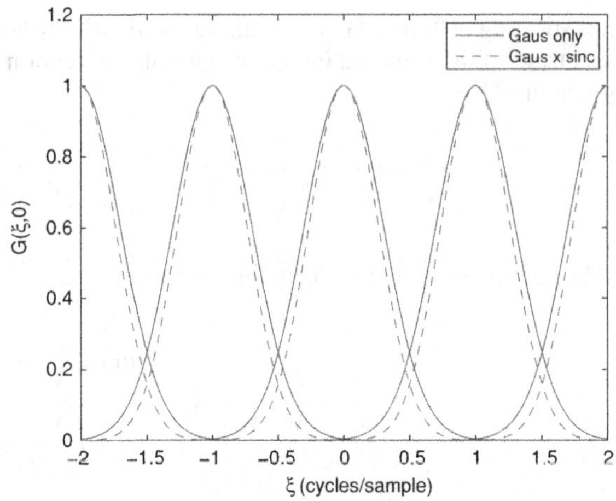


**Fig. 10** Spatial frequency responses of a sampled image

**Fig. 11** Example of aliasing in an image

shrink in size to maintain image fidelity. Second, it will require more of the smaller pixels to cover the same physical sensor area than it did the larger pixels. This results in a larger amount of pixel data that will require increased compute resources to process and store. Both these situations raise the cost of the imaging system, making an increased sampled rate solution to aliasing an expensive one. As a result, a second way to reduce aliasing is usually taken: *bandlimiting* the captured image. Through the use of an *antialiasing filter* [25], the image is optically low-pass filtered to eliminate the higher frequencies that produce aliasing. One of the most common antialiasing filters is the four-spot birefringent antialiasing filter. There are a number of different ways this filter can be constructed. Figure 13 is an exploded view of one of the more common filter configurations. A pencil of light from the object is first split into two pencils of light by a thin slab of birefringent crystal, typically quartz. This splitting is accomplished by the birefringent crystal having a different index of refraction depending upon the polarization of the incoming light and its orientation

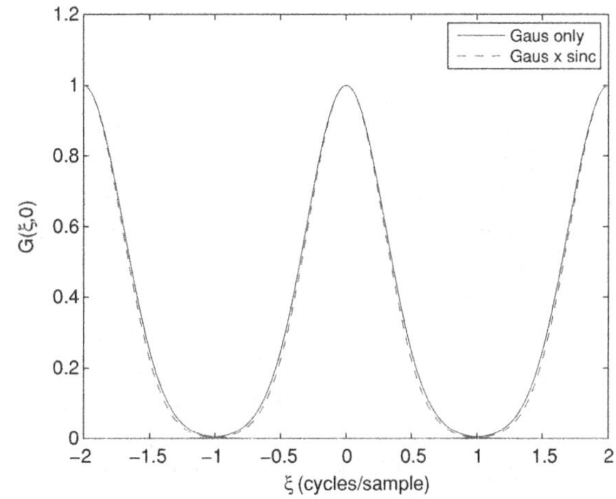


**Fig. 12** Spatial frequency responses of a sampled image with increased sampling rate

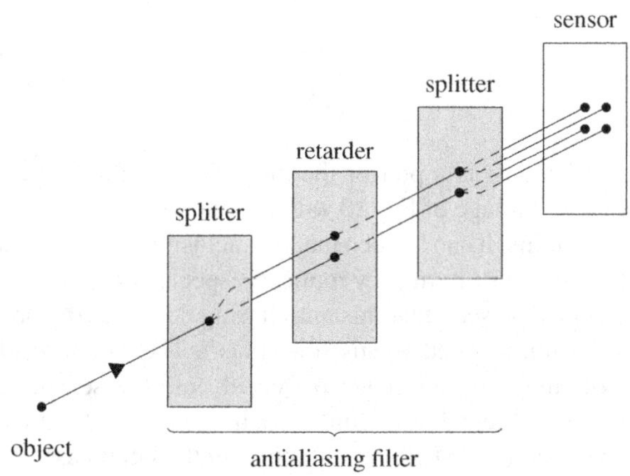


**Fig. 13** Four-spot birefringent antialiasing filter

with respect to the optical axis of the crystal material. In the case of the four-spot filter being described, an unpolarized pencil of light splits into two polarized pencils of light. These polarized pencils then pass through a retarder plate (another crystal slab) that effectively depolarizes the two pencils of light. Finally, a second splitter plate splits each of the incoming two pencils of light into two (polarized) pencils of light. In order to achieve the desired antialiasing effect, the thicknesses of the three crystal plates are adjusted so that the four spots are separated horizontally and vertically by the pixel pitch of the sensor. (This statement will be revisited when color filter arrays are discussed in Sect. 2.5.2.) Alternatively, one could think of the four-spot pattern as being the size of one pixel with a fill factor of unity. This is equivalent to the low-pass filter kernel given in (25).

$$h = \frac{1}{4}\begin{pmatrix} 1 & & 1 \\ & \bigcirc & \\ 1 & & 1 \end{pmatrix} \tag{25}$$

In (25) the center of the kernel, denoted by the $\bigcirc$, is taken to be a pixel vertex to indicate that the antialiasing filter can be shifted to minimize phase effects. The spatial representation and frequency response of this filter are given in (26) and (27), respectively.

$$h(x, y) = \delta\delta(2x, 2y) \tag{26}$$

$$H(\xi, \eta) = \cos(\pi\xi, \pi\eta) \tag{27}$$

The $\delta\delta$ function in (26) is defined below.

$$\delta\delta\left(\frac{x}{x_0}\right) = |x_0|\left[\delta(x - x_0) + \delta(x + x_0)\right] \tag{28}$$

$$\delta\delta\left(\frac{x}{x_0}, \frac{y}{y_0}\right) = \delta\delta\left(\frac{x}{x_0}\right)\delta\delta\left(\frac{y}{y_0}\right) \tag{29}$$

Figure 14 is a plot of the magnitude of the frequency response of the resulting sampled image of Fig. 10 with and without the four-spot antialiasing filter. From the dashed line it can be seen that the inclusion of the antialiasing filter largely eliminates the mixing of frequency replicas, especially at $\xi = n + 0.5$, $n \in \mathbb{Z}$. Unfortunately, it can also be seen that this antialiasing comes at the price of distorting the frequency spectrum (dashed versus solid lines). The visual result will be a low-pass filtering (softening) of the image projected onto the sensor. This loss of image fidelity is usually acceptable in consumer imaging applications, but can be anathema to the professional DSLR photographer. In the latter case, the only solution available is to try to compose the scene and the capture conditions to minimize the most grievous aliasing, e.g., changing the camera distance slightly so that the weave in the model's clothing is not at one of the worst aliasing frequencies.

It should be noted that for low-end imaging applications the cost of including an antialiasing filter in the camera can become objectionable. Since the image quality requirements are more relaxed, it is possible to simply use a taking lens that produces a lower quality (blurrier) image in the first place. While this may be a cost-effective solution, it imposes a hard upper bound on the possible image quality that can be produced by the imaging system. Still, for the low-cost solution, a lower quality taking lens will, itself, generally be lower in cost, and possibly smaller in size, which is usually another plus in this end of the market. The principle of antialiasing remains the same, regardless: eliminate the higher spatial frequencies that will mix together once the image is sampled by the sensor.

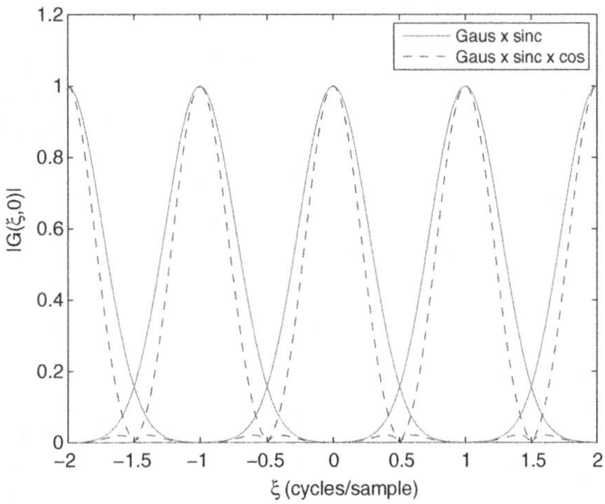


**Fig. 14** Spatial frequency responses of a sampled image with antialiasing filtering

## 2.3 *Infrared Cutoff Filter*

For all but the most specialized applications, digital cameras are used to capture images as they appear to the human visual system (HVS). As such, their wavelength sensitivity needs to be confined to that of the HVS, i.e., roughly from 400 to 700 nm [11]. The main photosensitive element of the digital camera imaging system is its silicon sensor which, unfortunately, does not match the photometric sensitivity of the HVS very well, as shown in Fig. 15. The most grievous difference appears in the near-infrared region of the spectrum. In order to address this an *infrared cutoff filter* or "IR cut filter" is incorporated into the digital camera optical path. This filter usually consists of a multilayer thin-film coating. Figure 1 shows the IR cut filter coated onto the antialiasing filter, although other optical surfaces in the optical path can also act as the substrate. Figure 16 shows typical IR cut filter and glass substrate spectral responses. It is noted that the glass substrate itself tends to be opaque in the near ultraviolet, so that both *stopbands* (the wavelength regions where the light is blocked) are addressed by the filter/substrate package. The key characteristics of the IR cut filter's spectral response are its *cutoff* wavelength and the sharpness of its cutoff. Foreshadowing the color filter array discussion of Sect. 2.5.2, the color sensing capability of the camera can be strongly distorted by the IR cut filter. In Fig. 17, the solid lines are the spectral responses of the camera color channels without an IR cut filter [14]. The dashed lines include the effects of adding an IR cut filter. The blue and green responses are left largely unchanged in their primary regions of spectral sensitivity. However, the red response from approximately 650 to 700 nm has been significantly suppressed, if not eliminated outright. This leads to two opposing issues: color accuracy and signal-to-noise. In terms of color accuracy, the HVS is highly sensitive to color variation in the orange-red-magenta region of color space. If the IR cut filter cutoff wavelength is too high and lets too much long-wavelength

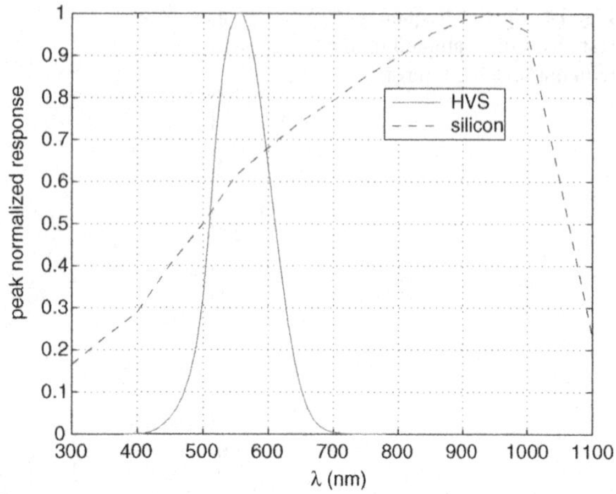


**Fig. 15** Peak normalized photometric sensitivities of the HVS and silicon

energy through (especially beyond 700 nm), color names can change very quickly. For example, the flames of a bonfire can turn from orange to magenta. Sunset colors can be equally distorted. If the IR cut filter cutoff wavelength is too low, the system can become starved for red signal and the corresponding signal gain needed to balance the red channel with the more photosensitive green and blue channels can lead to noticeable, if not unacceptable, noise amplification. Ideally, the proper IR cut filter spectral response is the one that produces three color channels that are linear combinations of the HVS's *color matching functions* [11]. By sensing color in the same way as the HVS (within a simple linear transform) maximum color accuracy is achieved. Of course, the camera's color channel spectral sensitivities are more strongly determined by the color filters in the color filter array. However, a poorly realized IR cut filter can significantly distort these responses. Since there are significant manufacturing limitations on producing the ideal spectral sensitivities in the color filters, the IR cut filter is usually used as the most "tunable" parameter in the digital camera's optical chain to compensate for any inaccuracies in the color filter spectral responsivities, at least to the degree possible.

In terms of controlling the cutoff response of the digital camera's IR cut filter, the usual considerations with multilayer thin-film coatings apply. The sharper the desired cutoff response, the more layers will generally be required. Each additional thin-film layer complicates the manufacturing process and adds to its expense. The choice of coating materials for durability and ease of manufacturing also significantly impact the cost and complexity of the multilayer stack. Recalling the discussion of Sect. 2.1.1, it is important that the angle of incidence be controlled when using a thin-film stack. The spectral response (such as the cutoff wavelength) will shift toward the blue end of the spectrum as the angle of the incident light moves away from the normal of the filter stack. A telecentric taking lens will achieve this control, but departures from telecentricity can lead to changes in color spectral sensitivity as a function of distance from the center of the image.

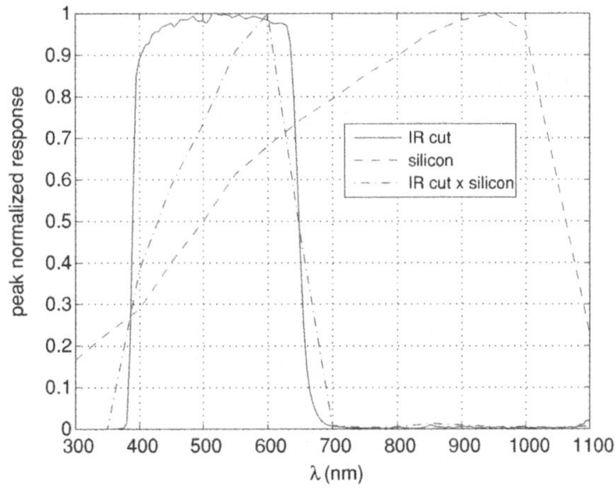


**Fig. 16** Peak normalized photometric sensitivities of IR cut filter and silicon

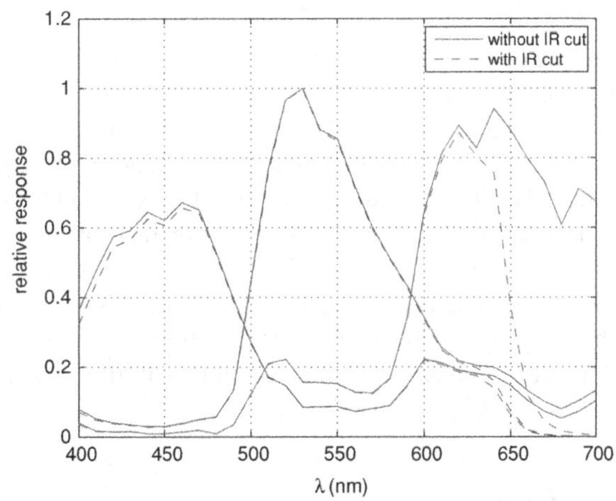


**Fig. 17** Relative color photometric sensitivities

There are simpler and less expensive filter technologies that can be used for producing IR cut filters. Perhaps the simplest is a filter made from IR absorbing glass [26]. These glasses absorb near-infrared radiation and re-radiate it as heat. A popular choice with some low-end consumer cameras, these glasses are characterized by more gradual cutoff responses, as shown in Fig. 18. As discussed above, these more gradual cutoffs can produce small amounts of IR leakage which, in turn, may produce color distortion. However, as with other engineering tradeoffs, the use of absorbing glass IR cut filters may be acceptable for certain imaging applications.

As with the other engineering considerations previously discussed, the professional DSLR market will tend to support the more expensive multilayer IR cut filter construction with additional thin-film layers to achieve a sharper cutoff at more precisely located cutoff wavelengths. Lower end consumer cameras will retreat from this position and strive only for more basic color control, e.g., preventing bonfires from

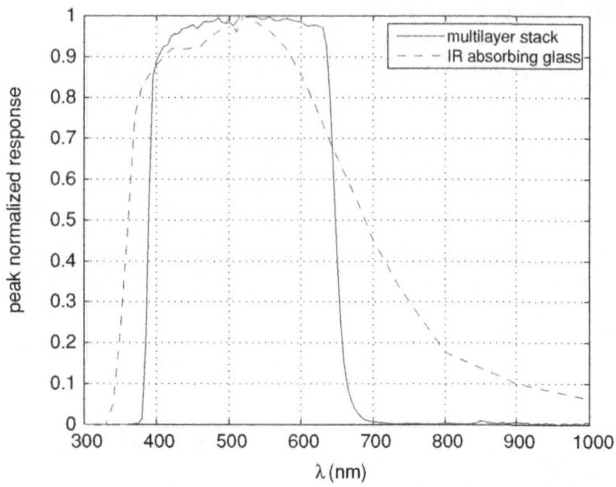


**Fig. 18** Peak normalized photometric sensitivities of multilayer and absorbing IR cut filters

turning magenta while letting other colors in the scene drift in fidelity. From a forensics perspective, the accuracy of the color reproduction, and in particular, significant color failures, can provide clues into the capturing camera's IR cut filter pedigree. Such clues are most easily revealed in scenes with illuminants with significant near-infrared energy, e.g., direct sunlight, firelight, and tungsten (incandescent) light. The reflectance of the objects in the scene in the near-infrared will also strongly influence the detection of undesirable IR cut filter designs, e.g., some flowers are notorious for *anomalous reflectance* such as blue morning glories, gentians, and ageratums due to their high reflectances in the near-infrared [15]. Significant departures from visual colors can indicate the nature of an IR cut filter's design.

## 2.4 Cover Glass

The cover glass is a piece of high-quality, defect-free optical glass that is placed on top of the sensor to prevent environmental contamination. Protection from oxidation and airborne dust is the primary purpose of this component. The cover glass is frequently bonded to the surface of the sensor to increase its effectiveness. In turn, the cover glass can also be physically combined with the antialiasing and IR cut filters into one optical "sandwich" element.

Due to its proximity to the photosensitive surface of the sensor, any dust or scratches on the cover glass will appear in the image in a manner similar to contact printing. In the case of digital cameras with nondetachable lenses, this is less of a concern as the camera is never opened up and the sensor exposed to the full environment. With the replaceable lenses of DSLR cameras, it is necessary to occasionally clean the cover glass to manage the dust that is introduced during lens exchange. During this cleaning process it is possible to inadvertently scratch the cover glass or expose it to cleaning fluids that may have adverse chemical reactions with the cover glass material. For these reasons, cleaning of the cover glass is usually performed as infrequently as possible and, instead, image processing techniques are used to compensate for dust and scratch-related artifacts during post-processing. From the forensics perspective, the implications of dust and scratches on the cover glass providing additional identifiers of a particular camera unit are fairly obvious. While dust may be a somewhat evanescent feature, scratches on the cover glass are more permanent since cover glass repair and replacement is a costly and delicate operation. Artifacts created by cover glass flaws will be revisited later in Sect. 3.1.

## 2.5 Sensor Optics

The heart of a digital camera imaging chain is the sensor. It is this device that converts the incident light image into a corresponding set of photocharges that, in turn, are read out and converted into a digital signal representation. Perhaps not surprisingly, it is also the most complex component in the imaging chain. Significant effort goes into creating a sensor that generates photocharges at precise locations in the image and then prevents the random dispersion of these charges before they have been read out of the device. Additionally, color information about the image must be acquired. A number of these issues and their technical solutions are discussed in the sections below.

### 2.5.1 Lenslet Array

Recalling Sect. 2.1.2, the fill factor of a pixel is generally less than unity to allow space for the non-imaging pixel components. (See Sect. 2.5.3.) Fortunately, this loss in potential light sensitivity can be mitigated by the use of *lenslets*, also known as *microlenses* [20]. Figure 19 illustrates two pixels, the left-hand one with a lenslet and the right-hand one without. In the right-hand pixel, light in the center of the pixel reaches the photosensitive area of the pixel. However, light at the edges of the pixel does not. In the left-hand pixel, all three rays of light reach the photosensitive area of the pixel due to the refractive nature of the lenslet. As a result, the fill factor is effectively increased. With the appropriate lenslet design, a fill factor approaching unity can be achieved. It can also be seen in Fig. 20 that having the rays of light normal to the pixel are essential to making the lenslets effective. Any significant angle off normal will cause most if not all the light to miss the photosensitive area of the pixel. As a result, the use of a telecentric taking lens becomes a requirement in order to achieve the best performance from lenslets.

In a digital camera, an array of lenslets is placed over the sensor so that each pixel has its own lenslet. There are many ways to fabricate lenslet arrays and research into new and improved methods is presently active. Perhaps the most common commercial methods involve a two-step process of first "carving" basic lenslets forms

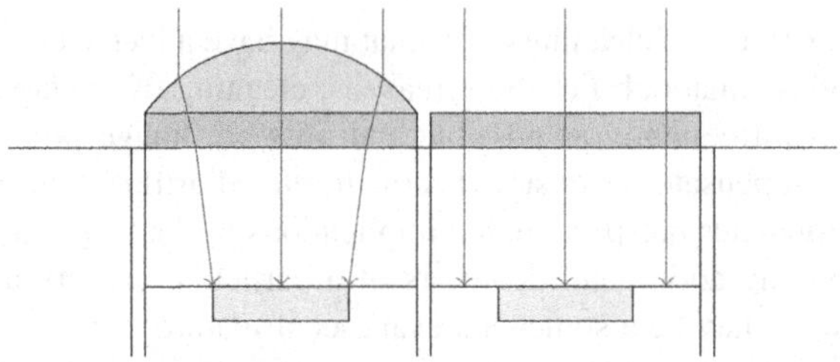

Fig. 19 Lenslet optics

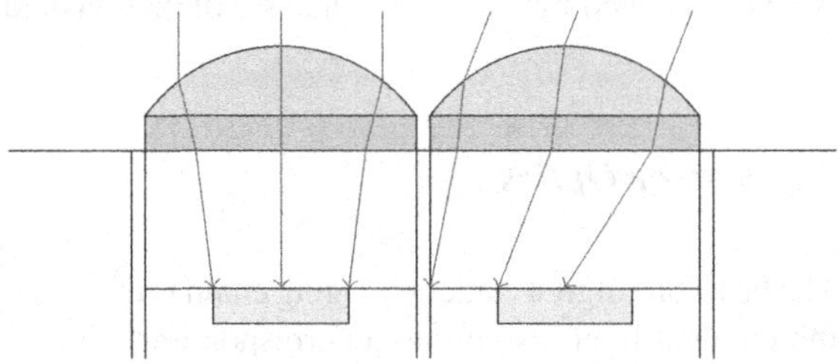

Fig. 20 Lenslet optics showing sensitivity to angle of incidence

(usually cylinders) and then melting ("reflowing") them into lens shapes [28]. In the case of lenslets made from photoresist materials, the carving process is usually done chemically through an etching process. When lenslets are made from glass or semiconductor materials, an optical carving can be achieved with a laser. The melting process can also be done with a laser or more conventional heating techniques. Alternatively, the lenslet array can be produced using a molding or stamping process, which tends to combine features of both carving and melting steps. Another fabrication approach is to use inkjet techniques to deposit lenslets made from UV curing materials. All of these approaches are directed at producing lenslet surfaces that are optically smooth and of the desired shape (radii). Departures from smoothness or proper shape will result in light scattering or simple misfocusing which, in turn, will reduce the potential gains in fill factor that the lenslets can achieve. These losses in fill factor, in turn, translate into reduced pixel photosensitivity, which will appear most readily as reductions in the signal-to-noise capability of the system.

It has been tacitly assumed that the curved surface of each lenslet in the array is spherical. However, additional reductions in the cost of manufacturing (with corresponding loss of performance) can be realized by departing from the spherical profile. One step away from a fully spherical surface is to control the curvature of the lenslet along only the horizontal and vertical axes and to allow the other orientations (i.e., diagonals) to vary. As a result, the profile of the lenslets assumes a more "pillow" shape. Finally, a fully cylindrical lenslet profile can be adopted so as to increase the fill factor in only one direction but to leave the optical path unaltered in the orthogonal direction. These kinds of lenslet profile compromises are generally found in low-end digital cameras, especially those with a small pixel size, i.e., less than 2 μ across.

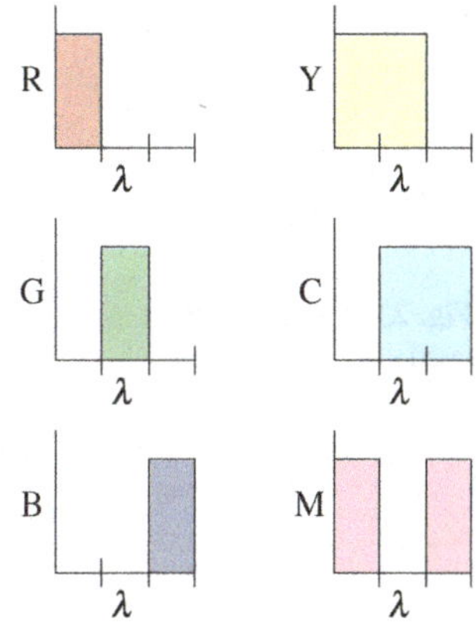


**Fig. 21** Idealized block spectral sensitivities

### 2.5.2 Color Filter Array

Perhaps the most distinctive and iconic element of the digital camera is the *color filter array* or CFA [1]. The CFA is a mosaic of pixel-sized color filters that are arranged over the surface of the sensor to enable the detection of full-color information from a single sensor. As indicated in Fig. 1, each pixel has its own color filter. These filters are usually fabricated from colored dyes, pigments, or photoresists that are deposited onto the sensor surface. Sometimes, a given color filter will consist of a single layer of colorant and at other times it may be composed of two or more layers of different colorants. The layers of colorants are usually thin enough that their effects on the direction and location of incident light can be safely ignored.

The colors chosen for the CFA usually compose one of two sets: red, green, and blue (RGB), or cyan, magenta, and yellow (CMY). Either set can be augmented with the inclusion of a non-colored pixel, usually denoted as panchromatic (P) or white (W). Therefore, one will also find RGBP and CMYW CFAs. It is possible to create hybrid color CFAs from these colors, such as CMYG. The decision to use RGB versus CMY is usually based on signal-to-noise issues. A simplistic theoretical analysis can suggest the overall signal-to-noise differences between RGB and CMY systems. In Fig. 21 the idealized block spectral sensitivities of these CFA colors are shown. As indicated by the plots, it is assumed that the area normalized CMY signals are $Y = (R + G)/2$, $C = (G + B)/2$, and $M = (R + B)/2$. Therefore, the CMY system is twice as light sensitive as the RGB system. However, the final image of the system is almost always required to be an RGB image, so a color rotation of the captured CMY image must occur as shown in (30).

$$\begin{pmatrix} R' \\ G' \\ B' \end{pmatrix} = \begin{pmatrix} -1 & 1 & 1 \\ 1 & -1 & 1 \\ 1 & 1 & -1 \end{pmatrix} \begin{pmatrix} C \\ M \\ Y \end{pmatrix} \tag{30}$$

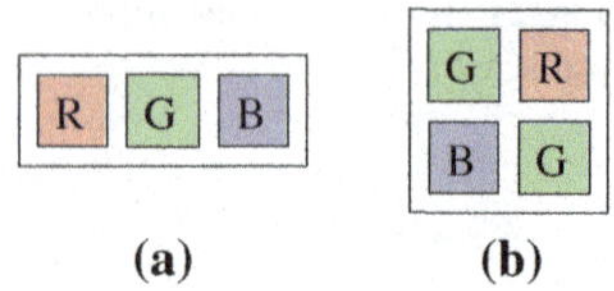


**Fig. 22** CFA minimum repeating patterns

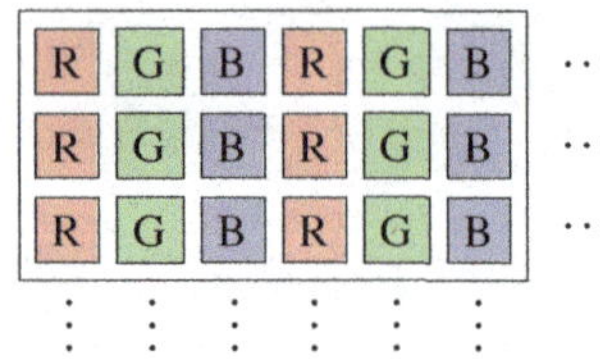


**Fig. 23** Fig. 22a MRP tessellated

If it is assumed that the noise in each of the RGB color channels is independent and its variance $\sigma_{RGB}^2$ is the same for all colors, then the variance of the noise in the CMY color channels is $\sigma_{CMY}^2 = \left(2\sigma_{RGB}^2\right)/4 = \sigma_{RGB}^2/2$. While this is a significant improvement in signal-to-noise performance, this improvement is lost by the color rotation operation: $\sigma_{R'G'B'}^2 = 3\sigma_{CMY}^2 = 3\sigma_{RGB}^2/2$. Therefore, in this idealized case the CMY system results in a lower signal-to-noise than does the RGB system. However, with more realistic spectral sensitivities and noise characteristics it is possible for CMY systems to have signal-to-noise responses comparable to RGB systems. The point is that there is no "free lunch" with respect to the set of the colors chosen for the CFA. Large improvements in signal-to-noise capability over RGB are not automatically achieved through the use of CMY color systems.

Once the CFA color set is chosen, the arrangement of the colors across the sensor needs to be determined. The CFA is usually composed of a *minimum repeating pattern* or MRP that is tessellated over the surface of the sensor. The MRP must tile the sensor surface completely with no gaps or overlaps, given the underlying pixel sampling grid. This underlying pixel sampling grid is almost always rectilinear, although hexagonal sampling grids have been investigated [23]. Because of the usual rectilinear pixel sampling grid, the MRP is almost always rectangular, although this is not the only possible shape. The simplest possible MRP is a $1 \times 3$ line of three colored pixels, as shown in Fig. 22a. When tessellated over the sensor surface, this MRP will produce red, green, and blue columns of pixels with columns of the same color separated by two pixels, as shown in Fig. 23. Each color channel is sampled with the same pattern and frequency. Recalling Sect. 2.2, the frequency spectrum of the sampled color channel is given in (31). A unit square pixel with a fill factor of unity is assumed.

$$G_s(\xi, \eta) = \sum_{m=-\infty}^{\infty} \sum_{n=-\infty}^{\infty} \operatorname{sinc}\left(\xi - \frac{m}{3}, \eta - n\right) G\left(\xi - \frac{m}{3}, \eta - n\right) \tag{31}$$

The frequency response along the $\eta$-axis has already been discussed and shown in Fig. 10. The corresponding frequency response along the $\xi$-axis is shown in Fig. 24.

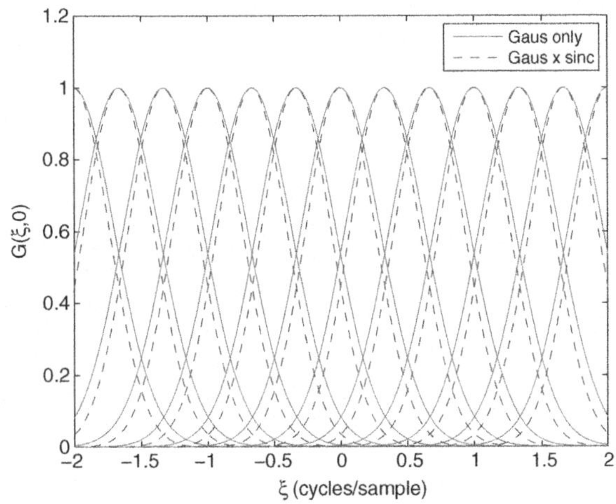


**Fig. 24** Spatial frequency responses of a CFA sampled image

It can clearly be seen that there is a high presence of aliasing corresponding to the horizontal direction in which the signal is sampled every three pixels. A four-point birefringent filter can be designed to produce spots that are horizontally separated by three pixel pitches and vertically separated by one. This is modeled as the convolution kernel given in (32). The corresponding spatial representation and frequency response are given in (33) and (34).

$$h = \frac{1}{4}\begin{pmatrix} 1 & 0 & & 0 & 1 \\ & & \bigcirc & & \\ 1 & 0 & & 0 & 1 \end{pmatrix} \tag{32}$$

$$h\left(x, y\right) = \frac{1}{3}\delta\delta\left(\frac{2x}{3}, 2y\right) \tag{33}$$

$$H\left(\xi, \eta\right) = \cos\left(3\pi\xi, \pi\eta\right) \tag{34}$$

A plot of the resulting filtered signal along the $\xi$-axis is given in Fig. 25. In Fig. 25 only the fundamental of the unfiltered sampled signal is shown to simplify the figure. It is clear that while the aliasing has been reduced by the four-spot antialiasing filter, it is far from eliminated. Also, comparing the unfiltered signal to its antialiased version shows a marked distortion of the frequency response. It is for these reasons that MRPs such as shown in Fig. 22a have presently fallen from favor with camera manufacturers.

Figure 22b shows the *Bayer pattern*, the most popular MRP in use in digital cameras today [2]. When tessellated over the surface of the sensor, the green pixels are sampled in a checkerboard manner and the red and blue pixels are sampled over a rectilinear grid with a pixel pitch of two in both horizontal and vertical directions, as shown in Fig. 26. The frequency response of the red and blue channels (ignoring phase shifts) is given in (35).

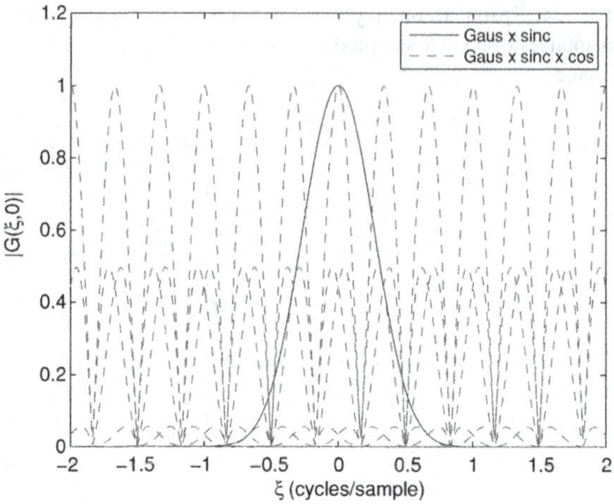


**Fig. 25** Spatial frequency responses of a CFA sampled image with antialiasing filtering

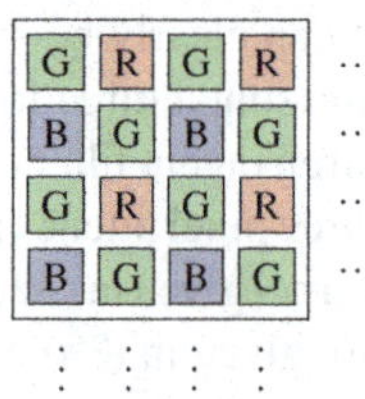


**Fig. 26** Bayer pattern tesselated

$$G_{rbs}(\xi,\eta) = \sum_{m=-\infty}^{\infty}\sum_{n=-\infty}^{\infty} \text{sinc}\left(\xi - \frac{m}{2}, \eta - \frac{n}{2}\right) G_{rb}\left(\xi - \frac{m}{2}, \eta - \frac{n}{2}\right) \tag{35}$$

Because of its checkerboard sampling nature, the frequency response of the green channel has a slightly different form, as given in (36).

$$\begin{aligned} G_{gs}(\xi,\eta) = \sum_{m=-\infty}^{\infty}\sum_{n=-\infty}^{\infty} & \text{sinc}\left(\xi - \frac{m+n}{2}, \eta - \frac{-m+n}{2}\right) \\ & \times G_g\left(\xi - \frac{m+n}{2}, \eta - \frac{-m+n}{2}\right) \end{aligned} \tag{36}$$

A commonly used intuitive interpretation for (35) and (36) comes from drawing the associated *Nyquist diagram*. Deriving this diagram from first principles, Fig. 27 shows the locations of the centers of the repeated frequency components for a sensor with no CFA pattern. Components appear at $m, n \in \mathbb{Z}$. The central square represents the boundary halfway between the fundamental at the origin and the nearest frequency sidebands. As such, this square describes the *Nyquist frequency* of the sensor. If the frequency spectrum of the fundamental is bandlimited (restricted) to

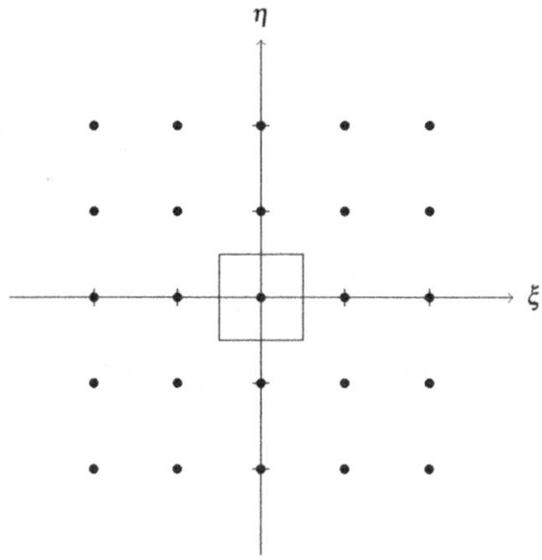


**Fig. 27** Sensor repeated frequency component locations

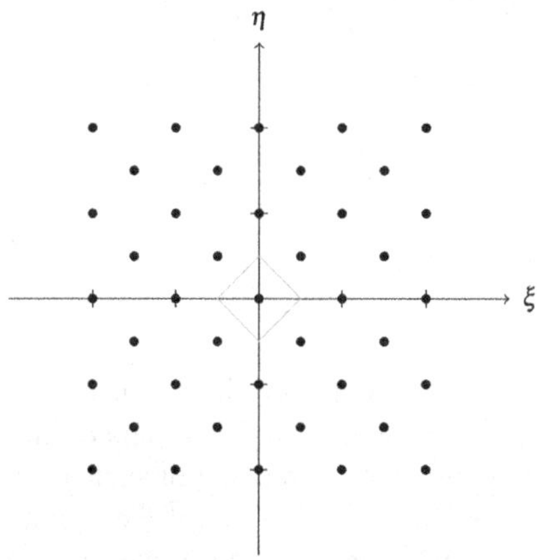


**Fig. 28** Bayer CFA green channel repeated frequency component locations

be entirely within the square, then aliasing is avoided. Figures 28 and 29 show the corresponding frequency component location plots for the green and red/blue channels of the Bayer MRP and their Nyquist frequencies. Amalgamating Figs. 27, 28, and 29, the resulting Nyquist diagram of Fig. 30 compares the frequency responses of the associated sampling patterns. It can be seen that along the horizontal and vertical frequency axes, the green channel has the same Nyquist frequency as the sensor without a CFA pattern. It is for this reason that the four-spot antialiasing filter is usually designed for a pixel pitch of unity in both horizontal and vertical directions, as described in Sect. 2.2. From the Nyquist diagram it can be seen that this strategy will break down for diagonally oriented edges, as the green channel Nyquist frequency pattern takes on a diamond shape that misses the corners of the sensor's Nyquist frequency square. The smaller square corresponding to the red and blue channel Nyquist frequency pattern is clearly smaller than the green channel and sensor patterns. These aliasing effects are demonstrated in Fig. 31. The effects

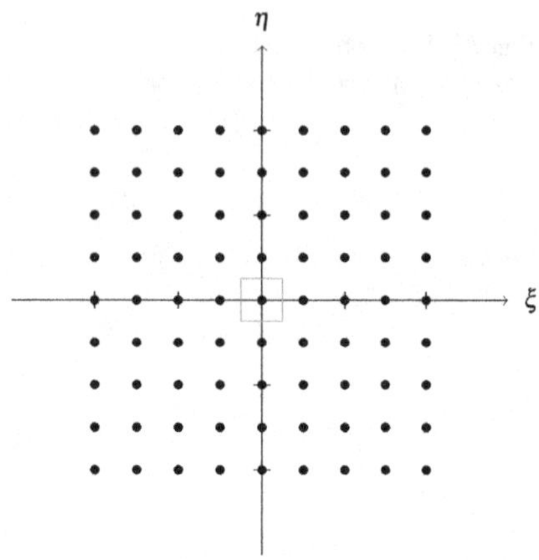


**Fig. 29** Bayer CFA red and blue channel repeated frequency component locations

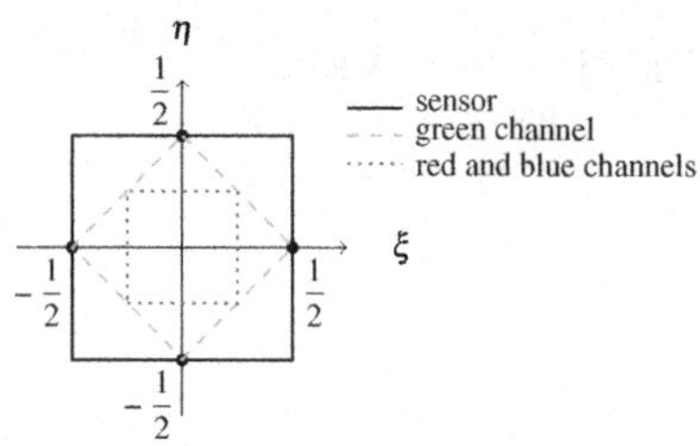


**Fig. 30** Bayer CFA Nyquist frequency diagram. Frequency axes are in cycles/sample

of exceeding the Nyquist frequency of the green channel can be seen in the green–magenta patterns in the corners of the figure. The effects of exceeding the Nyquist frequency of the red and blue channels can be seen in the cyan–yellow patterns at the middle of the edges of the figure. With the Bayer CFA MRP, the designer of the antialiasing filter must make a compromise. If the antialiasing filter is designed for the green channel, it will permit aliasing in the red and blue channels. However, an antialiasing filter designed for the red and blue channels will discard valid green channel high-frequency information. The usual solution is to design for the green channel, i.e., a one-pixel pitch, and to address the subsequent red and blue aliasing with sophisticated image processing operations after the capture.

### 2.5.3 Metal Wiring and Substrate

An unfortunate necessity in any sensor design is the metal wires and structures required to condition and transport the photocharges generated at the pixel sites to the subsequent signal processing electronics, e.g., the A/D converter. Since these wires and structures are generally collocated with the pixels, it is perhaps not unexpected that they can "get in the way" and block some of the light that might otherwise strike the photosensitive silicon surface. This has been previously discussed in Sect. 2.1.2 under the generalized concept of pixel fill factor. In the case of pixel sizes that are

**Fig. 31** Example of aliasing due to the Bayer CFA pattern

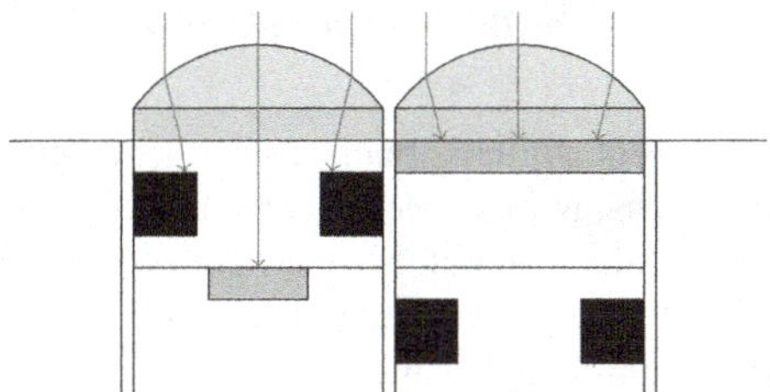


**Fig. 32** Idealized pixel designs: **a** front-illuminated pixel; **b** back-illuminated pixel

relatively large, i.e., $\geq 5\,\mu$, the amount of interference from these opaque objects is relatively small. However, in the case of small pixels, i.e., $\leq 2\,\mu$, where to "put" the wires can become a challenging design headache. Undesirable choices can lead to significant losses in photometric sensitivity.

Until very recently, the idealized standard design of a CMOS pixel was as given in Fig. 32a [4].

The design of Fig. 32a is referred to as a front-illuminated (FI) or a frontside-illuminated (FSI) pixel to distinguish it from the design of Fig. 32b, although the terminology came into being recently as the newer design was developed. In the FI pixel, light must pass through a region of metal wires and other opaque structures before it strikes the photosensitive silicon substrate. In Fig. 32a the metal wires are represented by two black regions at the sides of the pixel. In practice, these obstructions to light can occur anywhere within the cavity, perhaps even in the middle! As indicated by the gray region at the substrate surface, only a portion of the silicon substrate is able to receive light with the rest of the substrate lying in the shadows of the obscuring structures.

A seemingly obvious, although until very recently unmanufacturable, solution to the wire location problem is to invert the order of the silicon substrate and metal wire layers in the sensor, as shown in Fig. 32b [33]. This design is referred to as a *back-illuminated* (BI) or a *backside-illuminated* (BSI) pixel. (This terminology is derived from the light striking the "backside" of the standard pixel design in Fig. 32a). From the idealized representation in Fig. 32b it is clear that in a BI pixel nearly all of the photosensitive silicon surface can be used for photocharge generation, thereby increasing the photosensitivity over the FI pixel. The difficulty with the BI pixel design is its inherently high crosstalk due to its difficulty in keeping photocharges contained to the pixel locations where they were generated. Until recently, this crosstalk was sufficiently large enough to offset the advantages of BI technology. Within the past couple of years some manufacturers have discovered how to reduce the crosstalk while maintaining a significant portion of the increase in photometric sensitivity. With a fairly rapid transition from FI to BI technology occurring at the time of this writing, it is conceivable that BI technology will become the new standard pixel design in short order.

## 3 Image Sensor Operation and Artifacts

In present digital cameras, all image sensors use integrating sensing technology: each sensing pixel accumulates light-induced photocharge for a finite exposure time and is then read out. Most sensors convert 20 to 60% of the photons incident on the silicon into photocharge. The signal output from the sensor is usually linear with the accumulated charge.

Image sensors are commonly grouped, based on their fabrication processes, into Charge-coupled Devices (CCD) and Complementary Metal Oxide Semiconductor (CMOS) sensors. While design variations and terminology continue to develop, the key difference between these groups is that CCD sensors transport charge from each active pixel to an output gate, where charge is converted into a measurable signal. CMOS, or active pixel, sensors convert charge into voltage within each pixel and send that signal to the sensor output. Most sensors are illuminated from the front-side, but back-side illuminated sensors are becoming more common. The literature on sensor development and technology is rich; recent books discussing the field include [10, 22, 24]. The resentation here is at a very high level, providing only enough detail to help explain some of the common artifacts caused by sensor limitations. These artifacts are discussed in Sect. 3.3, after a brief overview of sensor operation in Sect. 3.1.

### *3.1 CCD Sensor Operation*

The operation of a CCD has three major elements: conversion of incident photons into photocharge and integration of the charge over time within each pixel, transport of charge to an output structure, and conversion of charge into a signal that can

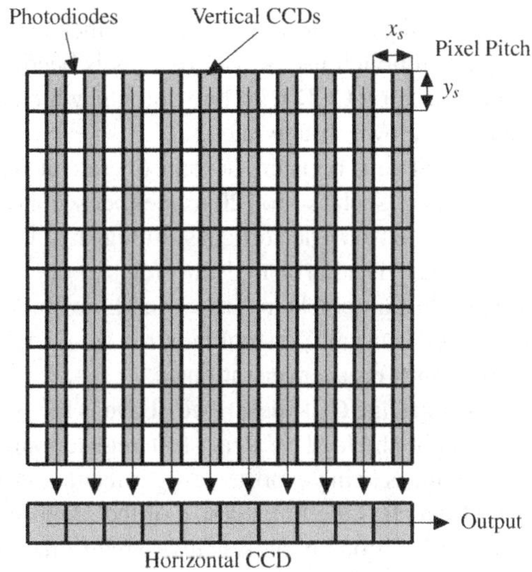


**Fig. 33** Diagram of area array interline CCD layout and charge flow

be digitized. These operations are illustrated in the diagram of an interline area CCD sensor in Fig. 33. In this figure, the white columns are photosensitive pixels for capturing photons and converting them into charge. The gray columns are light-shielded vertical shift registers for readout. They are used to shift charge for each pixel down the column to the horizontal shift register. The horizontal shift register at the bottom is also light shielded and is used to shift charge for each pixel across the row to the output. The black arrows on each shift register show the direction of charge flow during readout.

During operation, a reset signal can be used to drain all charge from the pixels to begin integration. After integrating charge for some time, charge is transferred from the photosensitive area of each pixel to the corresponding vertical shift register element. At this point, the photosensitive areas begin integrating charge again, while the image is being read out through the shift registers. During readout, pixels in each vertical shift register are shifted, one row at a time, into the horizontal shift register at the edge of the array. The horizontal shift register shifts each pixel in the row toward the output, where the charge for each pixel is converted into a signal that is digitized.

Because all pixels begin and end integration at the same time, this electronic exposure control is referred to as a global shutter. With this approach all pixels capture moving objects at the same moment, minimizing motion-generated artifacts in a captured image.

There are a number of options for setting up shift register structures, supporting different readout patterns. Many interline sensors intended for traditional video use support interlaced readout, in which the sensor is read out in two alternating interleaved fields. The key feature of a CCD is that within the design of the chip, readout is fundamentally serial—pixels are moved through the shift registers in series rather than supporting random access. It is possible to include multiple outputs, each for a portion of the area array, although this increases cost. Most CCDs support one or two outputs, although use of four outputs is becoming more common. Figure 34 shows the layout for a CCD with two outputs, which can also be read out through a single output at a lower frame rate [17].

Vertical shift registers usually operate at frequencies that are roughly the frame readout rate (such as 1 to 120 frames per second) times the number of rows read out of the image sensor. The vertical shift register clocking rate is usually in the range of 20 to 100 KHz. The horizontal shift register operates at much higher frequencies, since it must shift all pixels in a row each time a new row is shifted into it. The operating frequency is usually several hundred to several thousand times the frequency of the vertical shift register operation.

The interline CCD described above is the type used in most compact digital cameras. Light-shielded vertical shift registers allow readout of one image while the next image is being exposed, a significant advantage. There are some disadvantages, one of which is the limited size of the photosensitive area. Referring to Fig. 33, the distance from one photosensitive area to the next photosensitive area is known as the pixel pitch, labeled for a pixel in the upper right corner of the figure. The lower fill factor to make room for the vertical shift register leads to loss of sensitivity and aliasing unless mitigated by other factors such as lenslets and antialiasing filters. As the figure shows, the fill factor is greater in the vertical dimension than in the horizontal dimension, providing some motivation for the cylindrical lenslets mentioned in Sect. 2.5.1.

Some cameras, particularly those with larger formats (e.g., $24 \times 36$ mm or larger) intended for professional applications, use full frame CCDs, such as [16, 18]. This design dispenses with light-shielded vertical shift registers and uses mostly transparent material for circuitry on top of the photosensitive areas, using the same area for light capture and for the shift register. This allows nearly the full pixel area to be used for each photosensitive area (hence the term full frame), significantly increasing the signal that can be accumulated in each pixel. Because the vertical shift registers are not light shielded, an optical shutter is normally used to control exposure and shield the sensor from light during readout.

## 3.2 CMOS Sensor Operation

Sensors with active pixel designs, are referred to here as "CMOS" sensors. Most of them use Metal Oxide Semiconductor (MOS) diodes for accumulation of light-induced charge. Three notable variations on the CMOS design are the Foveon X3 [31], Nikon JFET [12], and 2PFC™ [30] sensors.

Commonly used CMOS sensors convert charge into voltage within the pixel, although variations exist. More significantly, the pixels in the array are addressed in a row and column fashion, providing greater flexibility in sensor readout than is practical with CCDs. Often, the sensor is read out in a progressive scan fashion, similar to a CCD sensor, but more use is being made of the readout flexibility. Because these

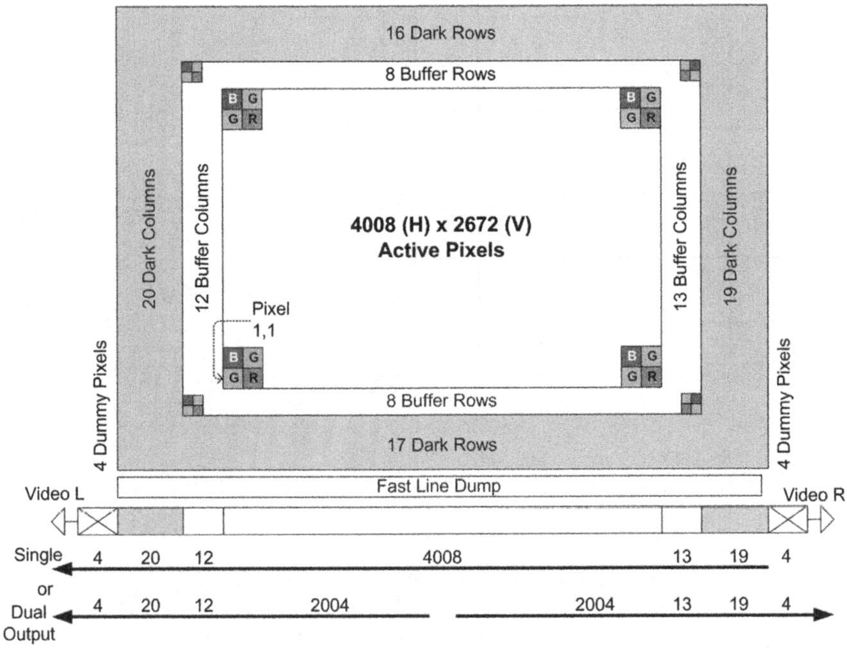


**Fig. 34** Example layout for a multiple output interline CCD

sensors lack shift registers, pixel binning is achieved with other circuitry, including shared photodiodes and other summing circuits [8, 13].

The readout of a CMOS sensor is illustrated very generally in Fig. 35. Each pixel in this array is addressed by a row select signal and a column sense signal. The gray filled areas roughly illustrate the region for each photosensitive area. The fill factor is usually relatively low, since most designs push the pixel pitch as small as possible. There are many design variations for the details of the pixel array and support circuitry, but the row and column addressing is common.

Because there is essentially no light-shielded storage in most CMOS sensors, charge cannot be stored for any significant time before readout. Because of this, most CMOS sensors use a *rolling shutter* readout scheme, unlike the global shutter of an interline CCD. With a rolling shutter, pixels in a row are reset to begin integration, and readout of that row ends the integration. Figure 36 illustrates this for a hypothetical sensor with $N$ rows reading out at $F$ frames per second. The bold diagonal lines indicate which row is being read out over time, starting with row 0 and getting to the last row ($N - 1$) of the sensor at the end of the frame readout time. In this example, the readout immediately cycles back to the top of the sensor, shown by the dotted lines, and readout continues. The thin diagonal lines indicate which row is being reset over time. In this example, the reset line precedes the readout line by $3/4$ of the frame time. The integration interval is shown by the gray fill between the lines—each row is integrated for $3/4\ F$, but each row integrates during a slightly different

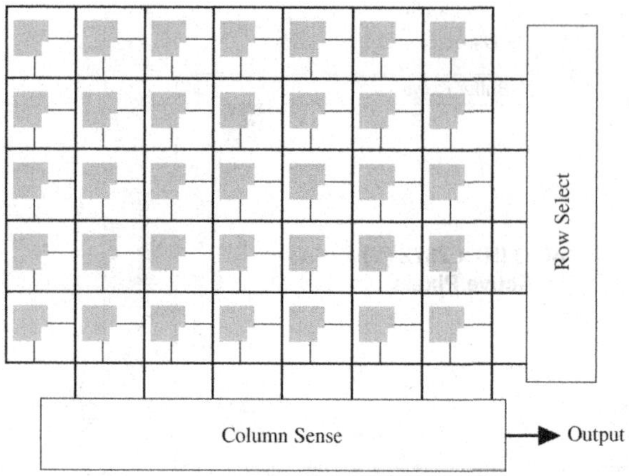


**Fig. 35** Simplified CMOS array addressing

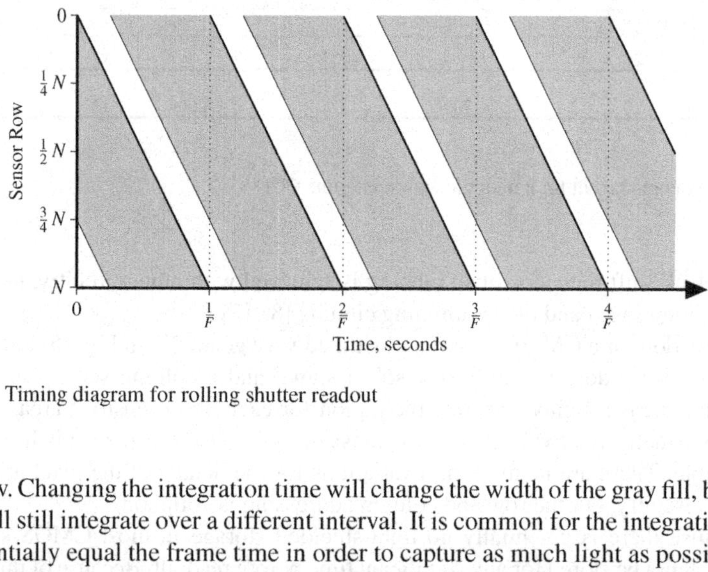


**Fig. 36** Timing diagram for rolling shutter readout

window. Changing the integration time will change the width of the gray fill, but each row will still integrate over a different interval. It is common for the integration time to essentially equal the frame time in order to capture as much light as possible, but this is not always the case, especially with bright illumination. This example shows the sensor being read out as quickly as possible; the frame time must be at least the time required to read out the required number of rows. If the frame time is chosen to be longer relative to the number of rows and the line readout time, the diagonal lines in Fig. 36 would become steeper and there would be gaps in the readout when readout is waiting for the next frame to begin.

## 3.3 Sensor Artifacts

Having presented a high-level flow of operation for the sensors, characteristic artifacts will be discussed next. Many of these artifacts may be used to help identify specific cameras or camera technologies used to form an image. Digital camera image processing chains, discussed in the following chapter, will normally mitigate these artifacts, but the correction may not be perfect, and concealment artifacts can be introduced in the image.

### 3.3.1 Sensor Defects

Defects are those pixels whose response is abnormal enough to be discarded after readout rather than used as image data. There are many causes of defects: high dark current arising from an impurity in the silicon crystal, a flaw in a filter or lenslet, a surface flaw on the sensor (dirt or scratch), an electrical fault, or even a flaw on the sensor cover glass. While manufacturers strive to minimize defects, they do occur and impact the yield of image sensors. In order to improve yield, limited quantities of certain types of defects are generally accepted in sensors. Details about what defects are acceptable are generally part of the specification negotiated between a sensor supplier and a camera manufacturer. Regardless of the details of the specification, defects are routinely concealed when processing images from cameras, although some defects are more difficult to conceal than others. The following discussion describes different classes of defects, while concealment of the defects is discussed in the next chapter.

Isolated single pixel defects are common; most sensors include dozens or even hundreds of these and they are easily concealed in the processing path. A flaw in a vertical shift register can introduce a single column defect, in which most or all of the pixels in a column are defective. This kind of defect is more difficult to conceal than an isolated pixel, but some camera processing paths support concealment of a few isolated column defects in order to improve image sensor yield and lower cost for the image sensor. Because CMOS sensors address pixels with row and column selection, row defects as well as column defects are possible. Defects on the surface of the sensor, such as dirt, often affect several or many pixels in a cluster. Sensors with these cluster defects are usually rejected during manufacturing, although some cameras have implemented defect concealment for small clusters of pixels. In some CMOS designs, multiple pixels (usually two or four) share circuitry for conversion of charge into voltage, so a failure in the shared circuitry will produce a cluster of defective pixels [8].

While most point defects in a sensor are present at manufacture and thus can be mapped during sensor or camera manufacture, some defects accumulate during the life of the sensor. Bright point defects can be caused by cosmic ray damage to the sensor, particularly when airborne, such as when being shipped or carried overseas [19].

A flaw, such as dirt, on the cover glass of the sensor will produce local variation in sensitivity that depends upon the f/number of the taking lens, distance from the exit pupil of the lens to the sensors, and the distance from the cover glass to the surface of the sensor. Generally, the shadow of the flaw cast onto the sensor subtends the same angle as the cone of light rays coming from the exit pupil of the lens. At lower f/numbers, the flaw will produce modest gain variations over a relatively large area, while at higher f/numbers, the flaw will produce a more compact region of more significant gain variations. Dirt on the cover glass is common in cameras with interchangeable lenses, as dust can easily get on the cover glass of the image sensor. Manufacturers of cameras with interchangeable lenses have introduced a variety of systems for cleaning the sensors and also for mapping and concealing the artifacts from dirt.

### 3.3.2 Dark Current

Dark current is signal accumulated within each pixel even in the absence of light. This happens because electrons are generated from thermal effects as well as from photon collection. The current is relatively constant, so the charge collected during integration depends linearly upon integration time. Dark current also varies exponentially with temperature [22]. Because dark correction must be precise over a wide range of temperatures and integration times, light-shielded (also known as dark, or optical black) pixels are placed around the periphery of the image sensor to provide accurate data on the dark current accumulated during each exposure. Figure 34 shows the layout for a CCD with two outputs. The dark pixels are shown on the left and right sides of the sensor as well as the top and bottom of the sensor. With small sensors, temperature variations across the sensor are usually insignificant, but larger sensors can have measurable variation in dark signal due to temperature non-uniformity.

In addition to temperature non-uniformities, some readout modes vary integration time slightly from row to row, with the last rows read out of the sensor being integrated longer than the first rows read out. This occurs more often with full-frame CCDs, since the exposure is usually controlled by performing a global reset of the CCD (draining all accumulated charge), opening the optical shutter for a time, then closing the optical shutter and reading out the CCD. The last rows read out have integrated dark current longer than the first rows read out, even though the optical shutter provides a global shutter for light collection. Depending upon the dark current and the readout timing, there may be measurable differences in noise from the first rows read out to the last rows read out, because shot noise occurs with dark signal as well as photocharge.

The statistical nature of dark current generation tends to blur the distinction between defects and the normal variation in dark response. Rather than two or three clear peaks (normal, dark defects, bright defects), a histogram of a low-light flat field image capture is more like that shown in Fig. 37. Because most defects have extremely high variance along with high dark current, the overall distribution looks more like a mixture of Gaussians, with a main peak and extremely wide tails. This

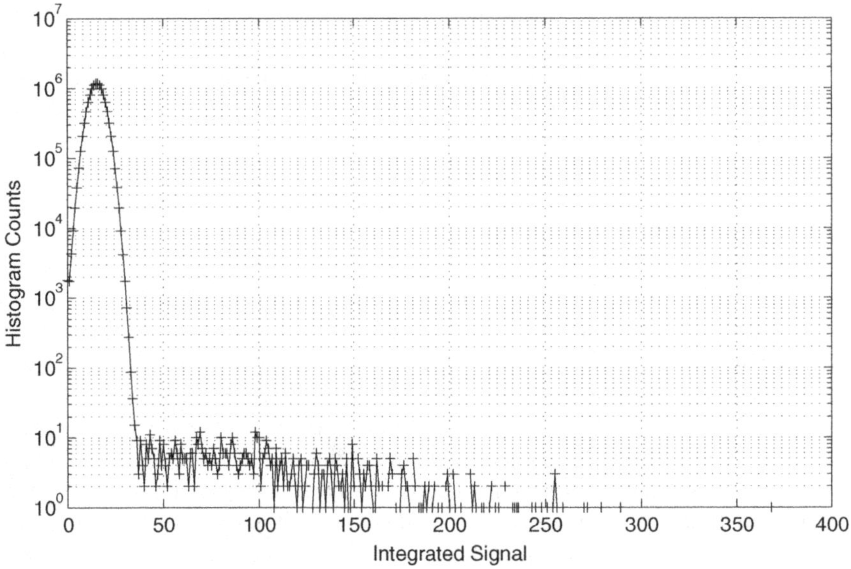


**Fig. 37** Typical histogram of low-light flat field exposure

makes classification of defective pixels nontrivial; the classification is improved by using a series of exposures rather than a single exposure.

### 3.3.3 Fixed Patterns

The dark floor of a captured image has two main contributing factors: dark signal, which is dark current integrated over the exposure time, and offset in the analog signal chain. Fixed pattern noise has two main components: variation in the dark floor, and variation in gain or sensitivity. Variation in dark signal is primarily driven by slight changes in dark current from pixel to pixel. With short exposure times and low dark current levels, the variation in dark signal is usually insignificant. Long exposure times can bring the dark signal up high enough to be visible, and summing or averaging multiple exposures can bring down temporal noise, making the fixed pattern noise more visible.

Fixed pattern noise is usually more significant and complex in CMOS sensors than CCDs because of the more complex circuitry in the array. In addition to more complex pixels, it is common to have signal variation for each row and for each column. The changes in dark signal fixed pattern with row or column are usually more significant than gain changes. Because these structured patterns are highly objectionable, they are usually corrected to subthreshold levels during processing.

Variation in sensitivity is most often caused by subtle non-uniformity in color filter arrays and lenslets, as well as interactions between the sensor and the taking lens.

These variations usually have a very low spatial frequency, although large sensors fabricated in several blocks, such as described in [18, 21], may have abrupt step features at block boundaries. Routine processing (discussed in the next chapter) rarely compensates very precisely for very gradual fixed offset or gain patterns, although partial correction is becoming more common.

### 3.3.4 Multiple Output Mismatch

In addition to pixel-based offset and gain variations, fixed patterns can also be introduced through the output electronics. This is especially true in sensors with multiple outputs. Even the most carefully matched designs are likely to have noticeable offset and gain variations from output to output. These are caused not only by variations in circuits leading to the output from the pixels on the sensor, but also in the analog circuitry leading from the output to the digitization step. This is especially true if there is an analog-to-digital converter for each output. Because real analog circuits and digitizers have measurable nonlinearities, it is also likely that multiple outputs will have measurable mismatches in the linearity of their response. Even if the linearity of each output is very good, there can be visible artifacts when pixels output and digitized through multiple paths are placed side by side in an image. The artifacts created by the mismatch will have a texture depending upon the geometric layout of the pixels routed through each output. In cases such as [17], the artifact will be a step boundary, because the two halves of the sensor are routed to the different outputs. In other cases, such as [21], columns are connected to the outputs in an interleaved fashion, so channel mismatches appear as patterned column variation.

### 3.3.5 Crosstalk

Most sensors suffer from crosstalk between pixels, where photons that should be collected in one pixel are collected in a neighboring one. Some crosstalk is optical, where light that passes through one lenslet and CFA filter ends up being collected in an adjacent pixel. As Fig. 32(a) shows, in a simplified way, the region between CFA and photosensitive pixel area is formed from layers of silicon and metal, with many opportunities for reflections. Crosstalk is also caused by charge diffusion, especially for longer wavelength light. Longer wavelength light penetrates farther into silicon before being absorbed and converted into charge and some photons get absorbed in the substrate rather than in the photosensitive pixel. The charge generated in the substrate can then diffuse to a nearby pixel.

With the presence of a CFA, crosstalk presents several artifacts. For example, in the Bayer CFA, the green channel is composed of two classes of pixels. Half of the green pixels have horizontal neighbors that are blue and vertical neighbors that are red. The other half have the converse—horizontal neighbors are red and vertical neighbors are blue. Because of this layout, crosstalk makes the effective spectral sensitivity of each red (or blue) pixel combine with a portion of the sensitivity of neighboring green pixels.

Another artifact is non-uniformity in response of pixels in a color channel, based upon their neighbors. Crosstalk in most sensors is asymmetric—either mostly vertical or mostly horizontal, due to asymmetry in pixel layout. In interline CCD sensors, vertical crosstalk is usually greater because of the vertical shift registers separating the photosensitive pixel areas. Thus, half of the green pixels have spectral sensitivity that has more of a red contribution, while the other half has more of a blue contribution. This change in sensitivity is usually a small percentage of the total sensitivity, but the highly patterned nature of the artifact can cause trouble for demosaicing. Any filter array pattern that places pixels within varying neighborhoods is vulnerable to this kind of pattern problem.

### 3.3.6 Saturation

Image sensors can exhibit artifacts in and around regions of the image where pixels are saturated, that is, where a pixel accumulates more charge than it can contain. When a pixel is saturated, charge can overflow from one pixel to adjacent pixels, creating what are known as blooming artifacts [10, 22]. Depending upon where the charge diffuses, this usually appears as increased signal in nearby pixels, especially in CCD sensors. Overflow drains are designed into sensors to drain excess charge to control blooming. These generally prevent blooming artifacts, although small pixels remain more vulnerable to them than larger pixels of similar design. In CMOS sensors, overflow to nearby pixels is less common; they are more likely to exhibit a signal inversion—the highlights go dark because the dark reference signal is corrupted by the excess photocharge.

### 3.3.7 Interline Smear

The design of an interline CCD also leads to an artifact known as interline smear. It is caused by excess charge from a photosensitive area leaking into the adjacent vertical shift register during readout, adding excess charge into every pixel shifted past the saturated pixel. This creates a vertical streak above and below a bright region in the image. This artifact is very rare in still captures when an optical shutter is used, because the photosensitive areas are shielded from light during most of readout, limiting excess charge generation. Smear is common in captures using only an electronic shutter, such as live preview or video capture.

### 3.3.8 Rolling Shutter

The rolling shutter used with most CMOS sensors creates artifacts when the scene changes significantly during capture of the image. This can be caused by motion in

**Fig. 38** Example of rolling shutter motion distortion

the scene, as shown in Fig. 38. In this figure, the school bus is moving to the left during the exposure, so the rows exposed later in time capture the bus farther to the left.

Another artifact caused by rolling shutter operation occurs when the illumination changes during exposure. This is most common with flickering illuminants, such as fluorescent lights, although CRT displays can create similar artifacts. The variation in illumination level during operation causes bands of varying exposure in the image. The spatial period of the bands depends upon the ratio of the readout rate and the flicker period. The magnitude of the bands depends upon the ratio of the integration time and the flicker period. Banding is eliminated when exposure time is an integer number of flicker periods and is maximized when exposure time is 1/2 of a flicker period or less. Poplin [27] discusses the problem of flicker detection in some detail.

## 4 Video Capture

For some time in the development of digital cameras, the somewhat different needs of camcorders and digital still cameras kept the technology distinct. Digital video cameras (DVCs)were developed using relatively small sensors with relatively low resolution, optimized for capturing just enough data for video under a wide range of illumination conditions. Digital still cameras (DSCs) were developed to capture the highest resolution images practical, usually using sensors with a larger area and electronic flash. More recently, technology has been converging: nearly all digital cameras are expected to acquire both stills and video. The differentiation between DSCs, DVCs, and other devices such as mobile phone cameras lies in the relative priority for cost, compactness, image quality, and so forth. Some current DVCs may use a sensor with roughly twice as many pixels as are provided in an output video stream, such as a five-megapixel sensor and a two-megapixel ($1920 \times 1080$) video stream. Most current DSCs use higher resolution sensors, such as a fourteen-megapixel sensor, while still providing a video stream limited to two megapixels.

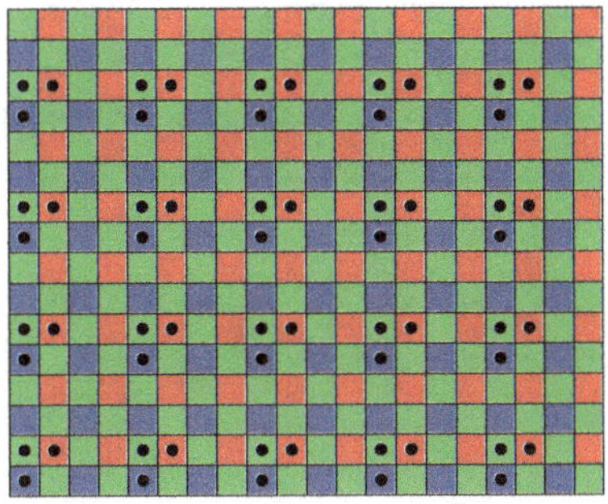

**Fig. 39** Example Bayer CFA sub-sampling pattern

While the sensor resolution can vary, both DVCs and DSCs require the ability to efficiently read a relatively low-resolution image from the sensor at video frame rates, such as 30 or 60 frames per second.

Key in this convergence has been the development of flexible binning readout architectures, allowing good-quality, low-resolution images to be read out from a high-resolution sensor. This was alluded to previously when discussing sensor readout schemes, but deserves more detail here. The need for good-quality video from still sensors has been driven by two needs: the feature of storing digital video files, and the feature of providing electronic live view functionality, either on a display or in an electronic view finder with an eyepiece. Both these features demand a rapidly updated image with relatively low resolution.

## *4.1 Subsampling*

Due to the computational cost and noise associated with very fast charge conversion and digitization, it is necessary to limit the readout rate for still and video capture. Using multiple outputs (and horizontal shift registers for a CCD) will help, but more common is the support of multiple readout resolutions at different frame rates. One approach is to flush several rows of pixels to a drain for every row that is read out, and flush several pixels in a row for every pixel that is read out. This easily supports subsampling in integer ratios. One example is illustrated in Fig. 39, which shows a Bayer pattern CFA with black dots superimposed on pixels that are sampled during operation in a low-resolution readout.

This approach was used for video in earlier DSCs that provided a live preview for scene composition. However, recalling the discussion from Sect. 2.2 about antialiasing filters, the subsampling approach to low-resolution readout can create severe aliasing problems because it lowers the effective sampling frequency. Accordingly, these cameras could not offer video capture that was nearly as good as devices with a sensor and optics optimized for video capture.

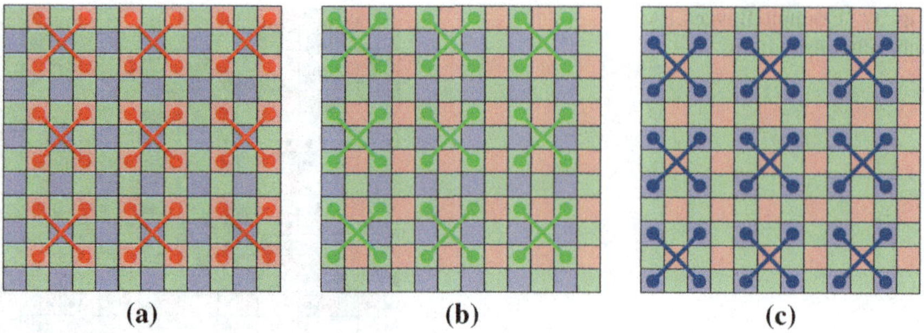


**Fig. 40** Example Bayer CFA binning pattern **a** *red* **b** *green* **c** *blue*

## *4.2 Binning*

A preferred, and now more common, approach to lower resolution is binning. In binning, charge from multiple pixels is combined before conversion into voltage, allowing a single pixel read out to represent the sum of multiple pixels on the sensor. This improves the signal-to-noise ratio for the summed pixel and also provides a great reduction in aliasing. Successful binning becomes more complex with a color filter array; the binning architecture and the CFA pattern must complement each other to provide the proper color response for binned pixels. Figure 40 illustrates a simple $2 \times 2$ binning pattern, showing each of three color channels separately to reduce clutter in a single figure. The diagonal lines joining the colored dots indicate which pixels are binned; the intersection of the colored lines conveniently shows the effective center of each binned pixel. The summing operation acts like a convolution with a sparse $2 \times 2$ kernel, applying an antialiasing filter as part of the subsampling.

The reader may notice that only half of the green pixels are being binned and read out. Binning and reading the other half of the green pixels is one of many additional options possible when designing binning and readout patterns for a sensor with a CFA. Other binning patterns are also possible, including $2 \times 1$, $1 \times 2$, and $3 \times 3$ patterns. Binning in CCDs is controlled by the design of shift registers and timing patterns, so some CCDs offer a number of binning options. Binning in CMOS sensors is controlled by specific interconnection hardware, so it tends to be less flexible.

## *4.3 Exposure Control*

In addition to the spatial sampling differences between full-resolution readout and low-resolution readout, the temporal sampling of video provides constraints on exposure time. Conventional video readout requires the integration time to be less than the frame time. Conversely, in the presence of motion, it is important for the integration time to be fairly close to the frame time. Excessively short exposures will cause an artifact sometimes known as *judder*, where individual frames seem to present a discontinuous motion. This is the temporal equivalent of a small pixel fill factor. When integration times approaching the frame time are used, moving objects have a motion blur in each frame that is perceptually (and physically) consistent with the displacement from frame to frame. A common rule of thumb is that integration times roughly 50% or more of the frame time limit temporal artifacts to a practical level. Animated videos specifically add motion blur that is consistent with frame displacements in order to present more convincing motion. An example from the field of stop-motion animation is [3], in which a sequence of frames is analyzed for motion displacements, and each frame is blurred with a locally varying PSF based on the displacements.

### 4.3.1 Rolling Shutter Exposure

The rolling shutter artifacts suffered with CMOS sensors due to scene motion tend to be less of a problem with video readout than still readout, simply because the frame time is generally shorter for the low-resolution readout. On the other hand, the flickering illuminant artifacts can be worse because of the periodic nature of the banding. Banding artifacts of low frequency and small magnitude (perhaps 5% or less) will often be masked by scene content in a single frame. The same artifact tends to be more visible in an image sequence shown as a video, because the bands will tend to roll slowly through the image.

**Acknowledgments** The authors gratefully acknowledge many helpful discussions with Mrityunjay Kumar and Aaron Deever in the development and review of this chapter.

## References

1. Adams J, Parulski K, Spaulding K (1998) Color processing in digital cameras. IEEE Micro 18:20–30
2. Bayer B (1976) Color imaging array. US Patent 3,971,065
3. Brostow GJ, Essa I (2001) Image-based motion blur for stop motion animation. In: Proceedings of the 28th annual conference on Computer graphics and interactive techniques, SIGGRAPH '01, ACM, New York, pp 561–566 http://doi.acm.org/10.1145/383259.383325
4. Chi MH (1998) Method of forming a new bipolar/CMOS pixel for high resolution imagers. US Patent 5,854,100
5. Four thirds standard. http://www.four-thirds.org/en/fourthirds/index.html
6. Gaskill J (1978) Linear systems, Fourier transforms, and optics. Wiley, New York
7. Goodman J (1968) Introduction to Fourier optics. McGraw-Hill, San Francisco
8. Guidash RM (2003) Active pixel sensor with wired floating diffusions and shared amplifier. US Patent 6,657,665
9. Guidash RM, Lee PP (1999) Active pixel sensor with punch-through reset and cross-talk suppression. US Patent 5,872,371
10. Holst GC, Lomheim TS (2007) CMOS/CCD sensors and camera systems. The International Society for Optical Engineering, Bellingham, WA, USA
11. Hunt R (1987) The reproduction of colour. Fountain Press, England

12. Isogai T (1996) Photoelectric conversion device utilizing a JFET. US Patent 5,528,059
13. Kim YC, Kim YT, Choi SH, Kong HK, Hwang SI, Ko JH, Kim BS, Asaba T, Lim SH, Hahn JS, Im JH, Oh TS, Yi DM, Lee JM, Yang WP, Ahn JC, Jung ES, Lee YH (2006) 1/2-inch 7.2mpixel cmos image sensor with 2.25/spl mu/m pixels using 4-shared pixel structure for pixel-level summation. pp 1994–2003
14. Kodak KAI-16000 image sensor. http://www.kodak.com/global/plugins/acrobat/en/business/ISS/datasheet/interline/KAI-16000LongSpec.pdf
15. Why a color may not reproduce correctly. http://www.kodak.com/global/en/professional/support/techPubs/e73/e73.pdf
16. KAF-50100 image sensor. http://www.kodak.com/global/plugins/acrobat/en/business/ISS/datasheet/fullframe/KAF-50100LongSpec.pdf (2010)
17. KAI-11002 image sensor. http://www.kodak.com/global/plugins/acrobat/en/business/ISS/datasheet/interline/KAI-11002LongSpec.pdf (2010)
18. Manoury EJ, Klaassens W, van Kuijk H, Meessen L, Kleimann A, Bogaart E, Peters I, Stoldt H, Koyuncu M, Bosiers J (2008) A 36 $\times 48mm^2$ 48m-pixel CCD imager for professional DSC applications. pp 1–4
19. McColgin WC, Tivarus C, Swanson CC, Filo AJ (2007) Bright-pixel defects in irradiated ccd image sensors. In: Materials research society symposium proceedings, vol 994, p 341
20. Meyers MM (1997) Diffractive/refractive lenlet array. US Patent 5,696,371
21. Meynants G, Scheffer D, Dierickx B, Alaerts A (2004) A 14-megapixel $36 \times 24 - mm^2$ image sensor. pp. 168–174, SPIE. 10.1117/12.525339. http://link.aip.org/link/?PSI/5301/168/1
22. Nakamura J (ed) (2005) Image sensors and signal processing for digital still cameras (Optical science and engineering). CRC Press, Boca Raton
23. Ochi S (1984) Photosensor pattern of solid-state imaging sensors. US Patent 4,441,123
24. Ohta J (2007) Smart CMOS image sensors and applications (Optical science and engineering). CRC Press, Boca Raton
25. Palum R (2009) Optical antialiasing filters. In: Lukac R (ed) Single-sensor imaging. CRC Press, Boca Raton
26. Pecoraro G, Shelestak L (1988) Transparent infrared absorbing glass and method of making. US Patent 4,792,536
27. Poplin D (2006) An automatic flicker detection method for embedded camera systems. IEEE Trans Consum Electron 52(2):308–311
28. Popovic Z, Sprague R, Neville Connell G (1987) Pedestal-type microlens fabrication process. US Patent 4,689,291
29. Stevens E, Komori H, Doan H, Fujita H, Kyan J, Parks C, Shi G, Tivarus C, Wu J (2008) Low-crosstalk and low-dark-current cmos image-sensor technology using a hole-based detector. pp 60–595
30. Tamburrino D, Speigle JM, Tweet DJ, LeeJJ (2010) 2PFC$^{TM}$(two pixels, full color): image sensor demosaicing and characterization. J Electron Imaging 19(2): 021103-1–021103-13
31. Turner RM, Guttosch RJ (2006) Development challenges of a new image capture technology: foveon X3 image sensors. In: International congress of imaging science, pp 175–181, Rochester, NY, USA
32. Watanabe M, Nayar S (1995) Telecentric optics for computational vision. In: Proceedings of the European Conference on Computer Vision, pp 439–451
33. Yamamoto Y, Iwamoto H (2006) Solid-state imaging device and method of manufacturing solid-state imaging device background of the invention. US Patent 7,101,726

# Digital Camera Image Formation: Processing and Storage

**Aaron Deever, Mrityunjay Kumar and Bruce Pillman**

**Abstract** This chapter presents a high-level overview of image formation in a digital camera, highlighting aspects of potential interest in forensic applications. The discussion here focuses on image processing, especially processing steps related to concealing artifacts caused by camera hardware or that tend to create artifacts themselves. Image storage format issues are also discussed.

## 1 Introduction

The hardware of a digital camera was discussed in the previous chapter. This chapter describes image processing operations used with digital cameras. Each operation can introduce characteristic artifacts under some conditions. The purpose of this chapter is to mention them so researchers can expect and recognize them when found.

These processing operations can take place either in a camera or on a computer, depending upon the chosen camera and workflow. Most processing steps are largely the same whether performed in a camera or on a computer, although increased complexity is often allowed when processing on a computer, rather than in a portable camera. The discussion here will describe the processing chain step by step, mentioning when computer processing is more likely to differ from processing in a camera.

Most of the discussion in this chapter focuses upon routine processing performed on most images from digital cameras. Naturally, more complex processing is occa-

A. Deever (✉) · M. Kumar · B. Pillman
Corporate Research and Engineering, Eastman Kodak Company,
Rochester, NY, USA
e-mail: aaron.deever@kodak.com

M. Kumar
e-mail: mrityunjay.kumar@kodak.com

B. Pillman
e-mail: bruce.pillman@kodak.com

H. T. Sencar and N. Memon (eds.), *Digital Image Forensics*,
DOI: 10.1007/978-1-4614-0757-7_2,

sionally used for some images and applications. These applications and processing techniques will be discussed in Sect. 4.

Images from digital cameras are stored in files, normally in one of several standard formats. The information included in the file can affect image use, especially in forensic applications, so image storage formats are briefly discussed in Sect. 3.

Finally, Sect. 5 discusses some of the characteristics of processing chains for video capture.

## 2 Nominal Image Processing Chain

To provide an order for the discussion of routine camera image processing, the processing steps are presented in the sequence shown in Fig. 1. The ordering shown in the figure is reasonable, although in practice steps are often moved, combined, or split up and applied in several locations in the chain. In Fig. 1, ellipses represent the image at key points of interest along the chain, while rectangles represent processing blocks. The blocks in this chain will be discussed in the following sections.

One reason for split or redundant operations is the tendency for noise to be amplified during the processing chain. White balancing, color correction, tone and gamma correction, and edge enhancement are all operations that tend to increase noise or at least increase visibility of the noise. Because noise is amplified during processing, it is common for several operations to take steps to reduce noise or at least limit its amplification. Also, the chain is often altered to meet different expectations for image quality, performance, and cost. For example, color correction usually works more accurately when performed on linear data, but lower cost image chains will often apply gamma correction fairly early in the processing chain to reduce bit depth, and apply color correction on the gamma-corrected data.

In general, image processing chain design is fairly complex, with tradeoffs in the use of cache, buffer memory, computing operations, image quality, and flexibility. This chapter will discuss some of the more common processing chain operations, but the reader is advised to consult [2] for further discussion of the design of camera processing chains.

### *2.1 Raw CFA Image*

The first image is the raw image as read from the sensor through the analog signal processing chain. It is a single channel image in which different pixels sense different colors through a color filter array (CFA), as discussed in the previous chapter.

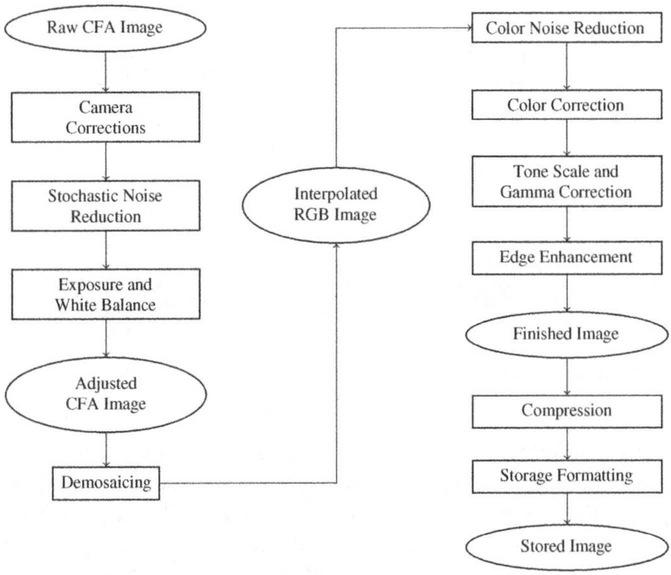


**Fig. 1** Nominal flow for a digital camera processing chain

## 2.2 Camera Corrections

Most of the processing blocks in the nominal processing chain are much simpler to implement if the incoming image has a known response to light exposure, a known offset for the dark signal level, and has no camera-induced artifacts. These ideal conditions are essentially never met with real hardware. At the very least, sensors essentially always have defective pixels and have a dark signal level that varies somewhat with integration time and temperature. Often, other artifacts are also present and require correction or concealment.

The first processing block in the chain of Fig. 1 is more precisely a collection of blocks, illustrated in Fig. 2, designed to convert the acquired raw CFA image into a more idealized "raw" image. The processing blocks required for a specific camera vary depending upon the hardware and the user expectations. Lower cost hardware typically leaves more artifacts in the raw image to be corrected, but the user expectations are often lower as well, so the choice of correction blocks used with a particular camera is the result of a number of system engineering and budget decisions. Few, if any, cameras use all of the processing blocks shown in Fig. 2. In some cases, users save images to a raw capture format and use sophisticated desktop software for processing, enabling a degree of control over the processing chain not usually exercised by the casual user.

While these blocks are presented in a specific order in this discussion, the ordering chosen for a specific camera is dependent upon the causes of the artifacts needing correction and interactions between the effects. Usually, the preferred order of cor-

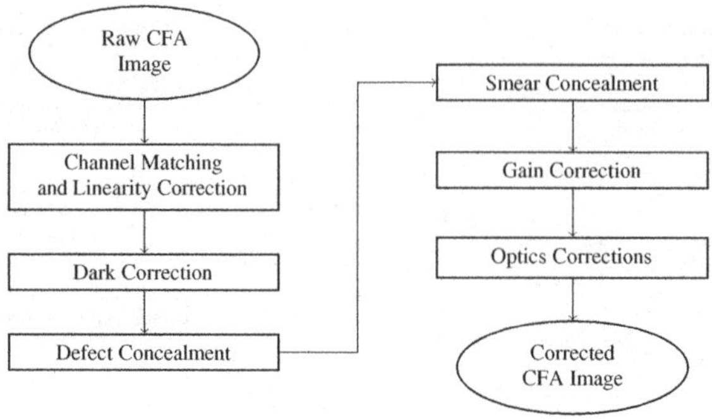


**Fig. 2** Typical flow for digital camera corrections

rection blocks is roughly the inverse of the order in which the artifacts are caused. For example, gain artifacts are mostly caused by optics and interactions between the taking lens and the sensor, while linearity problems usually occur primarily in the analog signal processing chain after collection of charge in the pixels. Therefore, linearity correction would typically be applied to the image before gain correction is applied. Each of these correction blocks is complicated by the artifacts that have not yet been corrected. For example, if a dark correction is computed before defect concealment is completed, care should be taken to avoid using defective pixels in calculation of statistics for dark correction.

### 2.2.1 Channel Matching and Linearity Correction

The first correction discussed here is to match the response of multiple outputs or analog signal processing chains, such as with the dual output sensor shown in Fig. 3. Because the artifacts due to channel mismatch are highly structured, usually a seam in the middle of the image or a periodic column pattern, the responses for the multiple outputs must match very closely. The most common form of this correction is to adaptively compute a dark offset correction for each output that will bring similar pixels from each output to match a common value, using reference dark pixels. More complex algorithms involve sampling pixels from the image area in order to perform gain or linearity matching as well as dark level matching. This requires either controlled capture of calibration images or the ability to estimate channel mismatches in the presence of scene content variation [68]. The key to successful matching of multiple output channels is to take advantage of the knowledge of which image pixels came from which output.

### 2.2.2 Dark Correction

Dark correction is always necessary, since the analog output from the image sensor is rarely precisely "zero" for a zero light condition. As mentioned in the previous

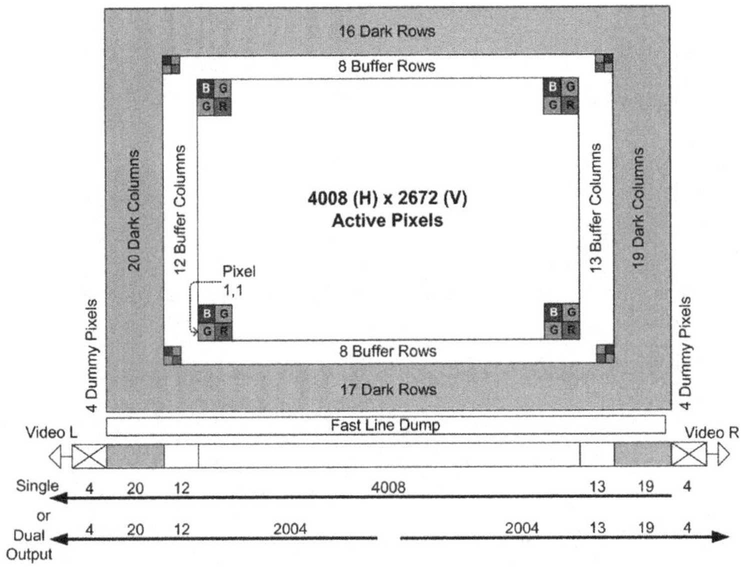


**Fig. 3** Example layout for a multiple output interline CCD

chapter, dark current is collected within pixels as well as light-induced charge. In addition, slight drifts in analog offsets mean the dark level for an image will usually drift by a small percentage of the full-scale signal. The nonlinearity of human perception and image tone processing means this kind of drift causes fairly obvious changes in the shadows of an image. This is illustrated in Fig. 4, showing a plot of the conversion from linear relative exposure to CIE L*, a standard perceptually even measure of lightness [18]. As shown in Fig. 4, the slope increases substantially in the shadows, below a midtone gray near 50 L*. To examine slope in the shadows more closely, Fig. 5 plots the change in L* due to a change in relative exposure of 0.01, plotting for a range from 0 to 0.2. A one-unit change in CIE L* is approximately one just-noticeable difference (JND). As shown in the diagram, a 0.01 change in exposure produces a much larger change in L* as relative exposure goes toward zero. Normal scene shadow exposures range from roughly 0.03 (corresponding to a typical diffuse black patch on a test chart) to 0.2 (corresponding to a midtone gray). In the figure, the dash-dotted line connects the black patch exposure of 0.03 with the slope at that exposure, roughly four $\Delta L*$, or JNDs.

Because the dark level must be controlled more precisely than low-cost analog electronics can provide, the analog dark level is chosen to be several percentage points into the range of the A/D converter, followed by a digital dark floor subtraction. If the dark floor of the image is uniform enough, dark subtraction is simply the subtraction of a global value from the image, usually based upon simple statistics from the light-shielded dark pixels during a capture.

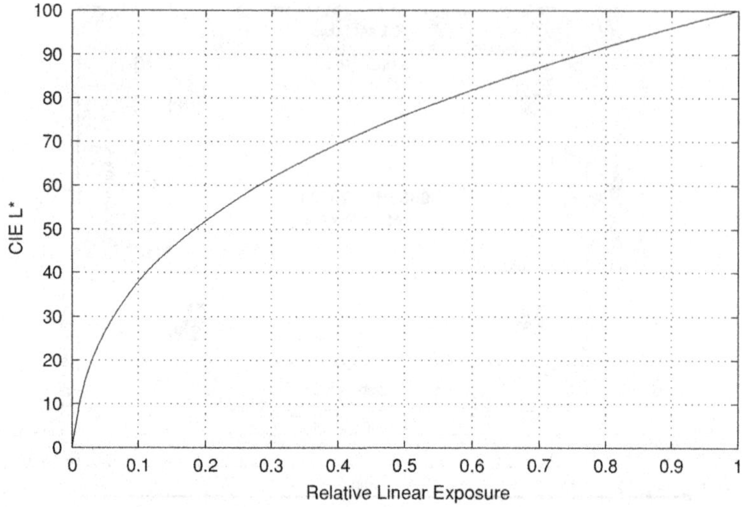


**Fig. 4** CIE L* versus relative exposure

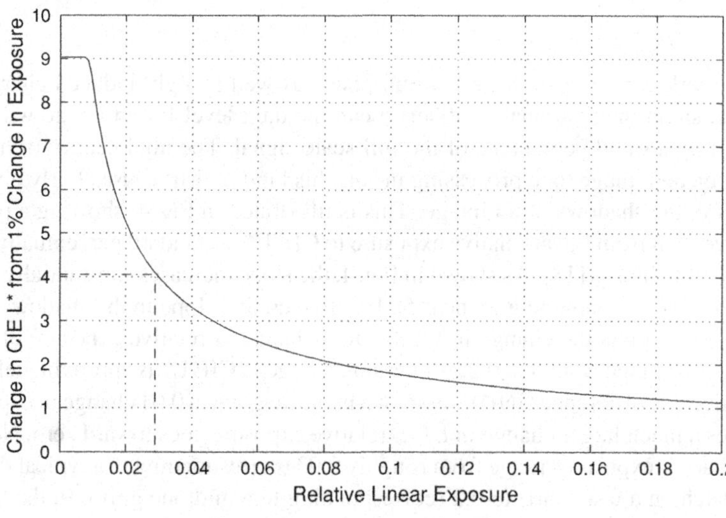


**Fig. 5** Change in CIE L* due to a 0.01 change in relative exposure

If the dark floor of the sensor is not uniform enough, then a full dark floor image is subtracted from a captured image to remove the fixed pattern. One extremely simple approach is to capture a second image immediately after capturing the scene image. The second image is captured with no exposure, providing a full image of dark pixels. This works most accurately if an optical shutter is closed and the integration time for the dark capture matches the integration time for the scene capture. This approach reduces fixed patterns in the dark floor of the image, but increases the random noise. Because of this, the dark capture technique is most often used with long integration times (such as 1/2 s or more), when the appearance of noise arising from the dark fixed pattern is clearly greater than the noise increase from the frame subtraction.

A second approach is to model the dark floor fixed patterns. Most often, this is done using light-shielded dark pixels from the border around the sensor. One common approach is to create a dark floor using dark pixels from the left and right of the image sensor, estimating a dark floor value for each row from the dark pixels for that row. This value can be corrupted by noise in the dark pixels, so some smoothing may be used to reduce the dark floor estimation error. Depending upon the sensor capabilities and the camera architecture, dark rows above and below the image may also be used. Dark pixels on the left, right, top, and bottom of the image provide enough data for a complete separable dark floor model. This one- or two-dimensional approach is very fast and is especially effective at correcting row and column patterns in a CMOS sensor.

In some cases, the dark floor is modeled using data from multiple dark captures. By averaging multiple dark captures, the impact of temporal noise on the dark floor estimate is minimized. This technique is still affected by changes in sensor temperature and integration time. Although changes in dark current caused by changes in temperature and integration time are well understood, other factors affecting the dark level may not be modeled as simply. This approach is quite rare in portable color cameras. Astronomical and other scientific applications, especially ones using a temperature-controlled sensor, routinely use this technique, made easier by the controlled temperature.

### 2.2.3 Defect Concealment

Sections 2.2.1 and 2.2.2 both referred to correction, because those artifacts can be essentially removed with no significant loss of data. Sensor defects are somewhat problematic, since they indicate lost data that simply was not sensed. Algorithms for treating defects interpolate the missing data. This process is referred to here as concealment rather than correction, to emphasize that it can only interpolate the missing data.

As mentioned in the previous chapter, the most common defects are isolated single pixel defects. Concealment of isolated pixels is usually done with a linear interpolation from the nearest adjacent pixels of the same color sensitivity. Some cameras at a low enough price point treat these defects with an impulse noise filter rather than maintaining a map of defective pixels. This tends to (inappropriately) filter out high-contrast details such as stars, lights, or specular reflections. With low-cost optics, the taking lens may spread fine detail from the scene over enough pixels to reduce confusion between scene detail and sensor noise.

Bright pixel defects caused by cosmic ray damage must be concealed without depending upon a map from the sensor or camera manufacturer. If at a low enough price point, the impulse removal approach works well. For higher user expectations, a camera can implement a dark image capture and bright defect detection scan in firmware, usually done at startup. New defects found in the dark image are added to the defect map. Because cosmic ray damage tends to produce bright points rather than marginal defects, detecting these defects is relatively easy.

Sensor column defects present a much greater concealment challenge, because the human visual system is extremely sensitive to correlated features such as lines. Concealment of defective columns has been approached as an extended CFA interpolation problem with some success [34]. As with single pixel defects, a broader taking lens point spread function (PSF) will prevent the highest frequency detail from being imaged on the sensor, making defect concealment easier. The general goal for column concealment has been to ensure that concealment artifacts are subtle enough to be masked by noise or scene content. Concealment algorithms may produce columns with slightly different noise characteristics even if the mean value is well estimated. Concealment of defects including two or more adjacent columns is much more challenging and is an area of current research, although it has been approached with some success [33].

As mentioned previously, some sensor flaws create a defect in several adjacent pixels, here termed a cluster defect. The difficulty of concealing one of these defects increases dramatically with the size of the defect. Methods usually involve filling in from the boundary of the defect with a variety of adaptive approaches. The advantage is that these defects are rare enough that complex processing for the defect concealment is relatively insignificant compared to the processing time for the rest of the image.

Dirt on the cover glass of the sensor creates a much more complex defect, since it varies in size depending upon the f/number of the lens and the distance from the exit pupil to the sensor. Concealment of these defects must take the variable size into account, usually using a combination of gain correction and pixel interpolation.

### 2.2.4 Smear Correction

Interline smear is a challenging artifact to correct or conceal because the artifacts vary with scene content. It is manifested as an offset added to some of the columns in the captured image. Since the added signal will usually vary from column to column, the effect will vary with the original scene content. If a small amount of charge is added to pixels that are well below saturation, the artifact is manifested as a column that is brighter and lower in contrast than normal. If the sum of scene charge and smear charge saturates the pixels in the column, then the column looks like a bright defective column. Smear usually affects several adjacent columns, so saturated columns become difficult to conceal well. Concealment approaches start with the use of dark rows or overclocked rows to estimate the smear signal that should be subtracted from each column. One example from the patent literature is [52], which subtracts a smear signal from each column and applies a gain adjustment after the subtraction. The gain adjustment prevents bringing saturated columns down below the maximum code value, but adds gain variations to each column. In order to control the gain variation, the amount of smear signal subtracted from each column is limited. In cases where the columns are saturated and thus effectively defective, Yoshida [90], and Kim [49] describe ways to approach concealment. Since the smear artifacts can be so wide, concealment may be of limited quality. Because smear artifacts are caused by the scene, imperfect concealment is usually more acceptable than with sensor defects. Because rapid movement can cause smear artifacts that appear as jagged diagonals, there is some recent art that seeks to correct even these artifacts [71].

### 2.2.5 Gain Nonuniformity Correction

As discussed in the previous chapter, gain nonuniformities are caused by several sources in digital cameras. The correction is essentially a multiplication of each pixel with a gain map. The variation in each implementation is the model used to determine the gain to be applied to each pixel. Early implementations, with very limited memory for storing gain corrections, used simple separable polynomials. Later implementations stored small images, with a gain value for each color channel for small tiles of the image, such as $4 \times 16$, $8 \times 8$, $16 \times 16$, and so forth. These maps were often created to make the sensor response to a uniform illumination completely flat, which left taking lens effects and interactions uncompensated.

With increasing adoption of CMOS sensors and evolution to smaller pixels, gain corrections now usually include lens interactions. For a camera with a fixed lens, these are relatively simple. For cameras with interchangeable lenses, this creates new overhead to combine a sensor gain map with a lens interaction gain map. When lens effects get too severe (such as 50% falloff in the corners of the image), gain correction is usually limited to minimize noise amplification. This results as yet another system optimization, trading off darkness versus noisiness in the corners. With suitable noise reduction algorithms, the increase in noise can be mitigated, although usually with an increase in noise reduction artifacts.

Banding artifacts can be caused by interaction of a flickering illuminant with a rolling shutter. These are noticed more often with video than still captures and are discussed in Sect. 5.

### 2.2.6 Optics Corrections

In addition to illumination falloff, the taking lens can also produce other effects, particularly chromatic aberrations. The main aberrations that are corrected are geometric distortion, longitudinal color, lateral color, and spatially varying PSF. Smith contains a more complete discussion of geometric distortion and chromatic aberrations [84].

Geometric distortion is caused by magnification varying across the image and is usually described as being pincushion or barrel distortion. Magnification can be modeled as a low-order polynomial (5th order or less) function of distance from the center of the image. Geometric distortion is corrected by warping the image to invert the change in magnification. The warping is usually done with a bilinear or other relatively simple interpolation, using a low-order function to represent a correction as a function of radial position [91]. The warping does not always completely correct the distortion, because variation during lens assembly causes the geometric distortion for specific cameras to vary slightly.

Lateral color occurs when magnification varies with color channel. Lateral color aberrations in the lens are corrected via a mechanism similar to correction of geometric distortion. Since the aberration is essentially a magnification that is slightly different for each color channel, resizing one or more color channels to achieve a common magnification can be folded in with the correction of geometric distortion.

Longitudinal color aberrations are caused when the different color channels are focused at different distances from the lens. For example, if the lens focus position is set so the green channel is in best focus, the red and blue channels may be slightly out of focus. Lens design can control this variation, but at increased lens cost. Usually, an autofocus system will bring the green channel into best focus and the red channel or blue channel will tend to be farther out of focus. The treatment for this artifact is to sharpen one or more color channels with a spatial filter. If this kind of correction is required, the geometric corrections can be included in the creation of these filters, by allowing use of asymmetric kernels and by allowing the filter kernels to vary with position in the image. Because distortion correction may spatially resample the color channels individually, it is often included in the processing chain after demosaicing.

Correction for a spatially varying PSF is similar to the correction for longitudinal color. Convolution with a spatially varying kernel is used, although usually only on a luma channel or on all three color channels.

Optics corrections can be particularly complex, especially if correcting for spatially varying PSF or longitudinal color aberrations. These corrections are especially common in more sophisticated desktop software for processing images from raw files, allowing the user to tune adjustments for the specific image and application.

## 2.3 Stochastic Noise Reduction

In most digital camera image processing chains, noise reduction is a critical operation. The main problem is that compact digital cameras usually operate with limited signal and significant noise. As mentioned at the beginning of Sect. 2, noise reduction is often addressed in several places in the processing chain. All noise reduction operations seek to preserve as much scene information as possible while smoothing noise. To achieve this efficiently, it is important to use relatively simple models to discriminate between scene modulation and noise modulation. Before the demosaicing step in Fig. 1, it is somewhat difficult to exploit inter-channel correlations for noise reduction. In the stochastic noise reduction block, grayscale techniques for noise reduction are usually applied to each color channel individually. There are many possible approaches, but two families of filtering, based upon different models of the image capture process, are discussed here.

The first of these models represents the random noise in the sensor capture and is used for range-based filtering, such as in a sigma filter [57] or the range component of a bilateral filter [85]. Early in a digital camera processing chain, a fairly simple model for noise variance is effective. There are two primary sources of random noise in the capture chain. The first is Poisson-distributed noise associated with the random process of photons being absorbed and converted into charge within a pixel. The second is electronic read noise, modeled with a Gaussian distribution. These two processes are independent, so a pixel value $Q$ may be modeled as $Q = k_Q(q + g)$, where $k_Q$ is the amplifier gain, $q$ is a Poisson random variable with mean $m_q$ and variance $\sigma_q^2$, and $g$ is a Gaussian random variable with mean $m_g$ and variance $\sigma_g^2$. Because $q$ is a Poisson variable, $\sigma_q^2 = m_q$, and the total variance for a digitized pixel $Q$ is

$$\sigma_Q^2 = k_Q^2(m_q + \sigma_g^2). \tag{1}$$

where $m_q$ is the mean original signal level (captured photocharge), and $\sigma_g^2$ is the read noise. This relationship, that signal variance has a simple linear relationship with code value and a positive offset, allows a very compact parameterization of $\sigma_Q^2$ based upon a limited number of tests to characterize capture noise. Noise reduction based upon the noise levels in captured images allows smoothing of modulations with a high probability of being noise, while not smoothing over (larger) modulations that have a low probability of being noise.

Range-based filtering can introduce two main artifacts in the processed image. The most obvious is loss of fine texture. Since the noise reduction is based on smoothing small modulations and retaining large modulation, textures and edges with low contrast tend to get over-smoothed. The second artifact is the tendency to switch from smoothing to preservation when modulation gets larger. This results in a very nonuniform appearance in textured fields or edges, with portions of the texture being smoothed and other portions being much sharper. In some cases, this can lead to contouring artifacts as well.

The second simple model is an approximation of the capture PSF. Any point in the scene is spread over a finite area on the sensor in a spatially bandlimited capture. Thus, single-pixel outliers, or impulses, are more likely to be caused by sensor noise than scene content. While range-based filtering handles signal-dependent noise fairly well, it is prone to leave outliers unfiltered and it tends to increase the kurtosis of the distribution since it smooths small modulations more than larger modulations. The likelihood that impulses are noise leads to use of impulse filtering noise reduction. A standard center-weighted median filter can be effective, especially with lower cost cameras that have a large enough PSF to guarantee any scene detail will be spread over several pixels in the capture, thus preventing it from appearing as an impulse. More sophisticated approaches may be used for cameras with smaller or variable PSFs, such as digital SLR cameras.

The characteristic artifact caused by impulse filtering is elimination of small details from the scene, especially specular reflections from eyes and small lights. When applying impulse filters to CFA data, the filtering is particularly vulnerable to creating colored highlights, if an impulse is filtered out of one or two color channel(s), but left in the remaining channel(s).

## 2.4 Exposure and White Balance Correction

The human visual system automatically adapts in complex ways when viewing scenes with different illumination. Research in the areas of color appearance models and color constancy continue to focus on developing models for how different scenes appear to human observers under different conditions. Because the human visual system generally has nonlinear responses and operates over a wide range of conditions, this process is extremely complex. A more restricted form of the problem is normally addressed in digital cameras. The goal is to capture neutral scene content with equal responses in all color channels (R=G=B), with the midtones being rendered near the middle of the tone scale, regardless of the illuminant or content of the scene.

White balance adjustment is accomplished by multiplying pixels in each color channel by a different gain factor that compensates for a non-neutral camera response and illuminant imbalance. In a digital camera, exposure adjustment is usually done primarily by controlling exposure time, analog gain, and f/number, but sometimes a digital exposure adjustment is used as well. This is done by scaling all three gain factors by a common factor. Application of the gain factors to the CFA data before demosaicing may be preferred, since some demosaicing algorithms may presume equal responses for the different color channels.

The other part of the exposure and white balance task is estimation or selection of appropriate gain factors to correct for illumination imbalance. Knowledge of the illuminant (and the camera's response to it) is critical. The camera's response to typical illuminants, such as daylight, incandescent, and fluorescent, is easily stored in the camera. Illuminant identification can be done manually through a user interface, which then drives selection of the stored gains for the illuminant. This is generally quite simple for the camera, but more tedious for the user. Another common approach is to allow the user to capture an image of a neutral (gray) under the scene illuminant and have the camera compute gain factors from that image.

The classic and most ill-posed form of the estimation problem is to analyze the image data to estimate the gains automatically, responding to the illuminant, regardless of the scene content. A classic difficult example is a featureless flat image with a reddish cast. Is it a gray card under incandescent illumination, or a reddish sheet of paper under daylight? Fortunately, normal scenes contain more information than a flat field.

Current cameras approach this estimation problem with different algorithms having different responses to scene content and illuminants. Camera manufacturers usually have somewhat different preferences, for example, biasing white balance to render images warmer or cooler, as well as different approaches to estimating the scene illuminant.

Most automatic white balance and exposure algorithms are based on some extension of the gray world model, that images of many different scenes will average out to 18% gray (a midtone gray). Unfortunately, this says very little about a specific image, and the algorithm must work well for individual images. Most extensions of the gray world model try to discount large areas of single colors, to avoid having the balance driven one way or another by red buildings, blue skies, or green foliage. There is also a tendency to more heavily weight colors closer to neutral more than colors far from neutral, but this complicates the algorithm, since the notion of neutral is not well-defined before performing illuminant estimation. This can be approached by applying a calibrated daylight balance to the image for analysis, then analyzing colors in the resulting image [25]. Another extension is to consider several possible illuminant classes and estimate the probability of each illuminant being the actual scene illuminant [24, 65].

Sometimes, highlight colors are given special consideration, based on the theory that highlights are specular reflections that are the color of the illuminant [56, 62, 64]. This breaks down for scenes that have no truly specular highlights. Some approaches also consider the color of particular scene content. The most common of these is using face detection and adjusting balance to provide a reasonable color for the face(s). This has other challenges, because faces themselves vary in color.

Using exposure control information such as scene brightness can help with the illuminant estimation. For example, a reddish scene at high illumination levels is more likely to be an outdoor scene near sunset, while the same scene with dim illumination is somewhat more likely to be indoor illumination. Another example is flash information. If an image is captured primarily with flash illumination, then the illuminant is largely known.

## 2.5 *Adjusted CFA Image*

The adjusted CFA image shown in Fig. 1 is still a single-channel image in which different pixels represent different color channels. It is now conditioned to more cleanly represent the scene, with few sensor-imposed artifacts. It also has less noise than the original capture, and is adjusted to represent the scene, with roughly equal red, green, and blue responses for neutral scene content, correcting out any illuminant imbalance.

## 2.6 *Demosaicing*

Capturing color images using a digital camera requires sensing at least three colors at each pixel location. One of the approaches to capture multiple colors at each pixel is to use a set of imaging sensors and project the scene onto each one of these sensors [58]. However, this increases the cost of the device and also requires careful alignment of

**Fig. 6** Bayer CFA pattern

the sensors to produce a visually pleasing color image. Therefore, to reduce cost and complexity, most digital cameras are designed using a single monochrome CCD or CMOS image sensor with a CFA laid on top of the sensor [31, 58]. The CFA is a set of color filters that samples only one color at each pixel location, and the missing colors are estimated using interpolation algorithms. The CFA interpolation algorithms are also widely referred to as CFA demosaicing or demosaicing algorithms [28, 51, 87].

Among many CFA patterns, the Bayer CFA pattern [15] in Fig. 6 is one of the most commonly used CFA patterns in digital cameras. Since the human visual system is more sensitive to the green portion of the visual spectrum, the Bayer pattern consists of 50% green filters with the remaining 50% filters assigned equally to red and blue colors. The red, green, and blue pixels in the Bayer pattern are arranged as a $2 \times 2$ periodic minimal repeating unit, where each unit consists of two green filters, one red, and one blue filter. An example of a Bayer CFA image and the corresponding CFA interpolated color image is shown in Fig. 7.

Because the quality of the CFA interpolated image largely depends on the accuracy of the demosaicing algorithm, a great deal of attention has been paid to the demosaicing problem. Although simple non-adaptive image interpolation techniques ( e.g., nearest neighborhood, bilinear interpolation, etc.) can be used to interpolate the CFA image, demosaicing algorithms designed to exploit inter-pixel and inter-channel correlations outperform non-adaptive interpolation algorithms as illustrated in Fig. 8. The original (ground truth) and the corresponding Bayer CFA images are shown in Figs. 8a, b, respectively. Three different demosaicing algorithms, namely, (i) nearest-neighborhood interpolation [32], (ii) bilinear interpolation [7], and (iii) directional linear minimum mean-square-error estimation (DLMMSE) [94] were applied to the CFA image and the corresponding demosaiced color images are shown in Fig. 8c–e. Both the nearest-neighborhood and bilinear interpolation algorithms are non-adaptive in nature and as a result, produced aliasing artifacts in the high-frequency regions. However, DLMMSE, due to its adaptive design, reconstructed the color image almost perfectly. For more details on the design and performance of various adaptive and non-adaptive CFA demosaicing algorithms, see [37, 60, 67, 93].

Placement of the CFA on top of the image sensor is essentially a downsampling operation. Therefore, the overall quality of color images (e.g., spatial resolution, color fidelity, etc.) produced by demosaicing algorithms not only depends upon

**Fig. 7** Example color image reconstruction from Bayer pattern. **a** Bayer CFA image, **b** full-resolution CFA interpolated color image

the accuracy of the demosaicing algorithm but is also influenced significantly by the underlying CFA layout. Careful selection of a CFA pattern and corresponding demosaicing algorithm leads to high-quality color image reconstruction.

Although the Bayer pattern is one of the most commonly used CFA patterns, many others such as GGRB, RGBE, CYMM, CYGM, etc., [36, 63, 88, 89] also have been suggested for consumer digital cameras. Primarily influenced by manufacturing constraints and implementation costs, these CFA constructions and the corresponding demosaicing algorithms have been researched extensively. However, a systematic research framework for designing optimal CFA patterns is still a fairly new research direction. Some of the state-of-the-art developments in this area include Kodak's panchromatic CFA [53, 77], second-generation CFA [38], etc. A detailed review of these and other similar CFA patterns is beyond the scope of this chapter and readers are encouraged to refer to [14, 16, 61, 66, 82] for more details.

## 2.7 Interpolated RGB Image

After demosaicing, the interpolated image has three channels, each fully populated. The demosaicing process may have amplified noise in some of the color channels. The white balance process almost assuredly has as well, since gain factors greater than unity are normally used.

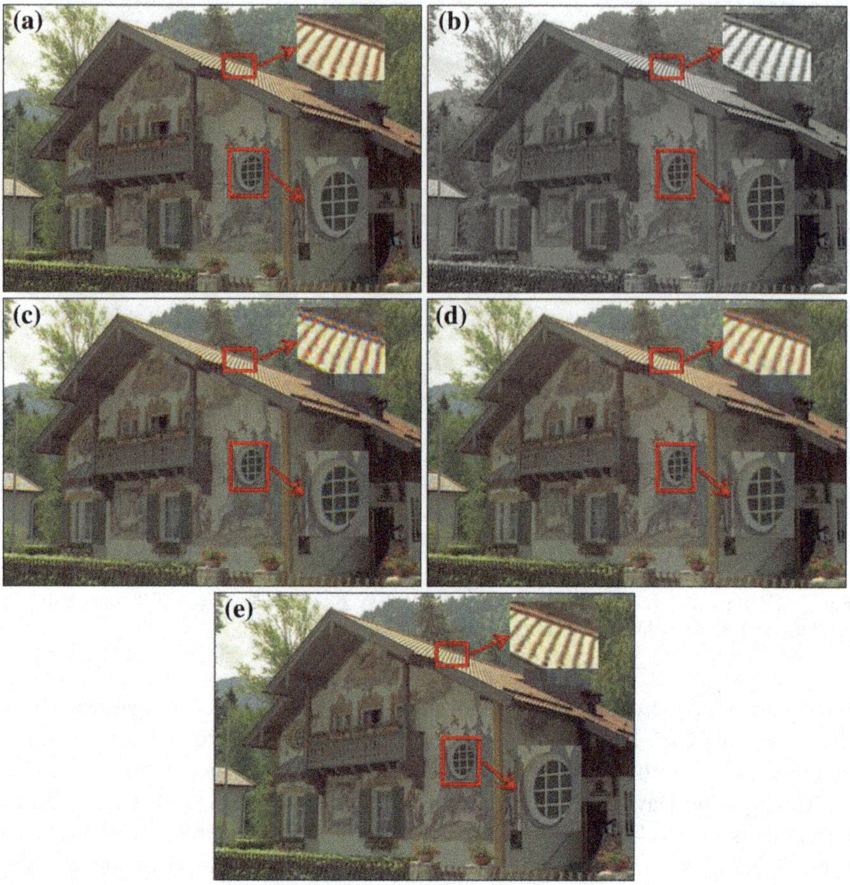


**Fig. 8** Demosaicing algorithm comparisons: **a** original (ground truth) image, **b** Bayer CFA image, **c** nearest-neighborhood interpolation, **d** bilinear interpolation and **e** DLMMSE [94]

## *2.8 Color Noise Reduction*

Section 2.3 discussed noise reduction based upon simple single-channel noise models. This section discusses exploitation of inter-channel correlation and knowledge of the human visual system for further noise reduction. As mentioned before, these concepts are easier to apply after demosaicing, although application of these concepts before and during demosaicing remains a research opportunity.

Once three fully populated color channels are present, it is straightforward to rotate the color image to a luma-chroma color space. Because color correction has not yet been performed, this luma-chroma space will typically not be colorimetrically accurate. Still, a simple uncalibrated rotation such as in (2), similar to one proposed by Ohta [70], suffices to get most of the scene detail into the luma channel and most of the color information into the two chroma channels.

$$\begin{bmatrix} Y \\ C_1 \\ C_2 \end{bmatrix} = \frac{1}{4} \begin{bmatrix} 1 & 2 & 1 \\ -1 & 2 & -1 \\ -2 & 0 & 2 \end{bmatrix} \begin{bmatrix} R \\ G \\ B \end{bmatrix} \tag{2}$$

Once separated, each channel is cleaned based upon the sensitivity of the human visual system. In particular, the luma channel is cleaned carefully, trying to preserve as much sharpness and detail as possible while providing adequate noise reduction. Digital cameras operate over a wide range of gain values, and the optimum balance of noise reduction and texture preservation typically varies with the input noise level. The chroma channels are cleaned more aggressively for several reasons. One reason is that sharp edges in the chroma channels may be color interpolation or aliasing artifacts left over from demosaicing [35]. The second is that viewers are less sensitive to chroma edges and especially sensitive to colored noise. The overall quality of the image is improved by emphasizing smoothness of the chroma channels rather than sharpness.

After noise reduction, this rotation is inverted as in (3).

$$\begin{bmatrix} R \\ G \\ B \end{bmatrix} = \begin{bmatrix} 1 & -1 & -1 \\ 1 & 1 & 0 \\ 1 & -1 & 1 \end{bmatrix} \begin{bmatrix} Y \\ C_1 \\ C_2 \end{bmatrix} \tag{3}$$

Chroma-based noise reduction can produce two signature artifacts, color blobs and color bleed. Color blobs are caused by chroma noise that is smoothed and pushed to lower spatial frequencies in the process, without being eliminated. Successful cleaning at lower spatial frequencies requires relatively large filter kernels or iterative filtering. Implementations with constraints on available memory or processing power tend to leave low-frequency noise unfiltered.

Desktop implementations, with fewer constraints, can use pyramid [5] or wavelet decomposition [17] to reach the lowest frequencies. Chroma-based noise reduction is also prone to color bleed artifacts, caused by substantial mismatch in edge sharpness between the luma and chroma channels. Adaptive techniques that smooth the chroma channels using edge detection techniques such as [1] reduce the color bleeding problem by avoiding smoothing across edges. The visibility of color bleed artifacts depends in part upon the luma-chroma color space used for noise reduction. If the color space is less accurate in separating colorimetric luminance from chrominance data, color bleed artifacts will also affect the lightness of the final image, increasing the visibility of the artifacts. Adaptive chroma noise cleaning that can clean to very low frequencies while avoiding significant color bleeding is an open research question.

Color moiré patterns are sometimes addressed in a processing chain, often treated as a form of color noise. The usual approach is a variation of chroma noise reduction, including an additional test to check for high-frequency textures [3, 4].

## 2.9 Color Correction

After suitable noise reduction, colors are corrected, converting them from (white balanced) camera responses into a set of color primaries appropriate for the finished image. This is usually accomplished through multiplication with a color correction matrix, as in (4).

$$\begin{bmatrix} R_S \\ G_S \\ B_S \end{bmatrix} = \mathbf{C} \begin{bmatrix} R_C \\ G_C \\ B_C \end{bmatrix} \tag{4}$$

In this equation, $\mathbf{C}$ is a $3 \times 3$ matrix with coefficients determined to convert from the camera's white balanced native sensitivity into a standard set of color primaries, such as sRGB [8], ROMM [81], or Adobe RGB (1998) [6].

One characteristic of color correction is the location of colors in the finished image color space. For example, the color of blue sky, skin, and foliage in the rendered image can vary from manufacturer to manufacturer. Different camera designers usually choose different objectives when doing this primary conversion; this can be thought of as combining color correction and preferred color rendering. Considerations affecting the determination of the color matrix are discussed in more depth by Hunt [40] and also Giorgianni and Madden [27]. If a standard color space with relatively small gamut, such as sRGB, is chosen, then many saturated colors will be outside the gamut. Depending upon the gamut-mapping strategy chosen, this can introduce clipping or gamut-mapping artifacts. The color correction matrix will usually amplify noise in the image. Sometimes, color correction is deliberately desaturated to reduce the noise in the rendered image.

If a more complex color correction transform is chosen, such as one that increases color saturation but preserves flesh tones at a less saturated position, then some nonuniform noise characteristics and even contouring may be observable.

## 2.10 Tone Scale and Gamma Correction

After color correction, a tone scale is applied to convert the image, still linear with respect to scene exposure, into a final color space. Sometimes, this is referred to as gamma correction, although most processing chains apply additional contrast adjustment beyond simple gamma correction. Along with color correction, the choice of the tone scale for rendering a reproduction of a scene is complex. As with color correction, issues involved in this selection are discussed in more depth by Hunt [40] and Giorgianni and Madden [27].

The most common processing chains simply apply a tone scale to all pixels in the image as a look-up table operation, regardless of scene content. Consumer preference is usually for a higher contrast tone scale, as long as no significant scene information is lost in the process. This is somewhat dependent upon scene and user expectation;

professional portraits are an example where the preference is for a lower contrast look.

More recently, processing chains are using adaptive tone scales that are adjusted for each scene. When such adjustments become aggressive (bringing up shadow detail, compressing highlight range), image texture becomes unnatural. If the contrast in the shadows is stretched, noise is amplified, leading to image quality degradation. If the highlights are compressed, texture in the highlights is flattened, leading to a different quality degradation.

The more sophisticated processing chains apply the adaptive tone scale with spatial processing. The approach is usually to use a multilevel decomposition to transform the image into a base image containing low spatial frequencies and a detail or texture image [26, 30]. Once the image is decomposed into base and detail images, the tone scale adjustments are applied to the base image with no changes or with controlled changes in detail. When the decomposition into base and texture images is imperfect, this approach can cause halo effects near high-contrast edges where the base image is adjusted extensively. In extreme cases, the image is rendered with an artificial decoupling of base image and texture, looking more like an animation rather than a photographic image. Image decomposition and derivation of spatially adaptive tone scales for optimal rendering of images are open areas of research [10, 20, 22].

## 2.11 Edge Enhancement

During image capture, edge detail is lost through optical and other effects. Most processing operations, especially noise reduction, further reduce high spatial frequency content. Display devices, printers, and the human visual system also attenuate system response. Edge enhancement, also known as sharpening, is a relatively simple spatial operation to improve the appearance of images, making it appear sharper and partially compensating for these losses.

The core of routine edge enhancement is a convolution operation to obtain an edge image, which is then scaled and added to the original image, as in (5).

$$\mathbf{A}' = \mathbf{A} + k\mathbf{E} \tag{5}$$

In this equation, $\mathbf{A}'$ is the enhanced image, $\mathbf{A}$ is the color-corrected image from the previous processing stage, $k$ is a scalar gain, and $\mathbf{E}$ is the edge enhancement image. The key variations in the process lie in the creation of $\mathbf{E}$. This is normally done with a standard spatial convolution, as in (6).

$$\mathbf{E} = \mathbf{A} * *\mathbf{h} \tag{6}$$

In this equation, $\mathbf{h}$ is a high pass convolution kernel. In other implementations, an unsharp mask formulation is chosen, as in (7).

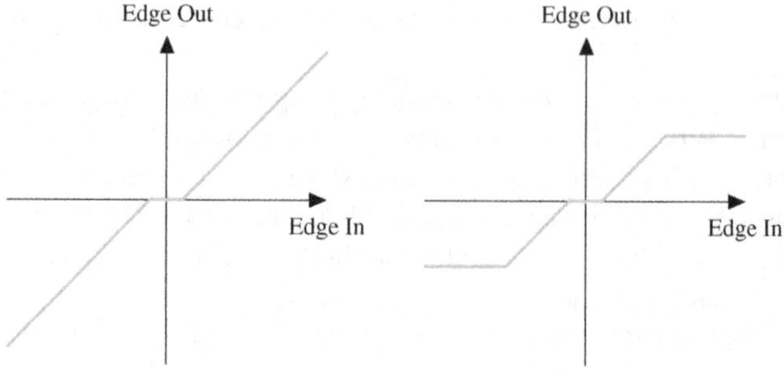


**Fig. 9** Example edge enhancement nonlinearities: **a** soft thresholding, **b** edge limiting

$$\mathbf{E} = \mathbf{A} - \mathbf{A} * *\mathbf{b} \tag{7}$$

In this equation, **b** is a low-pass convolution kernel. If **h** in (6) is chosen to be $\mathbf{h} = \mathbf{I} - \mathbf{b}$, then (6) and (7) are identical.

In both implementations, the design of the convolution kernel and the choice of $k$ are the main tuning parameters. The design of the kernel controls which spatial frequencies to enhance, while $k$ controls the magnitude of the enhancement. Often, the kernel is designed to produce a band-pass edge image, providing limited gain or even zero gain at the highest spatial frequencies. In most practical implementations, the size of the kernel is relatively small, with $5 \times 5$ being a common size.

Even with a band-pass kernel, this formulation amplifies noise and can produce significant halo or ringing artifacts at high-contrast edges. Both these problems can be treated by applying a nonlinearity to the edge image before scaling and adding to the original image, as in (8) or (9).

$$\mathbf{E} = L\left(\mathbf{A} * *\mathbf{h}\right) \tag{8}$$

$$\mathbf{E} = L\left(\mathbf{A} - \mathbf{A} * *\mathbf{b}\right) \tag{9}$$

Figure 9 shows two example nonlinearities, both based upon the magnitude of the edge value. The edge values with the smallest magnitude are most likely to be the result of noise, while larger edge values are likely to come from scene edges. The soft thresholding function shown in Fig. 9a reduces noise amplification by reducing the magnitude of all edge values by a constant, and is widely used for noise reduction, such as in [17]. Soft thresholding eliminates edge enhancement for small modulations, while continuing to enhance larger modulations. The edge-limiting nonlinearity shown in Fig. 9b also limits halo artifacts by limiting the edge value at the largest edge values, since high-contrast edges are the most likely to exhibit halo artifacts after edge enhancement.

Application of edge enhancement in an RGB color space will tend to amplify colored edges caused by any capture or processing artifacts earlier in the capture chain, as well as colored noise. For this reason, edge enhancement is often applied to the luma channel of an image rather than all three color channels, for the same reasons that noise reduction is often applied in a luma-chroma color space. With more aggressive edge enhancement, different artifacts can be caused by the choice of different color spaces. Selection of a carefully chosen luma-chroma space tends to minimize artifacts, although aggressive edge enhancement can still lead to color bleeding problems if luma edges are enhanced too much more than chroma edges.

Selection of a color space for edge enhancement can depend upon several factors. For chains planning to apply JPEG compression in a YCbCr color space, the Y channel is a natural choice. In other cases, the green channel may be an acceptable luma channel—it is often the channel with the lowest noise and highest captured spatial resolution. This choice can produce artifacts with edges of different colors, however, since some luma edges will not appear in the green channel.

## *2.12 Finished Image*

The finished image is ready for viewing, in a standard color space such as sRGB. It is possible to store this image in a file with no compression, such as in a TIFF file. In practice, this is very uncommon, since images from digital cameras are precisely the sort of natural photographic image for which lossy compression techniques such as JPEG were defined.

## *2.13 Compression*

Once an image is fully processed, it is often compressed to reduce the amount of physical storage space required to represent the image data. Image compression algorithms can be divided into two categories: lossy and lossless. Lossless image compression algorithms are reversible, meaning that the exact original image data can be recovered from the compressed image data. This characteristic limits the amount of compression that is possible. Lossy image compression algorithms allow some of the original image data to be discarded, and only an approximation to the original image is recovered from the compressed image data. Many image compression algorithms have been proposed both academically and commercially, but digital cameras predominantly utilize the JPEG image compression standard, and in particular a baseline implementation of the JPEG lossy image compression standard [75].

The fundamental building blocks of a lossy JPEG image compression algorithm are illustrated in Fig. 10. Decompression can be achieved by inverting the operations and performing them in the reverse order. These building blocks, along with a common preprocessing step, are described below.

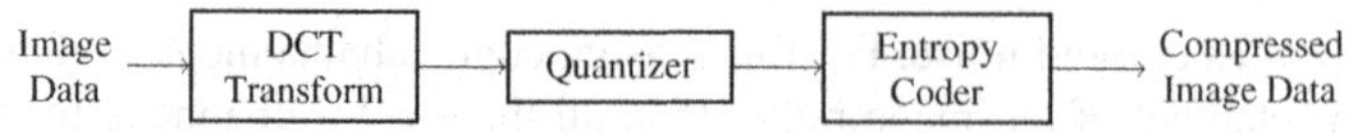


**Fig. 10** JPEG lossy encoder building blocks

### 2.13.1 Preprocessing

JPEG can be used for single-component (channel) images as well as multi-component images. Each component is compressed separately, however, and thus it is advantageous when compressing a three-component RGB image to first reduce the correlation among the components by converting to a luma-chroma space, such as YCbCr. This conversion allows for more efficient compression since the components have less redundancy among them. Additionally, the JPEG standard allows some variability in the size of each component, and commonly the Cb and Cr components are subsampled by a factor of 2 horizontally and vertically as a preprocessing step.

### 2.13.2 DCT Transform

Each component is divided into a grid of $8 \times 8$ blocks, and a two-dimensional discrete cosine transform (DCT) is applied to each $8 \times 8$ block of image data. This operation, shown in (10), converts pixel values into transform coefficients corresponding to spatial frequencies.

$$X(u, v) = \frac{C(u)C(v)}{4} \sum_{m=0}^{7} \sum_{n=0}^{7} x(m, n) \cos \frac{(2m+1)u\pi}{16} \times \cos \frac{(2n+1)v\pi}{16}, \tag{10}$$

where

$$C(u) = \begin{cases} 1/\sqrt{2} & u = 0 \\ 1 & 1 \le u \le 7. \end{cases} \tag{11}$$

In (10), $x(m, n)$ are the pixel values for a given block, $X(u, v)$ are the corresponding DCT transform coefficients, and $0 \le u, v \le 7$.

For smoothly varying natural imagery composed mostly of low frequencies, the DCT compacts the majority of the energy for a given block into a small subset of the transform coefficients corresponding to low spatial frequencies (small values of $u$ and $v$ in (10)). Given infinite data precision, the DCT is invertible and the inverse DCT is applied during decompression to recover pixel values from the transform coefficients.

The block-based nature of the DCT used in JPEG can lead to a specific type of artifact known as a blocking artifact. Each $8 \times 8$ block of image data is compressed separately, and at high levels of compression, artificial transitions can appear at the boundary between two blocks. Since the image data is partitioned into a uniform grid of $8 \times 8$ blocks, the location of potential blocking artifacts is known in a JPEG image. The presence of blocking artifacts at locations other than on the boundaries of $8 \times 8$ blocks suggests that an image has been modified in some way, such as by cropping. Blocking artifacts are a well-known characteristic of JPEG images, and many post processing techniques have been proposed to address them ([86] and references therein).

### 2.13.3 Quantization

Quantization is the lossy step of the JPEG algorithm. It refers to the "many-to-one" mapping of each input transform coefficient into one of a finite number of output levels. This is achieved by dividing each transform coefficient by the corresponding element of a quantization matrix (or quantization table) and rounding the result, as in (12).

$$Q(X(u, v)) = \text{round}\left(\frac{\text{X}(\text{u}, v)}{\text{q}(\text{u}, v)}\right). \tag{12}$$

In (12), $q(u, v)$ are the quantization table entries, and $Q(X(u, v))$ are the quantized transform coefficients. Larger values in the quantization table correspond to coarser quantization and greater compression. They also correspond to greater uncertainty and hence greater expected error when reconstructing the transform coefficients from their quantized values during decompression.

Quantization tables can be designed with regard to the human visual system. Because of the decreased sensitivity of the human visual system at high spatial frequencies, transform coefficients corresponding to high spatial frequencies can be quantized more aggressively than transform coefficients corresponding to low spatial frequencies. For multi-component images, JPEG allows multiple quantization tables. In particular, for YCbCr images, it is common to use one quantization table for the luma component (Y), and a separate quantization table for the chroma components (CbCr), exploiting the varying sensitivity of the human visual system to luma and chroma information.

The quantization tables used in the encoding process are included in an image file along with the compressed image data so that a decoder can correctly invert the quantization process. Quantization provides a rich source of information for forensic analysis of digital images. Different camera manufacturers use different quantization tables when generating JPEG images, and thus the values contained in the quantization tables can provide some information about the origin of a digital image. Additionally, requantization of an image, such as can happen when a compressed image is decompressed, modified, and recompressed, can result in unusual statistics in the compressed data [13].

### 2.13.4 Entropy Coding

The final main building block of a JPEG encoder involves entropy coding, which is a lossless process designed to efficiently encode a collection of symbols. In the case of JPEG, the symbols are related to the quantized transform coefficient values. Huffman codes are used as the entropy codes for the baseline implementation of JPEG lossy compression [39]. Broadly speaking, Huffman codes apply small codewords to represent symbols that occur frequently, and longer codewords to represent symbols that occur infrequently. The quantized transform coefficient values for an $8 \times 8$ block very often contain many zero values, in particular for coefficients corresponding to high spatial frequencies, and thus the Huffman codes are designed to efficiently represent these zeros (actually encoded as sequences, or runs, of zeros) with short codewords. During decompression, the entropy coding process can be exactly reversed to recover the quantized transform coefficient values.

# 3 Storage Formatting

Image storage formats are standardized means for organizing and storing digital images. They provide specifications for including image data and metadata in an image file. Metadata is any type of information that relates to the image, such as the camera model and manufacturer, the image size, and the date and time of the image capture. Widespread adoption of standardized image storage formats has ensured compatibility and interoperability among devices that capture and use digital images.

Current digital cameras almost uniformly use the Exif-JPEG image storage format to store JPEG-compressed image data [46, 47]. In some cases, however, JPEG-compressed data is not sufficient, and other standards, such as the TIFF/EP image storage format, are used to store raw image data [41].

## *3.1 Exif-JPEG*

The Exif-JPEG image storage format provides a standard representation of digital images that allows compatibility between digital cameras and the devices and software applications that use the digital images produced by the cameras. The metadata associated with Exif-JPEG files is defined in the Exif specification. This metadata includes general information about the digital camera, as well as specific information about the camera capture settings used for a particular image [73]. Some examples of this metadata are shown in Table 1.

It is possible to have proprietary metadata in an Exif-JPEG image file. Individual camera manufacturers use proprietary metadata to include private information in the image file, such as the serial number of the camera. It may also include information about image processing settings used to generate the final image. Often camera

**Table 1** Examples of metadata contained in Exif-JPEG image files

| Metadata field name | Description |
|---|---|
| Make | Name of the manufacturer of the digital camera |
| Model | Model name/number of the digital camera |
| DateTime | Date and time the file was last modified |
| Exposure time | The time duration that the image was exposed on the image sensor |
| Fnumber | The lens f/number used when the image was captured |
| Flash | Provides information about whether the camera flash was used at capture |
| DateTimeOriginal | The data and time the picture was taken by the camera |
| MakerNote | Different camera makers store a variety of custom information |

manufacturers use the "MakerNote" metadata field to store proprietary metadata, and encrypt the contents to protect the proprietary nature of the information.

In general, it can be difficult to identify the source digital camera from which a given Exif-JPEG image file was produced. The camera image processing is designed to remove camera-specific artifacts from the digital image itself. Metadata may indicate the camera make and model, but not necessarily the specific camera used.

Exif-JPEG image files have drawbacks, and are not appropriate in all scenarios. The image is restricted to 24-bit color, using only 8 bits for each color component. The lossy JPEG compression algorithm can introduce artifacts. Additionally, in-camera image processing algorithms used to generate the final image may have sacrificed image quality for computational efficiency.

In some applications, it may be desirable to retain the raw image directly from the image sensor. The raw image data often has greater bit-depth, with 12 or 16 bits per pixel. Additionally, if the raw image is subsequently processed in a more powerful computing environment than available in the digital camera, more complex image processing algorithms can be used to generate the final image.

### *3.2 TIFF/EP*

TIFF/EP was the first image format standardized for storing raw image data. In order to be able to properly interpret raw color image data, it is necessary to include metadata along with the image data that describes features such as the CFA pattern, and the color responses of the color channels. Table 2 shows some examples of the metadata associated with a raw color image included in a TIFF/EP file. In addition to the metadata specific to raw color image data, a TIFF/EP image file also contains metadata described previously with respect to Exif-JPEG image files.

Since the image data contained in a TIFF/EP image file has not been fully processed by the digital camera image processing path, it may still retain characteristics and artifacts described previously in this chapter that can be used to identify the specific camera from which the image was produced. Noise patterns and defec-

**Table 2** Examples of metadata contained in TIFF/EP image files

| Metadata field name | Description |
|---|---|
| Image width | Width of CFA image data |
| Image length | Length of CFA image data |
| Bits per sample | Bit-depth of each CFA sample |
| CFA pattern | The color filter array pattern |
| ICC color profile | Used to define RGB reference primaries, white point, and opto-electronic conversion function |

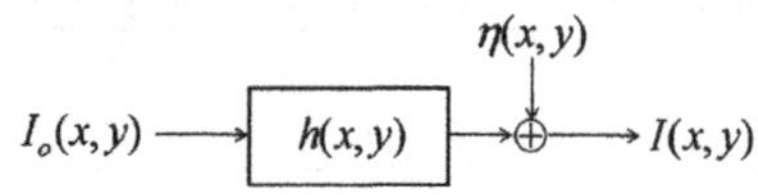


**Fig. 11** Image acquisition model for image restoration

tive data in the raw color image can be used to assess the likelihood that an image was generated by a particular camera.

One drawback of the TIFF/EP and other raw image storage formats is that while the format of the raw data is standardized, the image processing algorithms used to generate the final color image are not. Many possible finished images can be produced with different processing. Often, the image processing algorithms are proprietary, and the finished image may be of unknown quality if alternative processing algorithms are used.

# 4 Post-Processing Enhancement Algorithms

Despite significant advances in the areas of optics and semiconductors, digital cameras do not always provide *picture-perfect* images. Factors that affect image quality include sensor size, limited depth of field of the optical lenses, light condition, relative motion between the camera and the scene, etc. Image restoration techniques are commonly employed to restore the features of interest in the image [12, 29, 83]. However, image restoration techniques are (in general) computationally complex, which makes it impractical to use them during the image acquisition process. Therefore, image restoration is performed as a post processing step.

Classically, image restoration algorithms model the degradation process (such as motion blur, noise, out-of-focus blur, etc.) as a linear, space invariant system [29] as shown in Fig. 11. In this figure, $I_o(x, y)$ and $I(x, y)$ are (high-quality) unknown and (degraded) observed images, respectively, $h(x, y)$ is the system PSF and $\eta(x, y)$ is the measurement noise added during image acquisition.

The relationship between $I_o(x, y)$ and $I(x, y)$ can be expressed as a convolution operation as shown in ( 13) below,

$$I(x, y) = h(x, y) \ast\ast I_o(x, y) + \eta(x, y), \qquad (13)$$

where "$**$" denotes the 2-$D$ convolution operation. Typically, inverse techniques are used to estimate $I_o(x, y)$ from (13) and the most commonly used technique is deconvolution [9]. In cases where the degradation PSF $h(x, y)$ is known, a Wiener filter-based approach can be used [44]. In most practical applications, the only known quantity is $I(x, y)$. In such cases, a blind deconvolution technique [55] is necessary to estimate both $I_o(x, y)$ and $h(x, y)$. If the degradation process is spatially variant (e.g., out-of-focus blur, etc.), then (13) is applied locally, *i.e.*, the observed image $I(x, y)$ is divided into several regions and ( 13) is applied to each region.

In some cases, perfect reconstruction of $I_o(x, y)$ from $I(x, y)$ is difficult even if $h(x, y)$ is known completely. For example, in the case of motion blur, $I_o(x, y)$ cannot be reconstructed completely if zeros of $h(x, y)$ in the frequency domain coincide with the nonzero frequency components of $I_o(x, y)$. Estimation of $I_o(x, y)$ can be improved significantly by using a multichannel approach. In the multichannel setup, multiple images of the same scene with complementary characteristics (e.g., multiple exposures, flash/no-flash, etc.) are acquired and $I_o(x, y)$ is reconstructed by using all of the observed images simultaneously [54, 59, 80].

An example to illustrate the advantages of multichannel image restoration is shown in Fig. 12. The original "cameraman" image in Fig. 12a was blurred using three different types of motion blur filters—(i) motion filter of length 5 and angle 13 (capture 1 in Fig. 12b), (ii) motion filter of length 6 and angle 120 (capture 2 in Fig. 12c), and (iii) motion filter of length 6 and angle 90 (capture 3 in Fig. 12d). The results of the Richardson–Lucy blind deconvolution applied to each capture and the multichannel blind restoration proposed by Kumar et al. [54] are shown in Fig. 12e–h. Clearly, the proposed multichannel algorithm performs better than single-channel deconvolution.

Multichannel image restoration requires complementary images of the same scene. A detailed description of capturing multiple images and processing them simultaneously for various image processing tasks such as deblurring, denoising, etc., can be found in [19, 76, 79, 92].

Image restoration algorithms enhance salient image features but they are not capable of improving the spatial resolution of an image. In many image processing applications such as image forensics, automatic target recognition (ATR), remote sensing, etc., images with high resolution (HR) are desired and often required. Image super-resolution [21, 23, 69, 72, 74] is a signal processing technique to obtain an HR image (or sequence) from a number of observed low-resolution (LR) images or a LR video sequence captured from the same scene. Image super-resolution is technically possible because of the information contained collectively among the LR images. For example, the spatial misalignment of the LR images, due to spatial sampling on an integer lattice, introduces sub-pixel shifts, from which the lost spatial high-frequency components can be estimated. Additional information can also be incorporated, such as the prior knowledge of a scene and an imaging degradation model. The processed image has a higher spatial resolution and reveals more content details.

**Fig. 12** Cameraman image: **a** original, **b** capture 1, **c** capture 2, **d** capture 3, **e** restored image from capture 1 only, **f** restored image from capture 2 only **g** restored image from capture 3 only and **h** multichannel restoration from all the three captures [54]

## 5 Video Capture

Processing for video capture is normally a simplified form of the processing used for still image capture. The same basic processing blocks listed in the nominal processing chain in Fig. 1 are still required, although many of them are simplified. The primary reason for this simplification is the time constraint on the operations. Processing for a still capture can take from a small fraction of a second to many seconds, depending upon the size of the image and sophistication of the processing. Even when a rapid sequence of still images is captured, the normal practice is to buffer the raw images in memory and process the images after capture. In contrast, video capture is performed at a fixed frame rate, such as 30 frames per second (fps), for an indefinite period of time. This precludes buffering approaches such as used with still capture, and requires the processing path to keep up with the capture rate.

One advantage for a video processing path is the limited size of the frames, normally two megapixels (1920 × 1080) or less, compared with five megapixels or much more for still capture. Further, because each image in the video is normally viewed only fleetingly, the image quality of each frame need not be as high.

On the other hand, there are several challenges not present with full-resolution processing. One is the problem of aliasing and colored edge artifacts. Use of binning to limit aliasing in LR readout was discussed in the previous chapter. One side effect of binning is the readout of of color pixels whose spatial relationship (non-overlapping, partially overlapping) depends upon the the binning pattern. As a result, a video processing path must deal with raw data that can have more complex spatial relationships between pixels than is usual with full-resolution readout.

Several additional issues not relevant to a single image come to light when processing a sequence of images. One is the use of temporal noise reduction. Approaches such as [45, 50] can provide significant benefit, although processing budgets usually constrain them to be relatively simple. In addition, temporal compression techniques, such as MPEG or H.264, are normally used with video, although some processing chains simply use lossy JPEG compression on individual frames [42, 43]. This greatly increases the bit rate for a given level of quality, but saves a great deal of compression processing.

Another processing block of potential interest with video forensics is exposure control. Because exposure control is essentially a feedback control loop, the response of exposure control and automatic white balance to changes in the scene can vary with camera model.

The response of the camera to flickering illuminants can also be revealing, especially if a rolling shutter is employed. Bands caused by illuminant flicker interacting with a rolling shutter exposure are a form of gain artifact that is sometimes treated in the processing path. The standard flickering illuminants oscillate at two times the AC power line frequency. Because the AC frequency is one of two values, flicker detection can be approached as a classification problem, analyzing a video or preview stream for evidence of flicker-induced bands at either 120 or 100 Hz [48, 78].

The most common processing path treatment is really *avoidance* of flicker bands, by using an exposure time that is an integer multiple of the flicker period. If flicker is determined to be a problem and not avoidable through selection of exposure time, it can be mitigated using gain corrections to correct the banding, such as in [11]. The resulting variations in digital gain can lead to noise amplification in bands of the image. As with other gain corrections, suitable noise reduction algorithms allow mitigation of the noise, although usually with an increase in noise reduction artifacts.

**Acknowledgments** The authors gratefully acknowledge many helpful discussions with James E. Adams, Jr., in the development and review of this chapter. Ken Parulski also provided a great deal of assistance for the discussion of file formats.

## References

1. Adams JE Jr, Hamilton JF Jr (2003) Removing chroma noise from digital images by using variable shape pixel neighborhood regions. U.S. Patent 6621937
2. Adams JE Jr, Hamilton JF Jr (2008) Single-sensor imaging: methods and applications for digital cameras. Chapter, digital camera image processing chain design, 1st edn. CRC Press, Boca Raton, pp 67–103.
3. Adams JE Jr, Hamilton JF Jr, Hamilton JA (2004) Removing color aliasing artifacts from color digital images. U.S. Patent 6804392
4. Adams JE Jr, Hamilton JF Jr, Smith CM (2005) Reducing color aliasing artifacts from digital color images. U.S. Patent 6927804
5. Adams JE Jr, Hamilton JF Jr, Williams FC (2007) Noise reduction in color digital images using pyramid decomposition. U.S. Patent 7257271
6. Adobe RGB (1998) Color image encoding. Technical report Adobe Systems, Inc., San Jose. http://www.adobe.com/adobergb
7. Alleysson D, Susstrunk S, Herault J (2005) Linear demosaicing inspired by the human visual system. IEEE Trans Image Process 14(4):439–449
8. Anderson M, Motta R, Chandrasekar S, Stokes M (1995) Proposal for a standard default color space for the internet: sRGB. Fourth IS&T/SID color imaging conference, In, pp 238–245
9. Andrews HC, Hunt BR (1977) Digital image restoration. Prentice Hall, Englewood Cliffs
10. Bae S, Paris S, Durand F (2006) Two-scale tone management for photographic look. ACM Trans. Graph. 25(3):637–645 (2006). http://doi.acm.org/10.1145/1141911.1141935
11. Baer R (2007) Method and apparatus for removing flicker from images. U.S. Patent 7298401
12. Bando Y, Nishita T (2007) Towards digital refocusing from a single photograph. In: Proceedings of the 15th Pacific conference on computer graphics and applications, pp 363–372.
13. Bauschke HH, Hamilton CH, Macklem MS, McMichael JS, Swart NR (2003) Recompression of JPEG images by requantization. IEEE Trans Image Process 12(7):843–849
14. Bawolek E, Li Z, Smith R (1999) Magenta-white-yellow (MWY) color system for digital image sensor applications. U.S. Patent 5914749
15. Bayer BE (1976) Color imaging array. U.S. Patent 3971065
16. Bean J (2003) Cyan-magenta-yellow-blue color filter array. U.S. Patent 6628331
17. Chang S, Yu B, Vetterli M (2000) Adaptive wavelet thresholding for image denoising and compression. IEEE Trans Image Process 9(9):1532–1546
18. CIE publ. no. 15.2 (1986) Colorimetry. Techical report, CIE
19. Debevec PE, Malik J (1997) Recovering high dynamic range radiance maps from photographs. ACM SIGGRAPH, In, pp 369–378

20. Durand F, Dorsey J (2002) Fast bilateral filtering for the display of high-dynamic-range images. In: SIGGRAPH '02: Proceedings of the 29th annual conference on computer graphics and interactive techniques, pp 257–266. ACM, New York, NY, USA. http://doi.acm.org/10.1145/566570.566574
21. Elad M, Feuer A (1997) Restoration of a single superresolution image from several blurred, noisy, and undersampled measured images. IEEE Trans Image Process 6(12):1646–1658
22. Farbman Z, Fattal R, Lischinski D, Szeliski R (2008) Edge-preserving decompositions for multi-scale tone and detail manipulation. In: SIGGRAPH '08: ACM SIGGRAPH 2008 papers, pp 1–10. ACM, New York, NY, USA. http://doi.acm.org/10.1145/1399504.1360666
23. Farsiu S, Robinson MD, Elad M, Milanfar P (2004) Fast and robust multiframe super resolution. IEEE Trans Image Process 13(10):1327–1344
24. Finlayson G, Hordley S, HubeL P (2001) Color by correlation: a simple, unifying framework for color constancy. IEEE Trans. Pattern Anal. Mach. Intell. 23(11):1209–1221
25. Gindele E, Adams JE Jr, Hamilton JF Jr, Pillman BH (2007) Method for automatic white balance of digital images. U.S. Patent 7158174
26. Gindele EB, Gallagher AC (2001) Method for adjusting the tone scale of a digital image. U.S. Patent 6275605
27. Giorgianni EJ, Madden TE (2008) Digital color management encoding solutions, 2nd edn. Wiley, New York
28. Glotzbach J, Schafer R, Illgner K (2001) A method of color filter array interpolation with alias cancellation properties. In: Proceedings of the IEEE International Conference Image Processing, vol 1. pp 141–144
29. Gonzalez RC, Woods RE (2007) Digital image processing, 3rd edn. Prentice Hall, Englewood Cliffs
30. Goodwin RM, Gallagher A (1998) Method and apparatus for areas selective exposure adjustment. U.S. Patent 5818975
31. Gunturk B, Glotzbach J, Altunbasak Y, Schafer R, Mersereau R (2005) Demosaicking: color filter array interpolation. IEEE Signal Process Mag 22(1):44–54
32. Gupta M, Chen T (2001) Vector color filter array demosaicing. In: Proceedings of the SPIE, vol 4306. pp 374–382
33. Hamilton JF Jr (2004) Correcting for defects in a digital image taken by an image sensor caused by pre-existing defects in two pixels in adjacent columns of an image sensor. U.S. Patent No. 6741754
34. Hamilton JF Jr (2005) Correcting defect in a digital image caused by a pre-existing defect in a pixel of an image sensor. U.S. Patent No. 6900836
35. Hamilton JF Jr, Adams JE Jr (2003) Correcting for chrominance interpolation artifacts. U.S. Patent 6542187
36. Hamilton JF Jr, Adams JE Jr, Orlicki D (2001) Particular pattern of pixels for a color filter array which is used to derive luminanance and chrominance values. U.S. Patent 6330029B1
37. Hirakawa K, Parks T (2005) Adaptive homogeneity-directed demosaicing algorithm. IEEE Trans Image Process 14(3):360–369
38. Hirakawa K, Wolfe P (2007) Spatio-spectral color filter array design for enhanced image fidelity. In: Proceedings of the ICIP, pp II-81-II-84
39. Huffman DA (1952) A method for the construction of minimum-redundancy codes. Proc Inst Radio Eng 40(9):1098–1101
40. Hunt R (1987) The reproduction of colour. Fountain Press, England
41. ISO 12234-2:2001 (2001) Electronic still picture imaging—removable memory—part 2: TIFF/EP image data format
42. ISO/IEC 14496-10:2003 (2003) Information technology—coding of audio-visual objects—part 10: advanced video coding
43. ISO/IEC 14496-2:1999 (1999) Information technology—coding of audio-visual objects—part 2: visual
44. Jain A, Ranganath S (1981) Application of two dimensional spectral estimation in image restoration. In: IEEE International Conference Acoust., Speech, Signal Process (ICASSP), pp 1113–1116

45. Jain C, Sethuraman S (2008) A low-complexity, motion-robust, spatio-temporally adaptive video de-noiser with in-loop noise estimation. Proceedings of the ICIP, In, pp 557–560
46. JEITA CP 3451 (2002) Exchangeable image file format for digital still cameras: Exif version 2.2
47. JEITA CP 3451-1 (2003) Amendment 1 exchangeable image file format for digital still cameras: Exif version 2.21 (amendment to version 2.2)
48. Kaplinsky M, Subbotin I (2006) Method for mismatch detection between the frequency of illumination source and the duration of optical integration time for image with rolling shutter. U.S. Patent 7142234
49. Kim HS, Kim BG, Kim CY, Joo YH (2008) Digital photographing apparatus and method for detecting and correcting smear by using the same. U.S. Patent publication no. 2008/0084488 A1
50. Kim YJ, Oh JJ, Choi BT, Choi SJ, Kim ET (2009) Robust noise reduction using a hierarchical motion compensation in noisy image sequences. In Digest of Technical Papers of the ICCE, pp 1–2
51. Kimmel R (1999) Demosaicing: image reconstruction from color CCD samples. IEEE Trans Image Process 8(9):1221–1228
52. Kondo K (2009) Image sensing apparatus, image sensing method, and program. U.S. Patent 7545420
53. Kumar M, Morales E, Adams JE Jr, Hao W (2009) New digital camera sensor architecture for low light imaging. Proceedings of the ICIP, In, pp 2681–2684
54. Kumar M, Ramuhalli P (2005) Dynamic programming based multichannel image restoration. In: IEEE International Conference Acoustics, Speech, Signal Process (ICASSP), vol 2. pp 609–612
55. Kundur D, Hatzinakos D (1996) Blind image deconvolution. IEEE Signal Process Mag 13(3):43–64
56. Lee H (1986) Method for computing the scene-illuminant chromaticity from specular highlights. J Opt Soc Am A 3:1694–1699
57. Lee J (1983) Digital image smoothing and the sigma filter. Comput Vis Graph Image Process 24:255–269
58. Li X, Gunturk B, Zhang L (2008) Image demosaicing: a systematic survey. In: SPIE Conference on VCIP, vol 6822, pp. 68, 221J-68, 221J-15
59. Liu X, Gamal AE (2001) Simultaneous image formation and motion blur restoration via multiple capture. In: IEEE International conference acoustics, speech, signal process (ICASSP), vol 3. pp 1841–1844
60. Lu W, Tan YP (2003) Color filter array demosaicking: new method and performance measures. IEEE Trans Image Process 12(10):1194–1210
61. Lukac R, Plataniotis K (2005) Color filter arrays: design and performance analysis. IEEE Trans Consum Electron 51(4):1260–1267
62. McCann JJ, McKee SP, Taylor TH (1976) Quantitative studies in retinex theory. Vis Res 16:445–458
63. Miao L, Qi H, Snyder W (2004) A generic method for generating multispectral filter arrays. In: Proceedings of the ICIP, vol 5. pp 3343–3346
64. Miyano T (1997) Auto white adjusting device. U.S. Patent 5659357
65. Miyano T, Shimizu E (1997) Automatic white balance adjusting device. U.S. Patent 5644358
66. Moghadam A, Aghagolzadeh M, Radha H, Kumar M (2010) Compressive demosaicing. In: Proceedings of the MMSP, pp 105–110
67. Mukherjee J, Parthasarathi R, Goyal S (2001) Markov random field processing for color demosaicing. Pattern Recognit. Lett. 22(3–4):339–351
68. Multiple output sensors seams correction (2009) Technical report application note revision 1.0 MTD/PS-1149, Eastman Kodak image sensor solutions
69. Ng MK, Bose NK (2003) Mathematical analysis of super-resolution methodology. IEEE Signal Process Mag 20(3):62–74

70. Ohta YI, Kanade T, Sakai T (1980) Color information for region segmentation. Comput Graph Image Process 13(3):222–241
71. Okura Y, Tanizoe Y, Miyake T (2009) CCD signal processing device and image sensing device. U.S. Patent publication no. 2009/0147108 A1
72. Park SG, Park MK, Kang MG (2003) Super-resolution image reconstruction: a technical overview. IEEE Signal Process Mag 20(3):21–36
73. Parulski KA, Reisch R (2008) Single-sensor imaging: methods and applications for digital cameras. Chapter, digital camera image storage formats, 1st edn. CRC press, Boca Raton, pp 351–379
74. Patti AJ, Sezan MJ, Tekalp AM (1997) Superresolution video reconstruction with arbitrary sampling lattices and nonzero aperture time. IEEE Trans Image Process 6(8):1064–1076
75. Pennebaker WB (1993) Mitchell JL (1993) JPEG still image data compression standard. Van Nostrand Reinhold, New York
76. Petschnigg G, Szeliski R, Agrawala M, Cohen M, Hoppe H, Toyama K (2004) Digital photography with flash and no-flash image pairs. In: Proceedings of the ACM SIGGRAPH, vol 23. pp 664–672
77. Pillman B, Deever A, Kumar M (2010) Flexible readout image capture with a four-channel CFA. In: Proceedings of the ICIP, pp 561—564
78. Poplin D (2006) An automatic flicker detection method for embedded camera systems. IEEE Trans Consum Electron 52(2):308–311
79. Raskar R, Agrawal A, Tumblin J (2006) Coded exposure photography: motion deblurring using fluttered shutter. ACM Trans Graph 25:795–804
80. Rav-Acha A, Peleg S (2005) Two motion-blurred images are better than one. Pattern Recognit Lett 26(3):311–317
81. Reference output medium metric RGB color space (ROMM RGB) white paper (1999) Technical report version 2.2, accession number 324122H, Eastman Kodak Company
82. Roddy J, Zolla R, Blish N, Horvath L (2006) Four color image sensing apparatus. U.S. Patent 7057654
83. Shan Q, Jia J, Agarwala A (2008) High-quality motion deblurring from a single image. ACM Trans Graph 27:1–10
84. Smith WJ (1966) Modern optical engineering. McGraw-Hill, San Francisco
85. Tomasi C, Manduchi R (1998) Bilateral filtering for gray and color images. In: Proceedings of the 1998 IEEE international conference on computer vision. IEEE
86. Triantafyllidis GA, Tzovaras D, Strintzis MG (2002) Blocking artifact detection and reduction in compressed data. IEEE Trans Circuits Syst Video Technol 12(10):877–890
87. Trussell H, Hartwig R (2002) Mathematics for demosaicking. IEEE Trans Image Process 11(4):485–492
88. Yamagami T, Sasaki T, Suga A (1994) Image signal processing apparatus having a color filter with offset luminance filter elements. U.S. Patent 5323233
89. Yamanaka S (1997) Solid state camera. U.S. Patent 4054906
90. Yoshida H (2004) Image pickup apparatus and method of correcting deteriorated pixel signal thereof. U.S. Patent 6809763
91. Yu W (2003) An embedded camera lens distortion correction method for mobile computing applications. IEEE Trans Consum Electron 49(4):894–901. doi:10.1109/TCE.2003.1261171
92. Yuan L, Sun J, Quan L, Shum HY (2007) Image deblurring with blurred/noisy image pairs. ACM Trans Graph 26:1–10
93. Zhang F, Wu X, Yang X, Zhang W, Zhang L (2009) Robust color demosaicking with adaptation to varying spectral correlations. IEEE Trans Image Process 18(12):2706–2717
94. Zhang L, Wu X (2005) Color demosaicking via directional linear minimum mean square-error estimation. IEEE Trans Image Process 14(12):2167–2178

# Digital Image Formats

**Khalid Sayood**

**Abstract** Images contain a significant amount of information which translates to substantial memory for storage. In this chapter we briefly describe various formats in which images are stored. Of necessity this entails some description of image compression.

## 1 Introduction

A digital image is made up of pixels which are displayed as a rectangular array. Each pixel can be a single value or a vector of components. The single value can be the grayness level of a monochrome image or it can be an index into a colormap which maps a single value into a set of color components. The different kinds of images have been covered elsewhere in this volume. In this chapter we will look at different formats for the storage of digital images. In particular we will examine ways in which compressed representations of images are stored. There are two kinds of resolution associated with a digital image; the spatial resolution which depends on the number of pixels and the spatial extent covered by the image, and a pixel resolution which depends on the number of bits used to represent the color value of the pixel referred to as the number of bits per pixel, or the *bit depth*. This can can vary from between one in the case of bi-level images to 64 in the case of some color images. If images are stored without compression the amount of storage required can become prohibitive. A $1024x1024$ truecolor image with 24 bit resolution requires more than three megabytes to store. As we know nobody who is satisfied with a single image the storage requirement can easily become overwhelming. In order to reduce the storage most images are stored in a compressed form.

K. Sayood (✉)
University of Nebraska Lincoln, Lincoln, NE, USA
e-mail: ksayood@unl.edu

H. T. Sencar and N. Memon (eds.), *Digital Image Forensics*,
DOI: 10.1007/978-1-4614-0757-7_3,

Compression is the name given to all kinds of techniques for generating compact representations of information. These techniques make use of the particular kind of structure existing in the data. These might be relational structures, such as the relationship of neighboring pixels in an image, or they may be statistical structures such as the skewed distribution of pixel values. Often, as we shall see, both types of techniques are used within a compression algorithm in order to maximize the amount of compression. Compression algorithms can be divided into two broad classes; *lossless compression* algorithms and *lossy compression* algorithms. Lossless compression involves finding a compact representation for the image from which the original uncompressed image data can be recovered exactly. Lossy compression, as the name implies entails the loss of some information in some part of the compression procedure. This means that the the original pixel values cannot be recovered exactly from the compressed data. The tradeoff for the loss is that much higher level of compression can be achieved using lossy compression methods.

In this chapter we will examine a number of algorithms used for the compression of images. Our goal is to be able to get an understanding of the image formats which make use of these compression algorithms. We will first describe two approaches, Huffman Coding and Arithmetic Coding that make use of statistical asymmetries in the data to generate a compact representation of the information. We then describe two approaches, dictionary coding and predictive coding which use relational structures to generate compression. These approaches are used in the well-known image formats Portable Network Graphics (PNG), Graphic Interchange Format (GIF), and Tagged Interchange File Format (TIFF). We give a brief overview of these file formats. We then turn our attention to lossy compression. As lossy compression techniques involve the loss of information it is first necessary to represent the image in a way that allows us to select that information in the image which will be missed the least. We will look at two kinds of representations, one based on transforms and one based on wavelets which make it feasible to throw away information in a (more or less) controlled fashion. These two representations are the basis for the two lossy image compression algorithms, JPEG and JPEG2000 which we will describe next.

We should note that the final word for any format is the standards document. Information in this chapter should not be taken as definitive. It is only intended for use in understanding the standards.

## 2 Statistical Techniques

The idea of using statistical nonuniformity to create a compact representation of data has been around at least since Samuel Morse and probably earlier. In the Morse code common letters such as *E* and *I* are represented by short codes; *E* is represented by a dot and *I* by two dots. More unlikely letters like *Q* and *Y* are represented by much longer codes. *Q* is represented by dash-dash-dot-dash and *Y* is represented by dash-dot-dash-dash. Recalling that a dash takes as long as three dots this means that the unlikely letters take ten times as long to transmit as the more likely letters. However, as the more likely letters will occur more often, on the average, the length of time required to send a letter will be dominated by the length of the codes for these letters. Why not give a short code to each letter? Simply because there are only so many short codes.

Claude Shannon in his 1948 landmark paper [1] provided the theoretical underpinnings for the use of variable length codes. Shannon defined the amount of information contained in an event $A$ to be the logarithm of the inverse of the probability of the event. Furthermore, he showed that the fewest number of bits per symbol required to encode a random sequence $\{x_n\}$ which takes values over an alphabet $\mathcal{A}$ is given by the entropy of the source, where the entropy is defined in terms of the expected value of the self-information. Let

$$G_n = \sum_{x_1 \in \mathcal{A}} \cdots \sum_{x_n \in \mathcal{A}} P(x_1, x_2, \cdots x_n) \log_2 \frac{1}{P(x_1, x_2, \cdots x_n)}$$

Then the entropy $H$ is given by

$$H = \lim_{n \to \infty} \frac{1}{n} G_n$$

When the random sequence consists of independent identically distributed (*iid*) random variables the entropy becomes

$$H = \sum_{A \in \mathcal{A}} P(A) \log_2 \frac{1}{P(A)}$$

Even when the source is not *iid* this latter quantity is useful and is often referred to as the first-order entropy.

## 2.1 Huffman Coding

David Huffman [2] created a method for generating binary codewords which have the minimum average length for an alphabet with known probabilities of occurrence. His method for generating the codewords depends on two simple observations about optimal codes and a construction rule:

1. The code for a more likely symbol cannot be longer than the code for a less likely symbol.
2. The two least likely symbols have to be of the same length. To these two observations he added a constraint.
3. The least likely symbols differ in only the last bit.

to come up with an algorithm for the optimum variable length code. To see this algorithm in action consider the task of developing the optimum variable length code for a five-letter alphabet $\mathscr{A} = \{A_1, A_2, A_3, A_4, A_5\}$ with probabilities $\{0.3, 0.1, 0.5, 0.05, 0.05\}$. The two least likely symbols are $A_4$ and $A_5$. Therefore, by the third condition, the code for these two symbols will be a common binary string $s_{45}$ followed by 0 for one symbol and 1 for the other. It does not matter which symbol gets a 0 and which one gets a 1 as its last bit. Here, for convenience let us assume it is the codeword for $A_4$ whose last bit is a 0 and the codeword for $A_5$ whose last bit is a 1. Letting $*$ denote concatenation the codeword for $A_4$ is $s_{45} * 0$ and the codeword for $A_5$ is $s_{45} * 1$. The binary sequence $s_{45}$ is the codeword for $A_4$ $OR$ $A_5$. Let the symbol $A_{45}$ denote $A_4$ $OR$ $A_5$ and consider a new alphabet $\mathscr{A}_1 = \{A_1, A_2, A_3, A_{45}\}$. The probabilities associated with these symbols are $\{0.3, 0.1, 0.5, 0.1\}$ where the probability of the symbol $A_{45}$ is simply the sum of the probabilities of the symbol $A_4$ and $A_5$. In this set the two lowest probability symbols are $A_2$ and $A_{45}$; therefore, their codewords are the same except for the last bit. Let the codeword for $A_2$ be $s_{245} * 0$ and the codeword for $A_{45}$ be $s_{245} * 1$. Given that we previously declared the codeword for $A_{45}$ to be $s_{45}$ this means that the last bit of $s_{45}$ is 1. Continuing as before let us define a new symbol $A_{245}$ with codeword $s_{245}$ and a new alphabet $\mathscr{A}_2 = \{A_1, A_{245}, A_3\}$ with probabilities $\{0.3, 0.2, 0.5\}$. Now the two lowest probability symbols are $A_1$ and $A_{245}$. Let the codeword for $A_1$ be the binary sequence $s_{1245} * 0$ and the codeword for $A_{245}$ be the binary sequence $s_{1245} * 1$, where $s_{1245}$ can be viewed as the codeword for the symbol $A_{1245}$. Note that this means that the rightmost bit of the sequence $s_{245}$ is 1, and the two rightmost bits of $s_{45}$ are 10. Continuing as before we are left with a binary alphabet $\mathscr{A}_3 = \{A_{1245}, A_3\}$. Obtaining a binary code for a binary alphabet is easy; we code $A_{1245}$ with a 0 and $A_3$ with a 1. This means that $s_{1245}$ is the sequence 0, which in turn means the sequence $s_{245}$, which is the sequence $s_{1245} * 1$ is 01. As the sequence $s_{45}$ is $s_{245} * 0$ this means that the sequence $s_{45}$ is 010. Now we can put everything together to obtain the binary *Huffman code* for this alphabet.

| Symbol | Code |
|---|---|
| $A_1$ | 00 |
| $A_2$ | 010 |
| $A_3$ | 1 |
| $A_4$ | 0110 |
| $A_5$ | 0111 |

Looking at this process slightly differently, the construction of the Huffman code is simply the construction of a fully populated binary tree where the leaf nodes are the symbols and the probability of an intermediate node is the sum of the probabilities associated with its offsprings. At various points in the building of the Huffman code we make choices such as assigning a 0 or a 1 to a particular symbol or, where multiple symbols have the same probability, combining one pair over a different pair. These choices lead to different codes but they are all Huffman codes with the properties:

1. They have the minimum average length of any code for that particular alphabet.
2. They are comma-free codes.

By comma free we mean that a sequence of codewords can be uniquely parsed without the need for inserting delimiters between codes. Decoding simply means traversing the Huffman tree until a leaf is encountered at which point a symbol is decoded. This property of Huffman codes also provides us with a way of efficiently storing the code by simply storing the lengths of the codewords. For example, the *deflate* algorithm in *zlib* imposes the following conditions on a Huffman code [3]:

1. All codes of a given length have lexicographically consecutive values in the same order as the symbols they represent.
2. Shorter codes lexicographically precede longer codes.

In order to satisfy these conditions we could modify the code obtained above by making a different choice in the final step of the algorithm assigning a 0 to $A_3$ and a 1 to $A_{1245}$. This would give us the code:

| Symbol | Code |
|---|---|
| $A_1$ | 10 |
| $A_2$ | 110 |
| $A_3$ | 0 |
| $A_4$ | 1110 |
| $A_5$ | 1111 |

In order to store this code all we need to do is to store the lengths of the codewords, i.e., $\{2, 3, 1, 4, 4\}$. Knowing the constraints on the code it can be reconstructed from the lengths. Let us reconstruct the code for this example. We start with looking at all codes of length one. There is only one such code, the code for $A_3$. The lexicographically smallest codeword of length one is 0; therefore, the codeword for $A_3$ is 0. The codewords of length two have to be of the form 1x. There is only one codeword of length two, the codeword for $A_1$, therefore, has to be 10. The codewords of length 3 have to be of the form 11x. Again, there is a single codeword of length three, the codeword for $A_2$, which has to be 110. There are two codewords of length four. Therefore, the codeword for $A_4$ has to be 1110 and the codeword for $A_5$ has to be 1111. For Huffman codes of only five symbols the ability to store the code in this fashion is no great advantage. However, when the code size becomes large storing Huffman codes in this manner is efficient both in terms of storage as well as computations.

The Huffman coding procedure generates the optimal code in terms of average length for a given alphabet. We know that the lower bound on the average codeword length for a particular source is the entropy $H$ of the source. It can be easily shown that the average length $R$ of a Huffman code is within one bit of the entropy. Consider the example used above. Assuming a memoryless source the entropy of the source is

$$
\begin{aligned}
H &= -0.3\log_2 0.3 - 0.1\log_2 0.1 - 0.5\log_2 0.5 - 0.05\log_2 0.05 - 0.05\log_2 0.05 \\
&= 1.785 \text{ bits}
\end{aligned}
$$

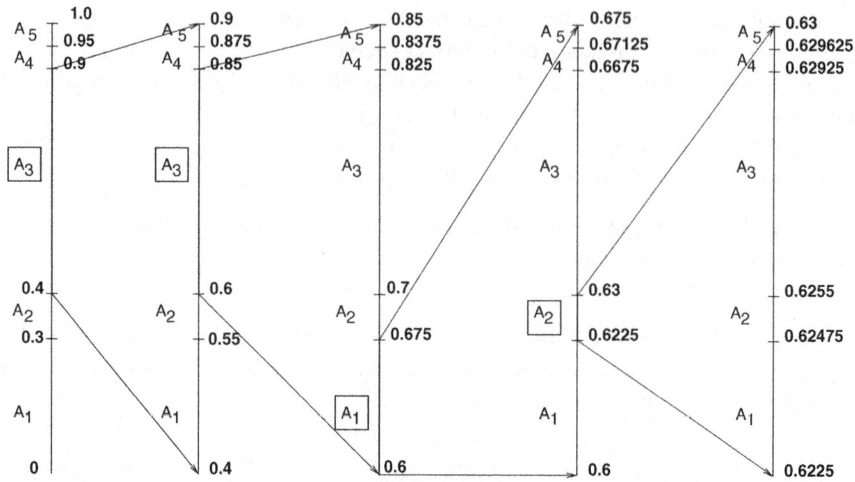


**Fig. 1** Construction of the interval containing the sequence $A_3A_3A_1A_2$

The average codeword length $R$ is

$$R = 0.3 \times 2 + 0.1 \times 3 + 0.5 \times 1 + 0.05 \times 4 + 0.05 \times 4 = 1.8 \text{ bits}$$

The average codeword length is quite close to the entropy. However, if we make the alphabet probabilities more skewed, for example $\{0.15, 0.1, 0.65, 0.05, 0.05\}$, the entropy goes down to 1.57 bits but the average codeword length for a Huffman code remains at 1.8 bits. The reason for this is that while the probability of encountering the symbol $A_3$ has gone up the codeword for $A_3$ cannot be smaller than one bit long. We can rectify the situation by encoding pairs of letters. We can construct an alphabet of 25 pairs $\{A_1A_1, A_1A_2, A_1A_3, \cdots, A_5A_5\}$ and generate a Huffman code for this alphabet. By using pairs of letters we open the possibility of assigning a fractional number of bits for a particular symbol thus allowing the rate to approach closer to the entropy bound. In fact, it can be easily shown that the average codeword length is bounded as:

$$H \leq R \leq H + \frac{1}{n}$$

where $n$ is the number of symbols we block together. By making $n$ large we can come arbitrarily close to the entropy bound. However, this implies an exponential growth in the codebook size which is not practical. An alternative approach is Arithmetic Coding which we explore in the next section.

## 2.2 Arithmetic Coding

Arithmetic coding can be viewed as the assignment of a label to a sequence of letters, where the label is a truncated representation of a number in the interval [0, 1). As there are an uncountably infinite number of values in this interval it is possible to obtain a unique label for a given sequence. While the ideas behind arithmetic coding have been around since Shannon's original work, practical arithmetic coding algorithms were developed by Rissanen [4] and Pasco [5] in 1976. Patents on the coding techniques initially prevented its wide acceptance; however, with the expiration of these patents arithmetic coding is becoming more and more widely used.

In order to find a label for a sequence the unit interval is progressively divided into smaller and smaller regions, each contained within the other. The label for the sequence is any number in the final interval. The process is shown in Fig. 1 for the sequence $A_3A_3A_1A_2$. Initially the entire unit interval is available for selection of the label. The unit interval is divided into subintervals where the size of the subinterval corresponding to a particular letter is proportional to its probability. Thus, initially the interval [0,0.3) is assigned to $A_1$ whose probability of occurrence is 0.3, an interval of size 0.1, [0.3, 0.4), is assigned to $A_2$ whose probability of occurrence is 0.1. In a similar manner the interval [0.4, 0.9) is assigned to $A_3$, the interval [0.0, 0.95) is assigned to the symbol $A_4$ and the interval [0.95, 1.0) is assigned to the symbol $A_5$. The first symbol in the sequence to be encoded is $A_3$. This means that the label corresponding to the sequence will be from the interval corresponding to $A_3$, that is, the interval [0.4, 0.9). We divide this restricted interval in proportion to the probabilities of the different symbols. The probability of $A_1$ is 0.3 so the interval corresponding to $A_1$ is from the lower end of the interval, that is, 0.4 to 0.3 times the size of the restricted interval, or $0.4 + 0.3 \times (0.9 - 0.4)$ which is 0.55. Continuing in this fashion we assign subintervals of the interval [0.4, 0.9) to the other letters in proportion to their probability as shown in Fig. 1. The second symbol in the sequence is $A_3$. This corresponds to the interval [0.6, 0.85) as shown in Fig. 1. This interval is again divided in proportion to the probability of the different letters to generate subintervals corresponding to the various letters. The appearance to the symbol $A_1$ further restricts the interval to [0.6, 0.675) and the appearance of $A_2$ gives us the final interval [0.6225, 0.63).

In this particular example the binary representation of any number in the interval [0.6225,0.63) suitably truncated and terminated will be a codeword for the sequence $A_3A_3A_1A_2$. The process is relatively simple to implement and several fast implementations of the coding procedure have been proposed.

The rate of the arithmetic code can be shown to be bounded by

$$H \leq R \leq H + \frac{2}{n}$$

which at first glance seems less efficient than the Huffman code. However, where in the Huffman coding procedure $n$ is limited in practice, implementations of the arithmetic code remove any limitations on $n$. Assuming the source is memoryless this allows the arithmetic coder to encode the source as close to the entropy as desired. Note the requirement that the source be memoryless. As it is generally not possible to guarantee that a source is memoryless it is more accurate to say that the average number of bits per symbol using an arithmetic coder is close to the first-order entropy of the source. As a source with a heavily skewed probability distribution has a lower entropy than a source with a more uniform distribution; it is advantageous to make the source distribution as skewed as possible. This is often done by using the context of the source symbol. Generally, knowing the context of a symbol gives us more idea of what the value of the symbol is, in other words knowing the context makes some values more probable than others thus lowering the first-order entropy.

## 3 Lossless Image Compression

Compression techniques attempt to exploit properties of the data to create compact representations. In images slated for human consumption there are two major features which can be exploited. There is generally a high degree of correlation between neighboring pixels in an image. This means that when representing a particular pixel the values of the neighboring pixels can be used to generate a more compact representation. This is the approach used in the *JPEG-LS* standard. The second feature is the existence of repeated sequence of pixel values. This feature can be exploited by building a dictionary containing often repeated sequences and then using this dictionary to provide a compact representation. The image compression formats GIF, PNG, and TIFF all use variations of dictionary-based methods for compression.

In the following we will briefly describe these compression approaches before detailing the particulars of the various image formats which use these approaches.

### *3.1 Runlength Coding*

The simplest type of repeated pattern is the repetition of the same pixel value. This kind of pattern is commonly founded in bi-level images. The first lossless image compression algorithms were developed for use with fax machines sending scanned bi-level images. The simplest of these is the *modified Huffman coding* scheme in which each row is represented by alternating runs of black and white pixels. A Huffman code is used to encode the length of the runs. As the number of possible runlengths is very large, the runlength $r$ is represented as

$$r = 64 \times m + t$$

To encode $r$ a code for $m$ and a code for $t$ are transmitted. These codes standardized in CCITT recommendations [6] are called *makeup* and *terminating* codes. If the length of the run is less than 64 only the terminating code is sent. In most implementations there is an assumption that the very first run is a run of white pixels. If this is not true in practice the terminating code for a runlength of zero is encoded.

The modified Huffman code makes use of the existence of runs in bi-level images to obtain compression. However, that is not the only obvious redundancy in bi-level images. Usually, runs in one row are heavily correlated with runs in the next. An algorithm called the modified Relative Element Address Designate (READ) makes use of this correlation to provide a more compact representation. In order to prevent errors from propagating through the document the modified READ algorithm requires that at specified intervals a line be encoded only using modified Huffman coding. Both the modified Huffman coding algorithm and the modified READ algorithm are part of CCITT recommendation T4. The CCITT also published recommendation T6 which contained the modified READ algorithm with one difference—the requirement for periodic use of modified Huffman was removed. The resulting algorithm is known as *modified modified READ* (MMR)!

Another runlength algorithm which we will refer to is Apple's PackBits algorithm [7]. In this algorithm each run is encoded as a packet. The packet consists of a one byte header, or flag-counter, followed by the data. If there are a sequence of $n$ bytes $1 \leq n \leq 128$, which do not contain any runs, the header byte is the value $n - 1$ followed by the $n$ bytes of data. However, if the data contains a run of $n$ identical bytes $2 \leq n \leq 128$, the header value is $1 - n$ followed by one copy of the repeated byte. The value $-128$ is used to signal no-operation and the following byte is treated as a header byte. For example, modifying an example from [7] the sequence

```
AA AA AA 80 00 2A AA AA AA AA 80 00
2A 22 AA AA AA AA AA
```

is encoded as

```
FE AA 02 80 00 2A FD AA 03 80 00 2A 22 FC AA
```

## 3.2 Dictionary Compression

### 3.2.1 LZ77

The dictionary compression techniques in use today all owe their existence to Jacob Ziv and Abraham Lempel, in particular to two papers published in the *IEEE Transactions on Information Theory* in 1977 [8] and 1978 [9]. In the 1977 paper they presented a technique, commonly referred to as LZ77, in which the dictionary for a sequence is simply the previous history of the sequence. The idea is very simple: when a sequence of symbols is encountered in the data which has occurred previously in a specified window, rather than encoding all the symbols separately, the encoder encodes the location of the previous occurrence and the length of the repeated pattern. The location is encoded as an offset from the current location in the datastream. Consider the following sequence [10]

$$\overline{c\ abracad\ abravrake}$$

where the search window contains the sequence *abracad* which has already been encoded. The next symbols to be encoded are *abra* which are contained in the search window. These symbols could be encoded by sending the offset 7 and match length of 4. What if there is no match for the current sequence of symbols in the search window. LZ77-based algorithms take two different approaches, each involving a third element to go with the offset and the match length. In the original version the third element is simply an encoding of the next symbol. In our example the triplet generated by the encoder would be $<c(7), c(4), c(v)>$ where by $c(\cdot)$ we mean the binary code for the argument. If the symbol to be encoded is not preceded by a match the first two elements of the triplet would be the code for zero. The other popular version was proposed by Storer and Syzmanski [11] (and hence this variant is sometimes referred to as LZSS). In this variant a flag bit is used to indicate when the encoding is of a match and when the encoding is of a symbol not present in the search window. Thus, in this example, we would have the encodings $<1, c(7), c(4)>$ and $<0, c(v)>$. In both case the actual codes used for offset and match length can be fixed or variable length. We will explore more implementation details when we discuss the various file formats.

### 3.2.2 LZW

In the 1978 paper [9] Ziv and Lempel presented an alternative approach which required the explicit construction of a dictionary. This approach was improved upon by Terry Welch and the resulting algorithm commonly known as LZW popularized the use of dictionary algorithms for compression. In this algorithm both the encoder and decoder begin with an initial dictionary which contains the alphabet of the source. Consider a sequence of two bit pixels, where for convenience we will refer to the gray levels by the letters $a, b, c$, and $d$. The initial LZW dictionary will consist of an indexed list containing these four values:

| Index | Entry |
|---|---|
| 1 | a |
| 2 | b |
| 3 | c |
| 4 | d |

The encoder then parses the sequence to be encoded and builds up the dictionary with unique words, i.e., words not already in the dictionary. Consider the sequence

$$abadbadba$$

The encoder would encode the first two elements of the sequence by sending the code for the indices 1 and 2 and add the entry *ab* into the dictionary with an index of 5. The next symbol to be encoded would be *a* and the sixth entry in the dictionary would be *ba*. Continuing, the next symbol *d* would be encoded using the index value of 4 and the entry *ad* would be added to the dictionary. The next two symbols are *ba*. There is an entry for *ba* in the dictionary so both symbols would be encoded with a single index value of 6. The next two symbols in the sequence are *db*. This pattern also exists in the dictionary and therefore these two symbols would be encoded using a single index 8. At this point the dictionary would look as follows:

| Index | Entry |
|---|---|
| 1 | a |
| 2 | b |
| 3 | c |
| 4 | d |
| 5 | ab |
| 6 | ba |
| 7 | ad |
| 8 | db |
| 9 | bad |

Note that as we continue the entries start becoming longer and if the sequence to be encoded has many repeating patterns, longer and longer patterns will be encoded with single index values leading to better and better compression. There are several implementation issues that need to be considered. The way the algorithm has been defined the dictionary continues to grow without end. In practice there has to a limit to the size of the dictionary. Furthermore the lower index values reflect behavior of the sequence that is nonlocal. In applications such as image compression keeping these entries may be detrimental to compression performance. We will address these issues when we deal with the various image storage standards.

### 3.2.3 Predictive Coding

Images destined for human consumption tend to be of objects, natural or man-made, which in a two-dimensional image are mostly homogeneous regions bounded by sharp transitions. The homogeneous regions generally take up significantly larger proportion of the image than do the transition regions. Image compression strategies that take this characteristic of images into account will tend to perform better at compression. Consider the *Sinan* and *Sena* images shown in Fig. 2. A histogram of the pixel values in the *Sena* image is shown in Fig. 3 while a histogram of the pixels in the *Sinan* image is shown in Fig. 4. Clearly the two images are different and have different statistical characteristics. However, if we examine the histogram of the difference between neighboring pixels as shown in Figs. 6 and 5 we can see that there are significant similarities between the two images.

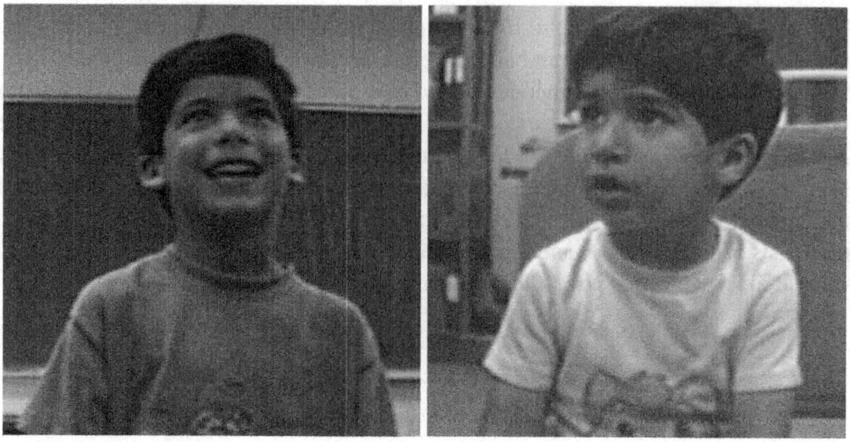

**Fig. 2** *Sena* and *Sinan* images

**Fig. 3** A histogram of the pixel values in the *Sena* image

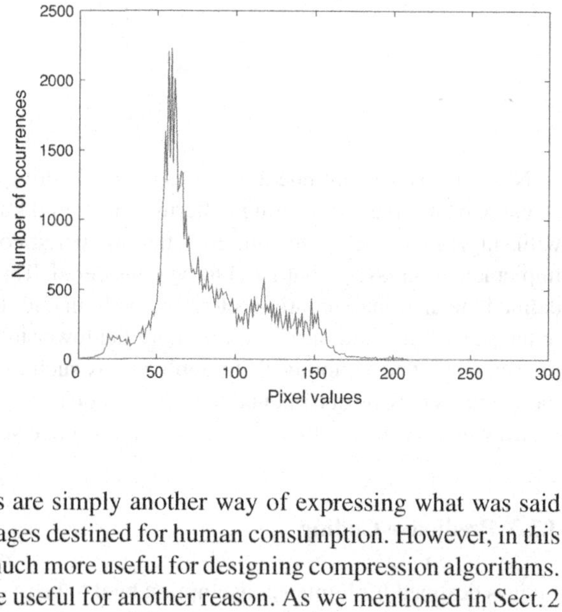


The difference histograms are simply another way of expressing what was said earlier about the nature of images destined for human consumption. However, in this form the characterization is much more useful for designing compression algorithms. The difference histograms are useful for another reason. As we mentioned in Sect. 2 the more skewed a distribution the lower the entropy. Clearly, the distribution of the differences is more skewed than the distribution of the pixels themselves, which means that the differences can be encoded using fewer bits than the pixels themselves.

The question then becomes can we do better? In that, can we obtain a representation that is more skewed? It turns out we can and the way to do that is predictive coding. Predictive coding techniques, as their name implies, attempt to predict the value of the pixel being encoded. The prediction is based on pixels that have already been encoded and therefore available to the decoder. Thus, the predicted value does not need to be encoded. The difference between the predicted value and the actual

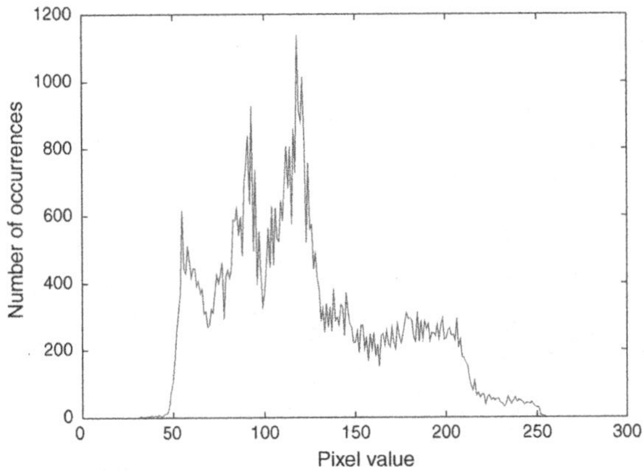


**Fig. 4** A histogram of the pixel values in the *Sinan* image

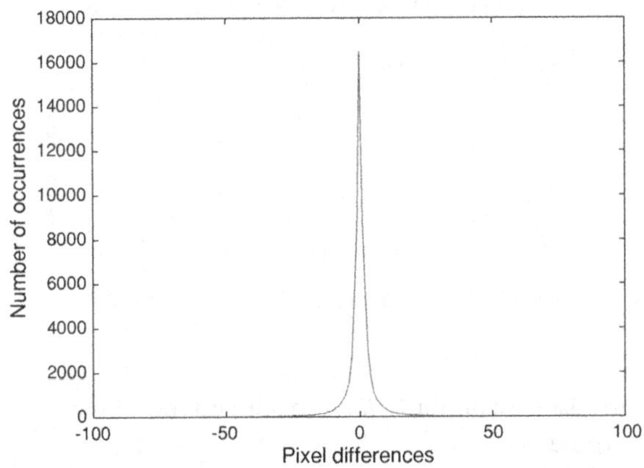


**Fig. 5** A histogram of the pixel value differences in the *Sena* image

value of the pixel is encoded and stored or transmitted. A simplified block diagram of a predictive encoder and decoder is shown in Fig. 7.

There are a number of predictive coding algorithms, the best known being CALIC [12]. The image compression standard JPEG-LS is also based on predictive coding. Predictive coding is also an option (under the name filtering) in the PNG format. We will describe the particular predictive scheme used in more detail when we discuss the related format.

### 3.2.4 Graphic Interchange Format

The GIF was one of the first lossless image formats, and one of the first standards to use dictionary compression. It was developed by CompuServe and rapidly became a popular method for storage of images on digital media. The standard organizes the

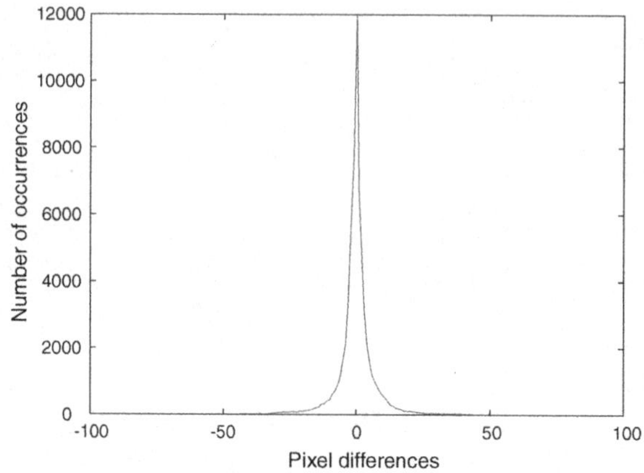


**Fig. 6** A histogram of the pixel values in the *Sinan* image

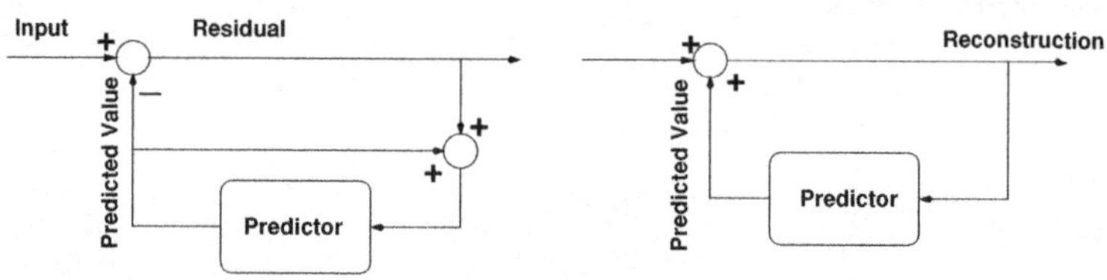


**Fig. 7** The block diagram of a predictive encoder and decoder

GIF datastream into *blocks* of three types: control, graphic rendering, and special purpose.

### Control Blocks

As evident from the name the control blocks define parameters used to set the stage for the image. The *Header* block is six bytes long with the first three bytes containing the signature and the next three bytes containing the version number. The signature bytes contain the values GIF. As might be expected the *Header* block is a required part of the GIF datastream. Each GIF datastream is required to have one and only one *Header* block.

The *Logical Screen Descriptor* block defines the physical context in which the image will be rendered. The first four bytes provide the dimensions of the display area with two bytes specifying the logical screen width in pixels and two bytes the logical screen height in pixels. The fifth byte contains a 1-bit flag which indicates the presence or absence of a global color map or palette, three bits to indicate the color resolution, a 1-bit flag to indicate whether the color map is sorted according to the frequency with which particular colors are used, and three bits to indicate the size of the global color table. In order to obtain the size of the global color table we take the decimal value of these three bits add one and raise two to this power. The sixth byte of the *Logical Screen Descriptor Block* is the background color index. This is an index into the color map which indicates the color of the pixels in the screen not covered by the image. The final byte contains a value used to compute an approximation to the aspect ratio of the pixels in the image. The aspect ratio is computed as

$$\text{aspect ratio} = \frac{\text{value} + 15}{64}$$

where value denotes the value of the seventh byte.

If the global color map flag in the *Logical Screen Descriptor* block is set the *Logical Screen Descriptor* block is immediately followed by a *Global Color Table* block. The size of the block is equal to $3 \times 2^{s+1}$ where $s$ the value of the three bits indicate the size of the global color table.

The last block in GIF datastream is the *Trailer* block. This block consists of a single byte with the fixed value 0x3B

## Graphic Rendering Blocks

The graphic rendering blocks are those blocks directly concerned with rendering the image on the display device. The *Image Descriptor Block* is the first of the graphic rendering blocks and follows the control blocks described above. The first byte is a separator byte which contains the fixed value 0x2C. This is followed by four bytes that specify the rendered image location relative to the logical screen. Two bytes denote the left position and two bytes denote the top position. These values are essentially offsets from the left and top boundaries of the logical screen.The next four bytes give the image width and height in pixels with two bytes devoted to each dimension. The tenth byte contains a one-bit flag indicating the presence or absence of a local color map, a one-bit flag indicating whether the image is interlaced, a one-bit flag indicating whether the local color table (if present) is sorted, two bits that are reserved, and three bits indicating the size of the local color map (if present).

If the flag for a local color map is present, the *Image Descriptor Block* is set and then the *Local Color Table* block immediately follows the *Image Descriptor Block*. As in the case of the *Global Color Map* the size of the *Local Color Table* block is equal to $3 \times 2^{s_l+1}$ where $s_l$ is the value of the three bits indicating the size of the local color table.

The actual data for the image is stored in the *Image Data* block which consists of a single byte containing the LZW initial index length followed by a sequence of subblocks which are almost 255 bytes in size. The first byte of each of the data subblocks contains a count of the number of bytes in the subblock. The initial dictionary size is given by $2^{b+1}$ where $b$ is the value contained in the first byte and is usually the number of bits per pixel in the image being encoded. The encoding procedure follows that described previously for the LZW algorithm with each index being represented by $b + 1$ bits. When the number of entries in the LZW dictionary reaches

$2^{b+1}$ the size of the indices is incremented by one bit. This procedure continues until the maximum of 12 bits per dictionary index. At this point the GIF encoder becomes a static dictionary encoder, no longer adapting itself to the changing image. An end of information code explicitly indicates the end of image data. This code is a byte with the value of $2^b + 1$.

### 3.2.5 Portable Network Graphics

Before describing the PNG format we need a brief description of some of the processing steps used by the PNG encoder to prepare a source image for encoding. The standard considers there to be five possible image types. A *truecolor with alpha* type where each pixel has four components, red, blue, green, and an alpha component which dictates the transparency of each pixel; *grayscale with alpha* image where each pixel consists of two components, the grayscale value and an alpha component; *truecolor* image with three components, red, green, and blue; grayscale image with one grayscale value per pixel; and an indexed-color image where the pixel value is replaced by an index into a palette or color map.

The PNG image can be scanned in one of the two ways. One is a simple raster scan reading pixels row-by-row from left to right. The second method decomposes the image into a sequence of seven sampled images where each successive image has more pixels and enhances the resolution further. The seven sampled images are generated according to the 8 sampling grid shown below repeated throughout the image.

```
1 6 4 6 2 6 4 6
7 7 7 7 7 7 7 7
5 6 5 6 5 6 5 6
7 7 7 7 7 7 7 7
3 6 4 6 3 6 4 6
7 7 7 7 7 7 7 7
5 6 5 6 5 6 5 6
7 7 7 7 7 7 7 7
```

Once the image has been scanned, instead of encoding the sequence of pixels the PNG standard allows for the encoding of the residuals resulting from prediction. The prediction and generation of residuals is at the byte level, even when the bit depth is less than 8. There are five different types of prediction or, in the language of the PNG standard, filtering. Type 0 is simply no prediction and the PNG image is directly encoded. In type 1 filtering the predicted value is the byte immediately before the pixel being encoded. In type 2 filtering the prediction is the byte above the byte containing the pixel being encoded. In type 3 the prediction is the truncated average of the byte above and the previous byte. The type 4 filter is somewhat more complicated. The first step is the generation of an estimate which consists of the sum of the byte above and to the left of the byte being processed minus the byte which is

**Table 1** Structure of the IHDR data field

| Information | Size |
|---|---|
| Width of image | 4 bytes |
| Height of image | 4 bytes |
| Bit depth | 4 bytes |
| Color type | 1 byte |
| Compression method | 1 byte |
| Filter method | 1 byte |
| Interlace method | 1byte |

diagonally above and to the left. The absolute distance of the bytes to the left, above, and diagonal from this estimate is computed and the closest of these three bytes to to the estimate is selected as the prediction.

The difference between the pixel and the predicted value is preserved as an eight-bit value; thus, the differencing is a modulo 256 operation. The differences are always between "like" pixels. That is, the byte containing a green value is always predicted using a byte containing a green value, and so on. Finally, each row can use a different filter. The filter type is added as byte preceding the row.

The filtered scan lines are then compressed using the *deflate* algorithm which is an implementation of LZ77 with maximum window size of 32,768. The offset, match length, and symbols are coded using Huffman codes which are stored using only their lengths as described in Sect. 2.

Now with these preliminaries we can begin to describe the PNG format. The PNG file is organized into *chunks* where each chunk consists of three or four fields. The first field is a four-byte integer which represents the number of data bytes $N$. The second four bytes are the chunk name. This is followed by $N$ data bytes. $N$ can take on values between 0 and $2^{31} - 1$. If the value of $N$ is 0 the data field is absent. The last field consists of a four-byte CRC computed over the data bytes. This allows PNG readers to detect errors without having to display the image.

There are two types of chunks, critical chunks and ancillary chunks. There are four critical chunks:

### IHDR

The IHDR chunk contains the the various image parameters as shown in Table 1. As shown, the IHDR chunk contains the dimensions of the image and the number of bits per pixel. The color type is as described before with type 0 being grayscale, type 3 being truecolor, type 3 being colormapped, type 4 being grayscale with an alpha channel, and type 6 being truecolor with an alpha channel.

PLTE

The PLTE chunk is mandatory only when the image is colormapped. In this case the PLTE chunk carries the lookup table used to map the color indices to the RGB values. There can only be one PLTE chunk and it has to precede the first IDAT chunk.

IDAT

This is the data chunk. The data is limited in size to $2^{31} - 1$ bytes. Thus, depending on the size of the image there can be multiple data chunks.

IEND

This chunk terminates the PNG file.

There are 14 ancillary chunks defined by the standard. Users can also define their own chunks. The naming convention of the chunks is used to discriminate between critical and ancillary chunks as well as to convey some more information. If the first letter in the chunk name is uppercase it is a critical chunk, otherwise it is an ancillary chunk. If the second character is uppercase the chunk is publicly defined, else it is privately defined. If the fourth letter is upper case the chunk is safe to copy, otherwise it is not safe to copy. The fourth letter is reserved. The public ancillary chunks contain transparency information (tRNS), colorspace information (cHRM, gAMA, sBIT, sRGB), time stamp (tIME), and other miscellaneous information (bKGD, hIST, pHYs, sPLT).

The first eight bytes of the PNG file is the PNG signature which is

| 0x89 0x50 0x4E 0x47 0x0D 0x0A 0x1A 0x0A |
|---|

This is followed by an IHDR chunk. The IHDR can chunk can be followed by ancillary chunks which contain colorspace information. These are followed, if needed, by the PLTE chunk. The PLTE chunk can be followed by other miscellaneous chunks such as the background (bKGD) which specifies the default background color or a chunk containing the image histogram (hIST) among others. These are followed by as many IDAT chunks as necessary. The last chunk is the IEND chunk.

### 3.2.6 Tag Interchange File Format

The TIFF is probably the most flexible of all the formats that use image compression. The format allows for a variety of image compression schemes to be used for a variety of image type. The format itself is straightforward using a series of information fields or tags organized in an *image file directory* (IFD) to specify the composition of the file. The image data is broken-up into strips, where the size of the strips is indicated in one of the fields. These strips can be distributed throughout the image file in any order desired as the offsets to the strips are contained in the information fields.

The first eight bytes of the TIFF file form the image file header. The first two bytes contain the values "II" (0x4949) or "MM" (0x4D4D) which indicate whether the byte order in the file is little endian or big endian. The "II" value indicates that the byte order is little endian while an "MM" value indicates that the file is big endian. The next two bytes contain the value 42 which is widely believed to be the answer to the problems of life, the universe, and everything [13]. The last four bytes contain the offset in bytes of the first IFD.

The IFD consists of sequence of 12-byte information fields. The structure of each entry is as follows:

| Parameter | Size (bytes) | Description |
|---|---|---|
| Tag | 2 | Identifies the field |
| Type | 2 | This identifies the type which includes byte, ascii, short (2 bytes), long (4 bytes), and rational (or two longs) (8 bytes) |
| Count | 4 | Number of values of the indicated type |
| Offset | 4 | May contain the offset to the values corresponding to this information or if the values can be accommodated within four bytes these bytes contain the values themselves |

## Baseline TIFF

There are some fields that are required for the baseline TIFF encoding of different image types. These include the number of rows and columns of the image, the physical dimensions of the image, the type of compression, and the interpretation of the pixel values as color.

There are many fields that are common between the four types of images, namely bi-level, grayscale, color mapped or palette color, and RGB full color images. We list these fields with values for the bi-level image in Table 2. The fields dealing with pictorial dimensions are Image Width and Image Length which describe the dimensions of the image in pixels per row and number of rows. The information required to translate the pixel-based dimensions to physical dimensions is provided in the Resolution Unit, Xresolution, and Yresolution fields. As mentioned previously the image is divided into strips. The size of the strips is provided in the Rows per Strip field. The number of strips can be obtained by dividing the Image Length by the Rows per Strip. The location of each strip is encoded in the TIFF file using two values, the Strip Offset and the Strip byte. The offset indicates where the data for the strip is located and the strip byte indicates how many bytes are stored for each strip. The offset and strip byte data consist of a sequence of values that are unlikely to fit

**Table 2** Required fields for baseline TIFF bi-level images. These fields are also required for the other image type; however, the values may be different

| Name | Tag | Type | Description |
|---|---|---|---|
| Image width | $0x100$ | Short or long | Number of pixels per scan line |
| Image length | $0x101$ | Short or long | Number of scan lines |
| Compression | $0x103$ | Short | compression strategy can be<br>1: no compression,<br>2: modified Huffman coding or<br>32773: PackBits compression |
| Photometric interpretation | $0x106$ | Short | The way a binary 0 or 1 is interpreted:<br>0: 0 is interpreted as white and the maximum value as black<br>1: 0 is interpreted as black and the maximum value as white |
| Strip offsets | $0x111$ | Short or long | Pointer to the sequence of stripe offsets<br>Each strip offset points to the beginning of the strip data |
| Rows/ strip | $0x116$ | Short or long | Number of rows in each strip |
| Strip byte count | $0x117$ | Long or short | Number of bytes of data encoded for each strip |
| Resolution unit | $0x128$ | Short | Resolution unit can be<br>1: none<br>2: inches<br>3: centimeters |
| Xresolution | $0x11A$ | Rational | Number of pixels per resolution unit in the image width |
| Yresolution | $0x11B$ | Rational | Number of pixels per resolution unit in the image length |

into the four-byte values field; therefore, the values field contains a pointer to where the sequence of offset values and the sequence of length values is stored.

For grayscale images there is an additional field called the BitsPerSample field with a tag of 0x102 and a type of short. The value indicates the number of bits per sample and in the baseline tiff images the number of bits per sample can be either four or eight. As modified Huffman coding is not defined for grayscale images the compression field is only allowed the values 1 and 32,773.

For color mapped or palette color images there is an additional field that provides the location of the color map. The tag for the color map field is 0x140 with a type of short. The value is a pointer to where the color map is located. In the TIFF color maps all the red values are stored first, followed by the green, and then the blue values. Black is represented by 0, 0, 0 and white is represented by 65535, 65535, 65535. For color mapped images the photometric interpretation value is set to 3 indicating the use of a color map.

In the RGB full color images there is one additional required field over those required for grayscale images. This is the Samples per Pixel field which indicates the number of components per pixel. The tag for this field is $0x115$. The photometric

**Table 3** A sampling of additional fields available for baseline TIFF images

| Name | Tag | Type | Description |
|---|---|---|---|
| Artist | 0x13B | ASCII | The individual who created the image |
| Copyright | 0x8298 | ASCII | Copyright notices including names, dates and statements of claims |
| DateTime | 0x132 | ASCII | pointer to a 20 byte date and time string with format YYYY:MM:DD HH:MM:SS |
| Make | 0x10F | ASCII | Make of scanner |
| Model | 0x110H | ASCII | Model number of scanner |
| MinSampleValue | 0x118 | Short | Minimum sample value for each component Default = 0 |
| MaxSampleValue | 0x119 | Short | Maximum sample value for each component Default = $2^{\text{BitsPerSample}} - 1$ |
| Orientation | 0x112 | Short | Orientation of the image with respect to rows and columns |

interpretation value is set to 2 to indicate a color image and the Bits per Sample value is set to 8, 8, 8.

While these fields are required in baseline TIFF there are not by any means the only fields available for users of baseline tiff. In Table 3 we present a sampling from [14] where a more complete set can be obtained.

### TIFF Extensions

The TIFF extensions, which may or may not be supported by all TIFF readers, mostly fall into three classes, though there are fields that do not fit into these classes, compression, storage, and display.

The number of compression formats allowed in the baseline implementation of TIFF is very limited. Essentially, the baseline TIFF allows for no compression or some variation of runlength coding, PackBits or, in the case of bilevel images, modified Huffman codes. The TIFF extensions allow for a much wider range of compression schemes and hence there are a number of additional fields which contain information required by these compression schemes. The additional schemes include those available in the CCITT T4 and T6 standards described earlier, an LZW-based scheme very similar to the one used in GIF and the JPEG compression scheme.

In the baseline TIFF there is not much attention given to how the data are to be stored with the only method being a raster scan. While the image can be partitioned the partitioning is only in the form of strips with fixed width. The extensions allow for tiling options which are particularly useful when storing large images. For the tiling option there are fields that indicate the dimensions of the tiles and their locations.

Finally, the extensions allow for enhancements in the display of the image including half-toning options and the inclusion of alpha channels. The kinds of images addressed by this format have also been extended to include YCbCr images, CMYK images, and CIE L*a*b* images.

# 4 Lossy Compression

Lossy compression is predicated on the assumption that there is information in the data, in images in this case, which is not essential, and can be traded off for compression advantage. The difficulty lies in identifying the information which can be discarded. In image compression there are two ways that are most often used for decomposing images to facilitate the process of selectively representing information. These are through transforms and through filtering. Though both can be viewed in a unified framework, for our purposes here it is most convenient to deal with them separately.

## *4.1 Transform Coding*

In transform coding the image is divided up into blocks and transform coefficients for each block are generated using a linear transform. Linear transforms, in particular the ones we are interested in, can be viewed in three ways. We can view them as rotations of the coordinate axes, as methods of decorrelating sequences, and as means for representing sequences in terms of a basis set. Consider a $2 \times 1$ block of pixels. Because pixel values tend to be correlated, if we treat the two pixel values as the $x$ and $y$ coordinates a of a point in two-dimensional space, we expect these points to cluster around a line at a 45° angle as shown in Fig. 8. In order to represent the value of the pixels in its original untransformed form requires two values. However, if we rotate the axes by 45°, we can see that most of the information about both the pixels is contained in the location along only one of the transformed axes. If we only kept this one value and then tried to reconstruct the two pixels we would introduce some distortion. However, it is clear that the amount of distortion would be significantly less than would have occurred if we simply deleted one of the pixels. Furthermore, and perhaps more importantly, the relationship between the pixels is, for the most part, preserved. This transformation can be accomplished by multiplying the vector of pixels with a transformation matrix. Given a block of pixels $X$ and a transformation matrix $A$ the transformed value, or transform coefficients are obtained as:

$$\Theta = X^T A X$$

where $\Theta$ is the matrix of transform coefficients. We can show [10] that the transform coefficients are weights associated with a set of basis matrices used to represent the image. In other words:

$$X = \sum_{i,j} \theta_{i,j} B_{i,j}$$

where $B_{i,j}$ are the set of basis images which are obtained by taking the outer products of the rows of the transform matrix. The particular transform that we are interested in

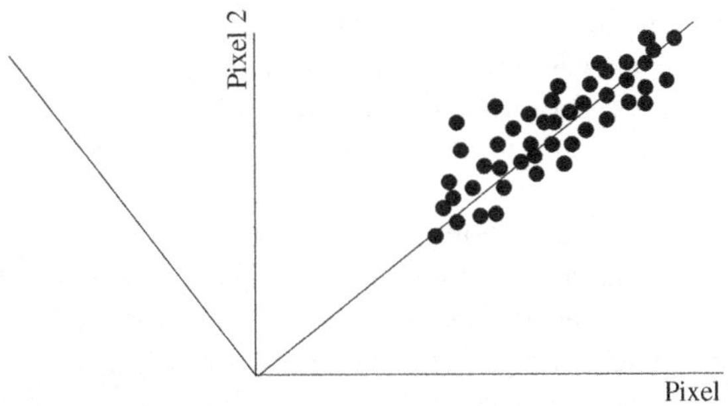


**Fig. 8** Plotting pairs of pixels in two dimensions

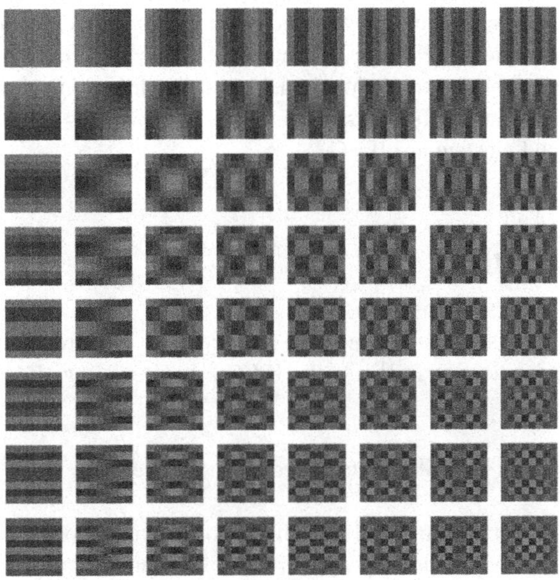

**Fig. 9** Basis images for the $8 \times 8$ discrete cosine transform

is the discrete cosine transform (DCT). The components of the $N \times N$ DCT matrix are cosine values.

$$[\mathbf{A}]_{i,j} = \begin{cases} \sqrt{\frac{1}{N}} \cos \frac{(2j+1)i\pi}{2N} & i = 0,\ j = 0, 1, \ldots, N-1 \\ \sqrt{\frac{2}{N}} \cos \frac{(2j+1)i\pi}{2N} & i = 1, 2, \ldots, N-1,\ j = 0, 1, \ldots, N-1. \end{cases}$$

The basis images for the DCT are shown in Fig. 9.

We can see that the basis images represent different levels of variation within the block of $8 \times 8$ pixels. The transform coefficients indicate the different amount of slowly varying or low frequency and fast varying or high frequency content. In particular, the coefficient of the basis matrix in the top left is called the DC coefficient because it is analogous to a zero frequency or direct current component.

The remaining coefficients are called AC coefficients. If the amount of low and high frequency was about the same in all images working with the transform coefficients would not give us a major advantage over working with the pixels themselves. There would be some advantage because the human visual system reacts differently to different frequency signals and a transformation to the frequency domain would allow us to use coding resources for each frequency coefficient according to its importance to the visual system. Luckily for us we do not need to turn to human visual system models to achieve coding gains. Because the images of interest to humans tend to be dominated by slowly changing regions the low frequency transform coefficients carry most of the information. Thus, we can devote most of the coding resources to low frequency coefficients. The DCT is especially good at compacting most of the information into a small number of coefficients when neighboring pixel values are highly correlated. This makes it an especially good transform to use for compressing images of interest to human beings. As in the case of the rotation example compression is achieved by discarding those coefficients which do not contain much information. We provide details with the description of the image coding standard JPEG.

## 4.2 JPEG

We first describe the JPEG compression standard and then continue with the format specification in JFIF—the JPEG File Interchange Format and Exif.

### 4.2.1 The JPEG Standard

The JPEG standard specifies two different modes, a sequential mode and a progressive mode. The former is most commonly used so we focus on the sequential mode. Because images consist of only positive valued pixels the DC coefficient of the transform can become very large. Therefore, for convenience the JPEG algorithm requires the pixel values be level shifted by subtracting $2^{p-1}$ from each pixel where $p$ is the bit depth. The algorithm then partitions the image into $8 \times 8$ blocks. If the width or height of the image is not a multiple of eight the algorithm replicates the last row or column until the dimensions are multiples of eight. These replicated rows or columns are removed during decoding. Each $8 \times 8$ block is transformed using the DCT. The resolution of the coefficients is generally higher than necessary for human viewing. This excess resolution costs bits to encode so the algorithm gets rid of the excess resolution through the process of quantization. In JPEG quantization is accomplished by dividing the coefficient by an integer value and then rounding up to the nearest integer. During decoding this value is multiplied by the integer value used to reduce the resolution. The integer used for resolution reduction is different for each coefficient and is selected based on the importance of the coefficient for visual perception. The standard allows user defined values for this set of integers, called

**Table 4** Sample quantization table for the $Y$ component

| | | | | | | | |
|---|---|---|---|---|---|---|---|
| 16 | 11 | 10 | 16 | 24 | 40 | 51 | 61 |
| 12 | 12 | 14 | 19 | 26 | 58 | 60 | 55 |
| 14 | 13 | 16 | 24 | 40 | 57 | 69 | 56 |
| 14 | 17 | 22 | 29 | 51 | 87 | 80 | 62 |
| 18 | 22 | 37 | 56 | 68 | 109 | 103 | 77 |
| 24 | 35 | 55 | 64 | 81 | 104 | 113 | 92 |
| 49 | 64 | 78 | 87 | 103 | 121 | 120 | 101 |
| 72 | 92 | 95 | 98 | 112 | 100 | 103 | 99 |

**Table 5** Sample quantization table for the Cb and Cr components

| | | | | | | | |
|---|---|---|---|---|---|---|---|
| 17 | 18 | 24 | 47 | 99 | 99 | 99 | 99 |
| 18 | 21 | 26 | 66 | 99 | 99 | 99 | 99 |
| 24 | 26 | 56 | 99 | 99 | 99 | 99 | 99 |
| 47 | 66 | 99 | 99 | 99 | 99 | 99 | 99 |
| 99 | 99 | 99 | 99 | 99 | 99 | 99 | 99 |
| 99 | 99 | 99 | 99 | 99 | 99 | 99 | 99 |
| 99 | 99 | 99 | 99 | 99 | 99 | 99 | 99 |
| 99 | 99 | 99 | 99 | 99 | 99 | 99 | 99 |

the quantization table; it also provides example tables. As these example tables work rather well, most JPEG applications use the example tables rather than generating their own. The quantization table for the $Y$ component is shown in Table 4.

Looking at the values in the quantization table it is clear that lower frequency coefficients, i.e., those in the top left corner are going to be represented with higher resolution than higher frequency coefficients. The quantization table for the Cb and Cr components is shown in Table 5. Comparing the two tables we can see a reflection of the fact that the vision system prefers higher resolution in the luminance component of images than in the chrominance components. Not only is less resolution required in higher frequency components, and the magnitudes of these components is generally smaller. Therefore, the quantized value of these coefficients is often zero.

To see this consider the following example from [10]. Table 6 is a typical $8 \times 8$ block of pixels taken from an image.

If we take the DCT of this block after level shifting we get the coefficients shown in Table 7. Notice

When quantized using the luminance quantization table we are left with the set of quantized values shown in Table 8. Most of the coefficients have been quantized to zero. In fact, there are only four non-zero values. The final step in the compression process is to encode these quantized coefficients. The JPEG standard encodes the DC coefficient and the rest of the coefficients differently. The DC coefficient is essentially the average value of the block. As such it usually will not differ substantially from the DC coefficient of the neighboring block. Therefore, instead of coding the DC coefficient directly the algorithm encodes the difference between the DC coefficients

**Table 6** An 8 × 8 block from an image

| | | | | | | | |
|---|---|---|---|---|---|---|---|
| 124 | 125 | 122 | 120 | 122 | 119 | 117 | 118 |
| 121 | 121 | 120 | 119 | 119 | 120 | 120 | 118 |
| 126 | 124 | 123 | 122 | 121 | 121 | 120 | 120 |
| 124 | 124 | 125 | 125 | 126 | 125 | 124 | 124 |
| 127 | 127 | 128 | 129 | 130 | 128 | 127 | 125 |
| 143 | 142 | 143 | 142 | 140 | 139 | 139 | 139 |
| 150 | 148 | 152 | 152 | 152 | 152 | 150 | 151 |
| 156 | 159 | 158 | 155 | 158 | 158 | 157 | 156 |

**Table 7** The DCT coefficients corresponding to the block of pixels in Table 6 after level shift

| | | | | | | | |
|---|---|---|---|---|---|---|---|
| 39.88 | 6.56 | −2.24 | 1.22 | −0.37 | −1.08 | 0.79 | 1.13 |
| −102.43 | 4.56 | 2.26 | 1.12 | 0.35 | −0.63 | −1.05 | −0.48 |
| 37.77 | 1.31 | 1.77 | 0.25 | −1.50 | −2.21 | −0.10 | 0.23 |
| −5.67 | 2.24 | −1.32 | −0.81 | 1.41 | 0.22 | −0.13 | 0.17 |
| −3.37 | −0.74 | −1.75 | 0.77 | −0.62 | −2.65 | −1.30 | 0.76 |
| 5.98 | −0.13 | −0.45 | −0.77 | 1.99 | −0.26 | 1.46 | 0.00 |
| 3.97 | 5.52 | 2.39 | −0.55 | −0.051 | −0.84 | −0.52 | −0.13 |
| −3.43 | 0.51 | −1.07 | 0.87 | 0.96 | 0.09 | 0.33 | 0.01 |

**Table 8** The quantizer labels obtained by using the quantization table on the coefficients

| | | | | | | | |
|---|---|---|---|---|---|---|---|
| 2 | 1 | 0 | 0 | 0 | 0 | 0 | 0 |
| −9 | 0 | 0 | 0 | 0 | 0 | 0 | 0 |
| 3 | 0 | 0 | 0 | 0 | 0 | 0 | 0 |
| 0 | 0 | 0 | 0 | 0 | 0 | 0 | 0 |
| 0 | 0 | 0 | 0 | 0 | 0 | 0 | 0 |
| 0 | 0 | 0 | 0 | 0 | 0 | 0 | 0 |
| 0 | 0 | 0 | 0 | 0 | 0 | 0 | 0 |
| 0 | 0 | 0 | 0 | 0 | 0 | 0 | 0 |

of neighboring blocks. The difference is encoded using a Huffman code. Because the number of possible values that need to be encoded is quite large it would be cumbersome to have a Huffman code for each value. The JPEG standard organizes the possible values into categories of variable size. Category 0 contains only the value 0. Category 1 contains two values −1 and 1. Category 2 contains four values −3, −2, 2, and 3, and so on. With each succeeding category containing twice as many elements as the preceding category, there is a Huffman code for the category followed by disambiguation bits to identify the particular value in the category. Thus, if the value fell in category 3 the encoder would generate the Huffman code for category 3 followed by three bits to differentiate between the eight members of category 3.

The rest of the coefficients are organized into a single sequence using a zigzag scanning pattern as shown in Fig. 10. Once the coefficients have been organized into

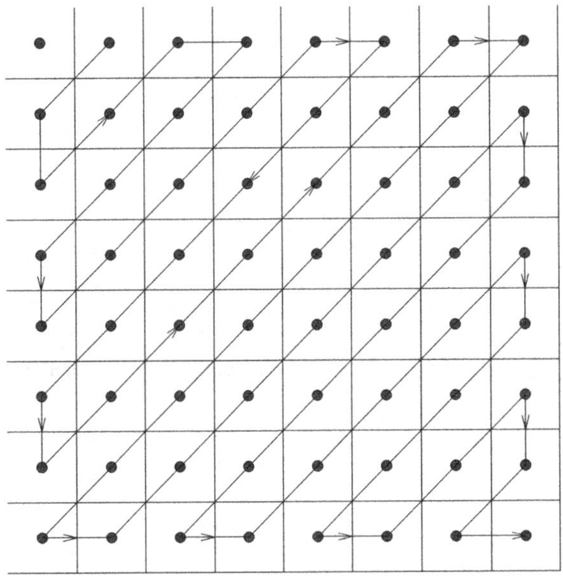

**Fig. 10** The JPEG zigzag scanning pattern for AC coefficients

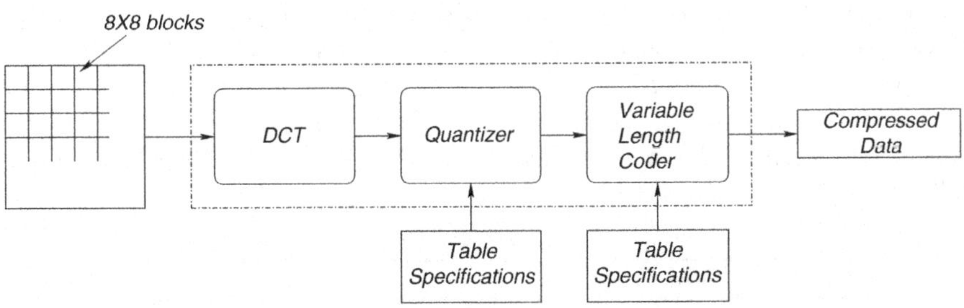


**Fig. 11** A block diagram of the JPEG compression algorithm

a vector the vector is scanned for nonzero. Each non-zero value is classified into a category. The category number and the number of zero values since the last non-zero value are used to generate a pointer into a code table. If a particular coefficient has the last non-zero value in the zigzag scan the code for that value is followed by a special code signaling the end of the block. A block diagram of the overall scheme is shown in Fig. 11.

### 4.2.2 JFIF

The JFIF format is the most common format used for storing JPEG encoded images. It uses a subset of the markers specified in the JPEG standard. These markers identify

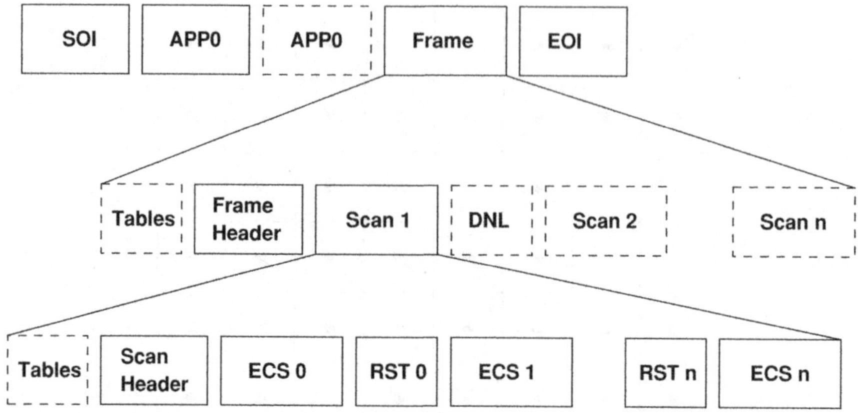


**Fig. 12** Syntax of the JFIF file

the various structural portions of the compressed data formats. Markers are assigned a two-byte code with the first byte being FF. The second byte of the marker code is neither FF nor 0. Markers can be standalone. Examples of these are the SOI (start of image) and EOI (end of image) markers which are FFD8 and FFD9. All JFIF files have to start with an SOI marker and end with an EOI marker. Other markers carry a payload. For these markers the first two bytes following the marker contains a length count. The length count includes the two bytes containing the count but not the two bytes that make up the marker. In JFIF the SOI marker is followed by the application-specific APP0 marker which has a code of FFE0. The APP0 format is as follows

| Field | Size | Content |
|---|---|---|
| Length | 2 bytes | Length of the header |
| Identifier | 5 bytes | 0x4A 0x46 0x49 0x46 0x00 (JFIF0) |
| Version | 2 bytes | MSB for major revision, LSB for minor revisions |
| Units | 1 byte | 0: no units, 1: dots per inch, 2: dots per cm |
| Xdensity | 2 bytes | Horizontal pixel density |
| Ydensity | 2 bytes | Vertical pixel density |
| Xthumbnail | 1 byte | Thumbnail horizontal pixel count |
| Ythumbnail | 1 byte | Thumbnail vertical count |
| $(\text{RGB})_n$ | $3n$ bytes | 24 bit RGB values for thumbnail<br>$n = \text{Xthumbnail} \times \text{Ythumbnail}$ |

A representation of the structure of the JFIF file is shown in Fig. 12. In this representation optional elements are shown as dashed boxes.

The APP0 header can be followed by an optional JFIF APP0 header that contains a thumbnail in a different format. The APP0 marker is followed by optional table markers. There are two table markers DHT (define Huffman table) with code FFC4 and DQT (define quantization table) with code FFDB. The syntax of the DHT marker

| FFC4 | Length | Class | Dest | # codes length 1 | # codes length 2 | • • • | # codes length 16 | Symbol Length Assignment |
|---|---|---|---|---|---|---|---|---|

**Fig. 13** Syntax of the Huffman table marker

| FFDB | Length | Prec. | Dest | Q[1,1] | Q[1,2] | Q[2,1] | Q[3,1] | Q[2,2] | • • • | Q[8,8] |
|---|---|---|---|---|---|---|---|---|---|---|

**Fig. 14** Syntax of the quantization table marker

is shown in Fig. 13. After the two bytes that contain the length the most significant four bits of the next byte indicates whether the Huffman code is for the DC coefficient or the AC coefficients. The number of Huffman codewords of lengths 1 through 16 is stored in the next 16 bytes. This is followed by a variable number of bytes which describe the Huffman codes in terms of their lengths.

The syntax of the quantization table marker is shown in Fig. 14. The most significant four bits specify whether the quantization table values are stored in one or two bytes. A value of 0 indicates a single byte is used for each quantization value while a value of 1 indicates 2 bytes are used for each quantization value. The next 64 bytes, or 128 bytes contain the 64 quantization table values stored in zigzag order.

Optional table marker segments are followed by the frame header which indicates the start of a frame. The first two bytes of the frame header make up the start of frame marker. While there are a number of possible start of frame markers the most commonly used marker is $SOF_0$ with value FFC0. The subscript 0 indicates that the frame is encoded using baseline DCT. The next two bytes specify the frame header length. The byte that follows gives the precision in bits per sample. The next four bytes specify the dimensions of the image; two bytes for number of lines followed by two bytes for number of samples per line. The following byte specifies the number of components in the image. For each component we get a component identifier, a horizontal sampling factor, and a vertical sampling factor. The latter two indicate if the chrominance components of the image have been subsampled. For each component we also have a pointer to the quantization table to be used with the component.

The frame header can be followed by a number of scans. A new scan begins with a DNL marker segment which has a value of FFDC. The DNL marker segment defines the number of lines in the next scan. This is used to reset the dimension of the image originally set in the frame header. The scan can begin with optional tables followed by the scan header. The first two bytes in the scan header is the start of scan (SOS) marker followed by the length, number of components, and the component parameters as in the frame header. The header is followed by the entropy coded data interleaved with restart markers. The restart marker specifies the number of minimum coded units (MCU) that are present between the Define Restart Interval (DRI) markers. These markers provide a degree of robustness to the JPEG datastream. One of the drawbacks of using variable length codes is that bit errors tend to propagate. The DRI markers prevent the error from propagating beyond a restart interval.

The file is terminated with an EOI marker.

### 4.3 Wavelet-Based Compression

While transforms are an efficient way of decomposing a signal their use in practice suffers from several drawbacks. A major source of these drawbacks is the fact that the transform treats the source as a stationary process. This not being the case it is necessary to break up the image into regions that are, for want of a better term, locally stationary. In practice we do this by breaking the image into square blocks, which in the case of JPEG are of size $8 \times 8$. While this helps both in terms of computational cost and compression there are unwanted side effects. The most ubiquitous of these is a blocking effect. As each block in a transform coder is treated independently there are often discontinuities at block boundaries in the reconstruction which are visually unpleasant. One way to get around this problem is through overlapping the blocks as in Lapped Orthogonal Transforms [15]. Another is by using alternative methods of decomposing the image.

Transforms are not the only way of decomposing a signal. A common approach, especially in audio applications is to use filters to separate the signal into different frequency bands. The frequency of a signal is a measure of the rate of change of the signal value. Thus, a high frequency signal varies more rapidly than a low frequency signal. The speed at which a signal varies also determines the rate at which it needs to be sampled in order to preserve its information content. A result usually attributed to Nyquist states that a signal with a bandwidth of $B$ Hz can be exactly reconstructed if it is sampled at more than $2B$ samples per second. These two ideas of decomposition and sampling are combined in a compression approach known as *subband coding* [10]. The signal to be compressed is decomposed into different frequency bands using linear filters. As the decomposed frequency bands have a lower bandwidth than the original signal they can be represented using fewer samples. Then, using the same argument as was used for transform coding the different frequency bands can be assigned different amount of coding resources depending on the amount of energy in each band and the perceptual importance of the particular band resulting in compression. Decomposition using linear filters usually implies a representation of the signal using a sinusoidal basis set. The use of wavelets provides the opportunity for selection of the basis functions used for decomposition from a much richer set. The implementation of the wavelet decomposition in the compression context is generally similar to that of subband coding and the wavelet decomposition can be implemented using linear filters. A block diagram of the subband/wavelet encoder and decoder for a one-dimensional signal is shown in Fig. 15.

As an example of the wavelet decomposition consider decomposition using the Haar wavelet. The signal $\{x_n\}$ is decomposed into two bands $\{y_{L,n}\}$ and $\{y_{H,n}\}$

$$y_{L,n} = \frac{1}{\sqrt{2}}(x_n + x_{n-1}) \tag{1}$$

$$y_{H,n} = \frac{1}{\sqrt{2}}(x_n - x_{n-1}) \tag{2}$$

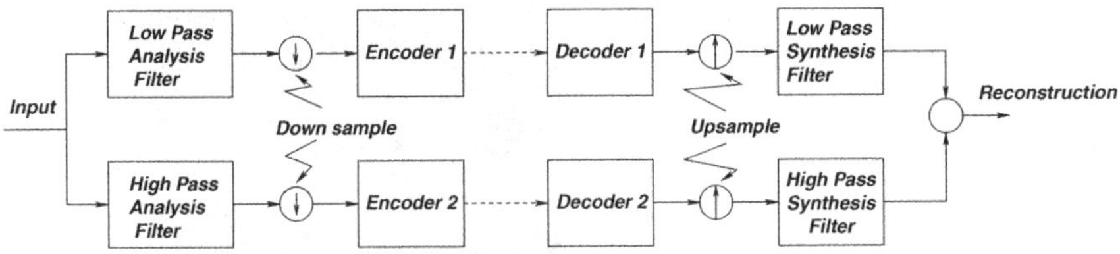


**Fig. 15** Subband or wavelet decomposition followed by downsampling and separate encoders for each downsampled sequence make up the encoder

where $\{y_{L,n}\}$ is the *low-pass* signal and $\{y_{H,n}\}$ is the *high-pass* signal. In this case clearly $\{x_n\}$ is a one-dimensional signal such as speech or music. In the case of two-dimensional signals such as images there is need for two-dimensional filters. However, most applications use a separable transform in which first each row of the image is filtered and then the filtered rows are again filtered as columns. Thus, one decomposition of an image results in four bands:

1. Low-pass filtering of the rows followed by low-pass filtering of the columns—the Low-Low (LL) band.
2. Low-pass filtering of the rows followed by high-pass filtering of the columns—the Low-High (LH) band.
3. High-pass filtering of the rows followed by low-pass filtering of the columns—the High-Low (HL) band.
4. High-pass filtering of the rows followed by high-pass filtering of the columns—the High-High (HH) band.

One of the results of the filtering operation is the reduction by half of the bandwidth. This means that the number of samples can also be decreased by half. Thus, for an $N \times N$ image the subband coefficient matrices are of size $N/2 \times N/2$. For display purposes the four sets of coefficients are represented as $N/2 \times N/2$ images which are usually displayed as four quadrants of an $N \times N$ image. As an example consider the Haar decomposition of the *Sinan* image shown in Fig. 16.

Note, for this example, that most of the energy is contained in the LL band. This is true of most images of interest. The degree to which the energy is compacted into the LL band depends on the efficiency of the wavelet basis. Although the Haar wavelet is not particularly good for compacting energy into the LL band we can see from this example that even this inefficient compaction results in most of the information being contained in the LL band. The LL band can be further decomposed into four bands using the same filters thus resulting in a seven-band decomposition. Of these seven bands, there will be four bands of $N/4 \times N/4$ coefficients and three bands of f $N/2 \times N/2$ coefficients. The decomposition can be continued by decomposing the lowest frequency band once more to give a 10-band decomposition with four bands of $N/8 \times N/8$ coefficients, three bands of $N/4 \times N/4$ coefficients, and three bands of $N/2 \times N/2$ coefficients. A conceptual drawing of a 10-band decomposition is shown in Fig. 17.

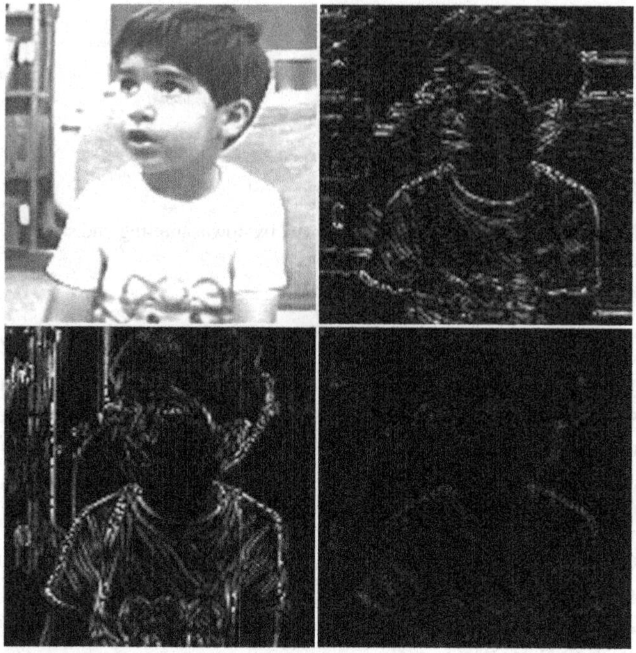

**Fig. 16** Haar decomposition of the *Sinan* image. The *top left* image is the LL image, the *top right* image is the LH image, and the *bottom left* image is the HL image while the *bottom right* image is the HH image

**Fig. 17** A 10-band decomposition of an image

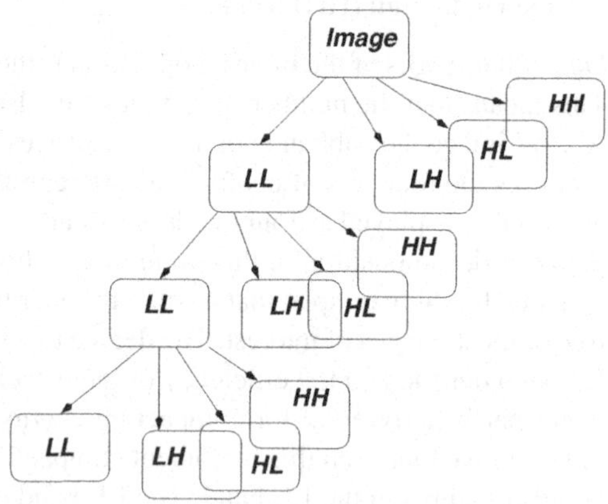


While the different bands contain different amounts of energy coefficients corresponding to the same spatial location will generally be related. In particular, if the coefficient in the lowest band has a value less than a threshold then it is quite likely that the corresponding coefficients in the higher bands will also have a value less than that threshold. This relationship allows for an efficient coding of the coef-

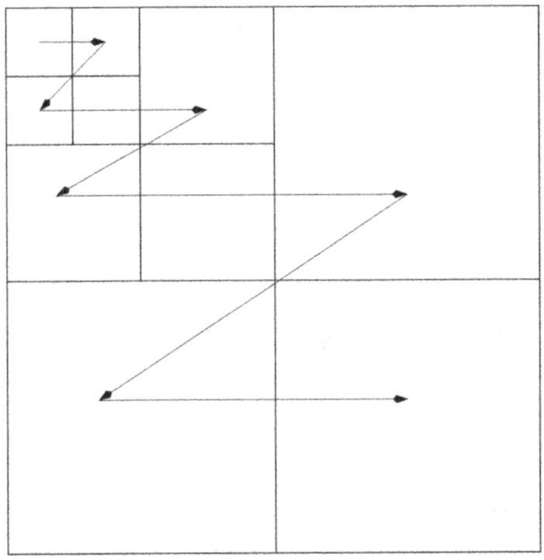

**Fig. 18** The coefficient scan used by the Embedded Zerotree Wavelet (EZW) coder

ficients. All coefficients corresponding to the same spatial location are organized as trees, where the root of the tree is the lowest band coefficient. The coefficient trees are encoded at different *significance* levels making use of the fact that when a coefficient is not significant at a particular level its offsprings in the tree are also likely not to be significant at that level. A coefficient is significant at a particular level if the magnitude of the coefficient is greater than a corresponding threshold. By using a sequence of progressively decreasing thresholds more and more of the coefficients become significant and a finer and finer resolution of the image can be reconstructed. Furthermore, the coding is *embedded* in which the bitstream can be truncated at any point resulting in a lower resolution reconstruction of the image. Instead of using a threshold to determine significance we can also use the most significant non-zero bit in the quantized representation of the coefficients to determine significance. Thus, an eight-bit quantization will result in eight levels of significance. In order for the coding to be embedded the encoder needs to scan the coefficients in a manner that will allow reconstruction from truncated bitstreams. This means that all coefficients need to be scanned in multiple passes with each pass filling in a bitplane. The particular scan used in one of the first wavelet compression algorithms is shown in Fig. 18.

This approach toward coding was first proposed by Shapiro [16] in the embedded zerotree wavelet (EZW) algorithm and refined by Said and Pearlman [17] in the set partitioning in hierarchical trees (SPIHT) algorithm. Both of these approaches resulted in an embedded bitstream that were resolution scalable. This approach was further extended by the layered zonal coding algorithm proposed by Taubman and Zakhor [18], and the embedded block coding with optimized truncation (EBCOT) algorithm proposed by Taubman [19]. The EBCOT algorithm is the compression algorithm used in JPEG2000 so we briefly review some of its characteristics.

While EBCOT uses the basic ideas of wavelet compression developed in EZW and SPIHT it trades off a small amount of the compression performance in order to

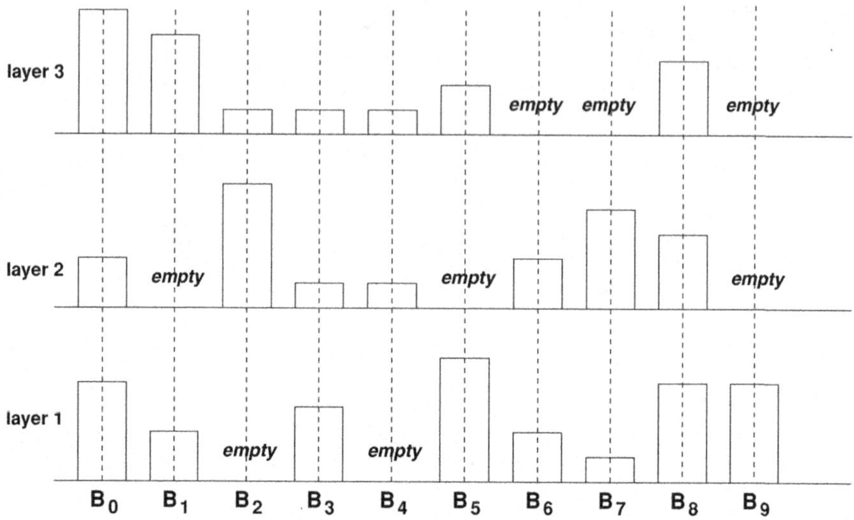


**Fig. 19** Three quality layers for a fictitious example using 10 blocks

get a much more feature-rich bitstream. One of the drawbacks of image compression is that it limits the ability to access parts of the image without decompressing the entire image. EBCOT provides random access capability by breaking up the image into *code-blocks* $B_i$ which are separately coded into embedded bitstreams. These bitsteams have truncation points $R_i^n$ which can be used to generate reconstructions at different resolution levels. Finally, the bitstream is divided into quality levels $Q_q$ which contain contributions from different blocks where some blocks may not contribute any bits to a particular quality level. An example of the distribution of bits for generating different quality levels is shown in Fig. 19.

Thus, the EBCOT algorithm provides a degree of random access, resolution scalability, and, through the use of quality levels, SNR scalability.

The actual bitstream representing the quantized coefficients is generated by an context-based arithmetic coder.

## 4.4 JPEG2000

The JPEG2000 standard is principally based on the EBCOT algorithm. While it provides arguably modest improvements in performance, the major contribution of this standard is its flexibility [20]. To the already feature-rich bitstream generated by EBCOT the standard adds some more flexibility through the definition of *tiles*. Tiles are rectangular partitions of the image, defined with respect to a high resolution grid, which are coded independently allowing for random access as well as editing functions such as cropping. The tile components at each resolution can be further partitioned into *precincts*. Recall that with EBCOT each subband of each component is partitioned into code-blocks. The subband coefficients are quantized using a mid-tread quantizer [10] and the bitplanes encoded using a three-pass procedure which generates an embedded bitstream. The bits are further encoded using a context-dependent adaptive arithmetic coder.

The data are packetized with a packet containing the binary codestream for each code-block within a precinct. The information about the data for each codeblock contained in the packet is contained in the packet header. The packet is indexed by layer, resolution, precinct, and component. These packets are combined into a *layer*. Thus, the layer contains contributions from all the code-blocks in a tile. The data from a tile can be broken down into *tile parts* which can be interleaved with other tile parts to provide flexible access.

As in the case of JPEG the bitstream is organized using markers. As with JPEG the markers are two bytes long with the first byte being 0xFF. There are six kinds of markers, delimiting markers, fixed information markers, functional markers, pointer markers, in bitstream markers, and informational markers.

### Delimiting Markers

There are four delimiting markers which are used to frame the header and the data. The start of codestream (SOC) marker with code 0xFF4F appears at the beginning of the codestream and is required as the first marker in the main header. The start of tile part (SOT) marker with code 0xFF90 is the first marker of each tile and is required at the beginning of each tile part header. The start of data (SOD) marker with code 0xFF93 is the last marker of the tile part header and separates the header from the data. Finally, the end of codestream (EOC) marker with code 0xFFD9 terminates the codestream. This is required as the last marker in the codestream.

### Fixed Information Marker

The only fixed information marker is the image and tile size marker (SIZ) with code 0xFF51. The segment following this marker contains information about the uncompressed image. The structure of this segment of the codestream is as follows:

A graphical representation of these parameters is shown in Fig. 20.

The relationships of the tile parameters with respect to the reference grid can be seen in Fig. 21.

### Functional Markers

There are six functional marker segments. If these markers and corresponding segments are found in the main header they describe the functions used to code the entire image. If they are found in the tile header they describe the functions used to code

| Parameter | Size (bytes) | Description |
|---|---|---|
| SIZ | 2 | Marker (0xFF51) |
| Lsiz | 2 | Length of segment in bytes |
| Rsiz | 2 | |
| Xsiz | 4 | Width of reference grid |
| Ysiz | 4 | Height of reference grid |
| XOsiz | 4 | Horizontal offset from grid origin to left of image area |
| YOsiz | 4 | Vertical offset from grid origin to top of image area |
| XTsiz | 4 | Width of one reference tile with respect to grid |
| YTsiz | 4 | Height of one reference tile with respect to grid |
| XTOsiz | 4 | Horizontal offset from grid origin to left side of first tile |
| YTOsiz | 4 | Vertical offset from grid origin to top side of first tile |
| Csiz | 2 | Number of components in the image |
| $Ssiz^i$ | 1 | Precision in bits and sign of the $i$th component |
| $XRSiz^i$ | 1 | Horizontal separation of $i$th component with respect to grid |
| $YRSiz^i$ | 1 | Vertical separation of $i$th component with respect to grid |

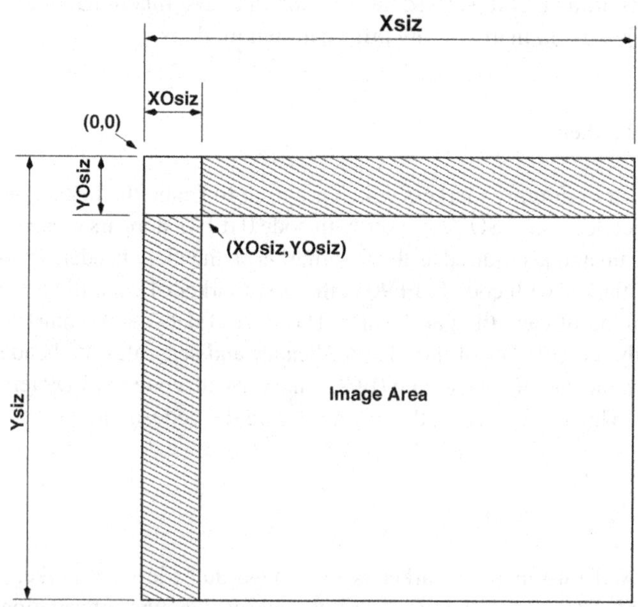


**Fig. 20** Some of the parameters in the SIZ marker with respect to a reference grid [21]

the tile. These include the coding style marker (COD), the coding style component (COC) marker, the region of interest (RGN) marker, the quantization default (QCD) marker, the quantization component (QCC) marker, and the progression order change default (POD) marker.

The coding style (COD) marker with code 0xFF52 is followed by a segment that describes the type of entropy coding used, the number of decomposition levels with

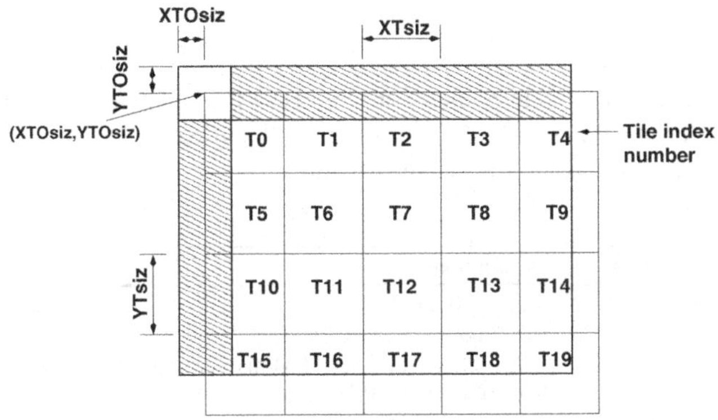


**Fig. 21** Some of the tiling parameters in the SIZ marker with respect to a reference grid [21]

zero implying no transform, the progression order, and the number of layers. Other parameters include the code-block dimensions, the wavelet transform used, and the multiple component transform used. As noted above this segment can appear in the main header or in the tile header depending on the scope of its application.

The COD marker can be overridden for a particular component by the coding style component (COC) marker. The marker coded as 0xFF53 is followed by a segment applicable to one or more components which contain information about the decomposition level, code-block dimensions, transform, and packet partition size.

The ability to display regions of interest with higher accuracy than other regions is a very useful aspect of the JPEG 2000 standard. The region of interest (RGN) marker has a value of 0xFF5E and describes how many components region of interest coding will be applied to as well the method of coding. Currently, the only available method is to shift up the coefficient forcing the significant bits into higher bitplanes thus allowing them to enter the bitstream earlier and allowing for more refinement of the coefficients in the region of interest.

The quantization default (QCD) marker with a value of 0xFF5C specifies the quantization style for all components and the quantizer stepsize values for all subbands. The QCD marker can be overridden for a particular component or set of components by the quantization component (QCC) marker. The marker value is 0xFF5D and the parameters are the same as those in the QCD marker segment.

The final functional marker is the progression order change (POD) marker which overrides the progression field of the COD marker segment. Note that JPEG 2000 allows for five different orders of progression including by increasing resolution or increasing quality.

Pointer Markers

The pointer markers provide pointers or lengths in the tile parts. The TLM marker (0xFF55) describes the length of every tile part. This segment optionally occurs in the main header and has the following structure:

| Parameter | Size (bytes) | Description |
|---|---|---|
| TLM | 2 | Marker (0xFF55) |
| Ltlm | 2 | Length of segment in bytes |
| Ztlm | 1 | Index relative to other TLM marker segments in the header |
| Stlm | 1 | Size of Ttlm and Ptlm parameters |
| $\text{Ttlm}^i$ | 0,1, or 2 | Tile number of the $i$th tile part |
| $\text{Ptlm}^i$ | 2 or 4 | Length in bytes from beginning of SOT marker to end of data for the $i$th tile part |

The packet length (PLM) marker is an optional marker that can appear in the main header. The PLM segment provides a list of packet lengths in the tile parts. The structure of the PLM segment is as follows:

| Parameter | Size (bytes) | Description |
|---|---|---|
| PLM | 2 | Marker (0xFF57) |
| Lplm | 2 | Length of segment in bytes |
| Zplm | 1 | Index relative to other PLM marker segments in the header |
| $\text{Nplm}^i$ | 1 | Number of bytes of Iplm information for $i$th tile part<br>One value for each tile part |
| $\text{Iplm}^{i,j}$ | variable | Length of $j$th packet in the $i$th tile part |

The counterpart to the PLM is the PLT marker which also indicates the list of packet lengths but appears in the tile part header rather than the main header. The PLT and PLM markers are both optional and can be used together or separately. The structure of the PLT marker segment is similar to that of the PLM marker segment albeit containing less information.

| Parameter | Size (bytes) | Description |
|---|---|---|
| PLT | 2 | Marker (0xFF58) |
| Lplt | 2 | Length of segment in bytes |
| Zplt | 1 | Index relative to other PLT marker segments in the header |
| $\text{Iplt}^i$ | variable | Length of $i$th packet. |

The packet headers can be organized in one of three ways. All the packet headers can be placed together using a PPM marker. This segment can occur only in the main header and is organized as follows:

| Parameter | Size (bytes) | Description |
|---|---|---|
| PPM | 2 | Marker (0xFF60) |
| Lppm | 2 | Length of segment in bytes |
| Zppm | 1 | Index relative to other PPM marker segments in the header |
| $Nppm^i$ | 4 | Number of bytes of Ippm information for $i$th tile part<br>One value for each tile part |
| $Ippm^{i,j}$ | variable | Packet header for every packet in order in the tile part The component number, layer and resolution determined from method of progression or POD. One value for each packet in the tile part |

The second approach is to place packed packet headers in the tile part header using the marker PPT. The structure of the segment is the same as that of the PPM except only the headers of the packets in the tile part are included.

| Parameter | Size (bytes) | Description |
|---|---|---|
| PPT | 2 | Marker (0xFF61) |
| Lppt | 2 | Length of segment in bytes |
| Zppt | 1 | Index relative to other PPT marker segments in the header |
| $Ippt^i$ | variable | Packet header for every packet in order in the tile part. The component number, layer and resolution determined from method of progression or POD. One value for each packet in the tile part |

Finally the packet headers can be distributed in the bitstream.

## In Bitstream Markers

There are two marker segments that are found in the bitstream rather than in the headers. They are the start of packet (SOP) marker segment and the end of packet (EPH) marker. The SOP segment is organized as follows:

| Parameter | Size (bytes) | Description |
|---|---|---|
| SOP | 2 | Marker (0xFF91) |
| Lsop | 2 | Length of segment in bytes |
| Nsop | 2 | Packet sequence number starting from 0. the numbering is modulo 65,536 |

The end of packet marker has a value of 0xFF92.

## Informational Marker

There is only one informational marker "Comment and Extension" (CME) which can be used an unrestricted number of times in the main or tile part header. The segment consists of the marker followed by the length of the segment in two bytes, followed by two bytes of registration value and an undefined number of bytes of unstructured data.

## 4.5 The JP2 File Format

The JPEG 2000 codestream contains provisions for only a single informational marker. Instead of storing metadata within the codestream the standard suggests the use of the JP2 file format for storing application-specific metadata with the JPEG 2000 codestream. The JP2 file format is a binary container for the JPEG 2000 codestream and uses the concept of boxes to organize the information. A schematic representation of this organization is shown in Fig. 22.

In this schematic representation required data structures are shown in solid boxes while optional structures are shown in dashed boxes. Note that some boxes contain other boxes. These boxes are sometimes referred to as super boxes. Each box consists of four fields. The first four bytes make up the **LBox** field which indicates the length of the box either directly or with a pointer to the **XLBox** field. The next four bytes make up the bf TBox field which indicates the type of the box. The third field is the extended length or **XLBox** field which specifies the length of the box if the value of the **LBox** field is 1. In this case this field consists of eight bytes. The final field, the **DBox**, is made up by the contents of the box.

While the function of most of these boxes is self-evident, in the interest of completion we briefly describe the contents of some of these boxes. The JPEG signature box is a 12-byte header with the first four bytes indicating the length (0×0000 000C). The type of the box is given by the string $jp$ (6A50 2020). And the content of the box is the four-byte character string $<CR><LF><0\times 87><LF>$ (0D0A 870A).

The file-type box specifies the version and compatibility information. They type of the file is $ftyp$ (0×6674 7970) and the contents include the recommendation defining the file, the minor version number, and a compatibility list.

The JP2 header box is a super box containing boxes which specify the image header, the number of bits per component, and the colorspace of the decompressed image. There can be multiple color specification boxes providing for multiple compatibility options for color processing. The super box can also include a number of optional boxes including a box containing a colormap, and a box specifying the capture and default grid resolutions of the image.

The Contiguous Codestream box contains the complete JPEG2000 codestream. The type of this box is $jp2c$ (6A70 3263).

This can be followed by a number of optional box including the IPR box which contains intellectual property rights information and the XML box containing vendor-specific information in XML format. Vendor-specific information can also be included in the UUID box.

**Fig. 22** Schematic representation of a JP2 file [21]

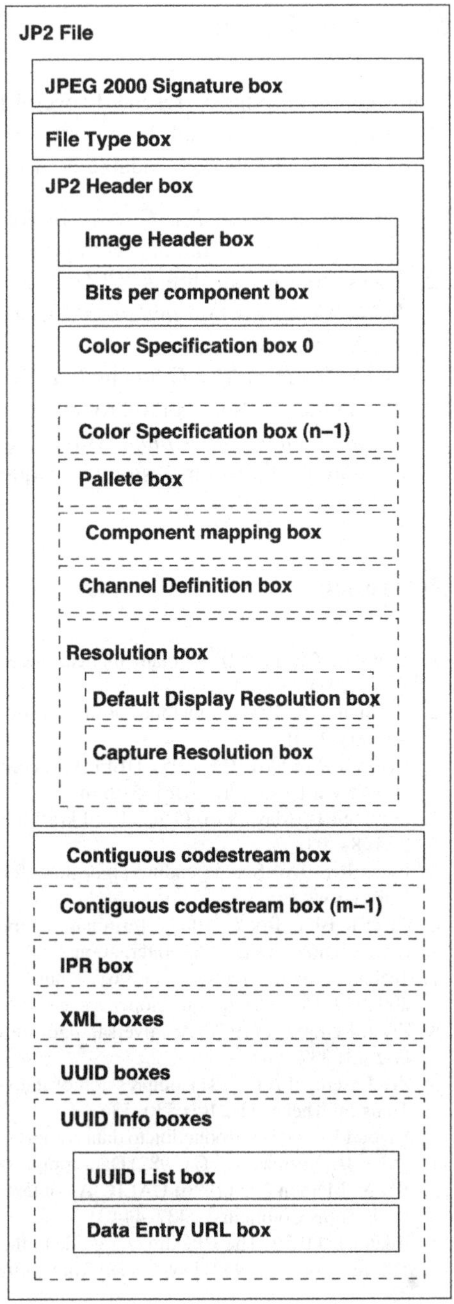

## 5 Further Reading

In this chapter we have presented several lossless and lossy compression formats. The ultimate source for all this information is the standards themselves. However, there are also a number of readable books available on the different formats.

1. *Compressed Image File Formats* by John Miano covers the JPEG, PNG and GIF formats in some detail and with great clarity.
2. The source for all things JPEG 2000 is the comprehensive book *JPEG2000: Image Compression Fundamentals, Standards and Practice* by David Taubman and Michael Marcellin.
3. *PNG: The Definitive Guide* by Greg Roelofs is exactly what it says—the definitive guide to PNG, and fun to read besides.
4. *Introduction to Data Compression* by this author provides an introduction and a bit more for the compression techniques described here.

## References

1. Shannon CE (1948) A mathematical theory of communication. Bell Syst Tech J 27 (379–423):623–656
2. Huffman DA (1951) A method for the construction of minimum redundancy codes. Proc IRE 40:1098–1101
3. Deutsch P (1996) RFC 1951-DEFLATE compressed data format specification version 1.3. http://www.faqs.org/rfcs/rfc1951.htm
4. Rissanen JJ (May 1976) Generalized kraft inequality and arithmetic coding. IBM J Res Dev 20:198–203
5. Pasco R (1976) Source coding algorithms for fast data compression. Ph.D. thesis, Stanford University
6. CCITT Blue Book (1989) Terminal equipment and protocols for telematic services—recommendations t.0–t.63. International Telecommunication Union, Geneva
7. Technical note TN1023 (1996) Understanding packBits. http://web.archive.org/web/20080705155158/http://developer.apple.com/technotes/tn/tn1023.html
8. Ziv J, Lempel A (1977) A universal algorithm for data compression. IEEE Trans Inf Theory IT-23(3):337–343
9. Ziv J, Lempel A (1978) Compression of individual sequences via variable-rate coding. IEEE Trans Inf Theory IT-24(5):530–536
10. Sayood K (2005) Introduction to data compression, 3rd edn. Morgan Kauffman, San Francisco
11. Storer JA, Syzmanski TG (1982) Data compression via textual substitution. J ACM 29:928–951
12. Wu X, Memon ND (1996) CALIC-A context based adaptive lossless image coding scheme. IEEE Trans Commun 45:437–444
13. Adams D (1979) The Hitchhiker's guide to the galaxy
14. Tiff, revision 6.0 (1992) Dowloaded from partners.adobe.com/public/developer/en/tiff/TIFF6.pdf
15. Malvar HS (1992) Signal processing with lapped transforms. Artech House, Norwood
16. Shapiro JM (1992) An embedded heirarchical image coder using zerotrees of wavelet coefficients. In: Proceedings DCC '92, IEEE, pp 215–223
17. Said A, Pearlman WA (1996) A new fast and efficient coder based on set partitioning in hierarchical trees. IEEE Trans Circuits Syst Video Technol 6:243–250

18. Taubman D, Zakhor A (1994) Multirate 3-D subband coding with motion compensation. IEEE Trans Image Process IP-3:572–588
19. Taubman D (2000) High performance scalable image compression with EBCOT. IEEE Trans Image Process IP-9:1158–1170
20. Taubman D, Marcellin M (2001) JPEG2000: image compression fundamentals, standards and practice. Kluwer Academic Press, Boston
21. ISO/IEC JTC1/SC29 WG1 (2004) JPEG 2000 image coding system: core coding system

# Searching and Extracting Digital Image Evidence

**Qiming Li**

**Abstract** Searching and extracting digital images are important tasks in disaster recovery and digital forensics. Many modern file systems arrange data in complex ways, and files are often fragmented, which makes these tasks very challenging, especially when the file systems or the underlying storage media are damaged or corrupted. To design efficient and accurate data recovery systems, we require not only the understanding of the underlying file systems, but also the nature of the data in question. In this chapter, we will discuss the difficulties in searching and extracting digital images from possibly corrupted file systems, and introduce some recent advances.

## 1 Introduction

Over the years, more and more digital devices are used to store important information, including personal computers, mobile phones, thumb drives, and optical discs. This brings new challenges to forensics investigators since now critical evidence may be only in digital form and requires a lot of efforts to be extracted and analyzed. The main difficulties come from the huge and ever-increasing volume of the data that needs to be analyzed, as well as the increasing complexity of the computer systems that are used to store the data.

For example, when a computer is suspected to be involved in the production or distribution of child pornography, a forensic investigator may need to examine the storage devices used by the computer and search for digital images in the disk volumes to see if they contain illegal content. The investigator would have to figure out a way to bypass the security mechanisms on the target computer if there are any, and extract the data from the disk volumes that may have been damaged.

Q. Li (✉)
Institute for Infocomm Research, Singapore, Singapore
e-mail: qiming.li@ieee.org

H. T. Sencar and N. Memon (eds.), *Digital Image Forensics*,
DOI: 10.1007/978-1-4614-0757-7_4,

After that, the file system on the disk volume may be checked for suspicious content. There is a chance that the file system has been damaged or the files deleted, either accidentally during the acquisition of the data, or purposely by the suspect before the computer was captured. It is also possible that the disk storage may contain hundreds of thousands of pictures among which only a few can serve as the evidence of a crime, thus making it very difficult to locate critical information from the given data.

Besides crime fighting, digital forensics and data recovery technologies are also important in disaster recovery. Due to the increasing popularity of digital cameras and built-in cameras on mobile phones, it is not surprising that important multimedia data are often stored in digital forms on flash memories, hard drives, or optical disks. Since these storage devices can be unreliable and become corrupt, it becomes important for users of these devices to be able to recover the original files in unfortunate events.

To find and extract useful image evidence from unreliable or corrupt file systems, we will need to have in-depth understanding of the file systems, as well as the syntactical and semantic structures of the files that we intend to extract and analyze.

Here we will give a brief introduction of commonly used file systems and how data can become difficult to recover due to fragmentation. After that we will see how people have been tackling the problem of data recovery using different techniques and under different assumptions, and how these techniques evolve during the years.

## 2 File Systems and Fragmentation

In many modern file systems, the physical storage media is organized into data blocks. For example, in magnetic and optical disks, the smallest data unit is referred to as a *sector*, and the size of each sector is typically 512 bytes or a multiple of 512 bytes. File systems, on the other hand, further groups contiguous sectors into *clusters*, where the size of a cluster, in terms of the number of sectors that is contained in a cluster, is typically configurable during formatting, and its default value is determined by the size of the volume. A cluster is the smallest unit for data storage and retrieval in a file system.

The main job of a file system is to organize *user data* so that operating systems can locate the data when it is needed. Different file systems may handle data in different ways, but would typically store some structural data about the user data it needs to organize, and such structural data can be used to perform operations on the user data, such as allocation of space for new data to be stored, and marking of certain space as *free space* (which becomes available for future data allocation).

Among many file systems, we are going to describe in more detail two widely used file systems, namely, File Allocation Table (FAT) and its variants, and the more recent New Technology File System (NTFS).

## 2.1 The FAT File Systems

The File Allocation Table (FAT) file system was introduced by Microsoft in 1980. Since then, it has been extended and updated throughout the years, which resulted in FAT-12, FAT-16, and now FAT-32, where the number indicates how many bits are used for each entry in the file allocation table. Although most recent Microsoft Windows operating systems use NTFS by default, FAT and its variants are still widely used in flash memory cards, thumb drives, and other removable devices.

The center of an FAT file system is a data structure called file allocation table (hence the name of the file system). In this table, among other things, there is a list where each entry maps to an actual data cluster. The value stored in each entry can be 0, which indicates that the cluster is available for use, or a special value to indicate that the cluster is a bad cluster, or the index of another cluster, or a special symbol EOC that stands for end-of-chain.

The clusters that are used to store a file are represented as a chain of clusters in the file allocation table. If we know the first cluster of the file, we can find the entry in the file allocation table that corresponds to the first file cluster, whose value would indicate the index of the next cluster that belongs to the file. We can then repeatedly look up the next cluster, until we encounter the EOC symbol.

### 2.1.1 File Allocation

The file allocation process in FAT-32 consists of two steps. In the first step, each file to be allocated is assigned to a *directory* (also known as a *folder*). In FAT file systems, each directory is represented by a special file called *directory entry*. Hence, the entry of the directory to which the file is assigned is located, and a file entry is added to it.

The file entry contains information about the file so that the file system can locate the file later. This includes the name of the file, starting cluster number, file attributes, creation and modification time stamps, together with possibly other metadata about the file. When the file needs to be retrieved later (e.g., by specifying its path), the file systems locate its file entry, and the index of the first cluster of the file can be read from the file entry. After that, we can traverse the file allocation table as mentioned above to locate the chain of clusters of the file.

For example, when we create a file named `evidence.jpg` under the root directory, which requires 5 clusters to store its content, the file system may decide that cluster number 300 is where the first cluster of the file should be stored. The file system would write an entry in the root directory entry to indicate that the file with the name `evidence.jpg` starts from cluster 300. The file system further stores the remaining content of the file into clusters 301, 302, 305, and 306, and make appropriate changes to the file allocation table as shown in Fig. 1. Note that clusters 303 and 304 are used to store another file `hello.txt`.

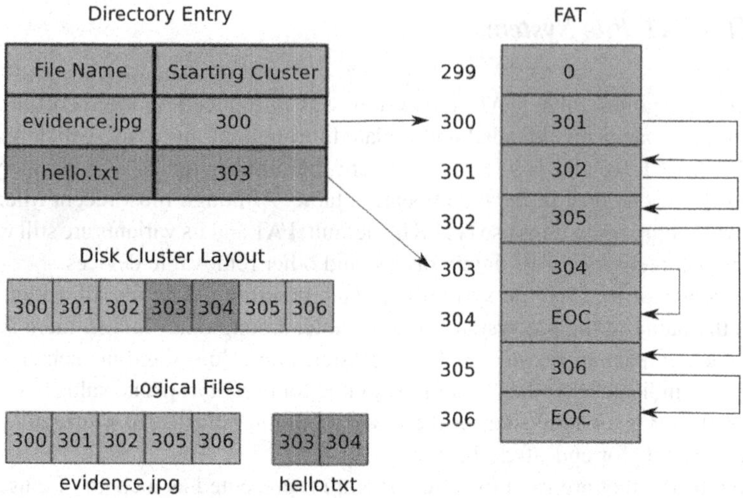


**Fig. 1** FAT file allocation of `evidence.jpg` and `hello.txt`

When we needs to retrieve `evidence.jpg`, we would look into the directory entry, find its first cluster number, which is 300, and then find the 300th entry in the file allocation table, which contains the value 301, and we can then follow the chain until we hit the EOC at the 306th entry.

### 2.1.2 Deletion

When a file is to be deleted, the file system does not actually erase the content of the file from the underlying storage media. Instead, all the file system will find the cluster chain belonging to the file in the file allocation table and mark all the clusters in the chain as "available" by setting their entry values to 0. Interestingly, the file entry in the directory is not removed either, but the first byte of the entry is replaced by the value of $0xE5$ (Meta-e) to mark the entry as "deleted" and available for use.

Figure 2 shows what happens to the file allocation table when `evidence.jpg` is deleted. The entries in the file allocation table that correspond to the clusters used by the file are assigned the value of zero to indicate that they are available for future file allocations. However, the content of the actual clusters has not been removed as yet. The first byte of the file name is also changed in the directory entry to represent that the file was deleted.

### 2.1.3 Recovery

Since the actual file content is not erased when the file is deleted from the file system, it becomes possible to recover a deleted file provided that the clusters that contain the content of the file have not been recycled and used by other files.

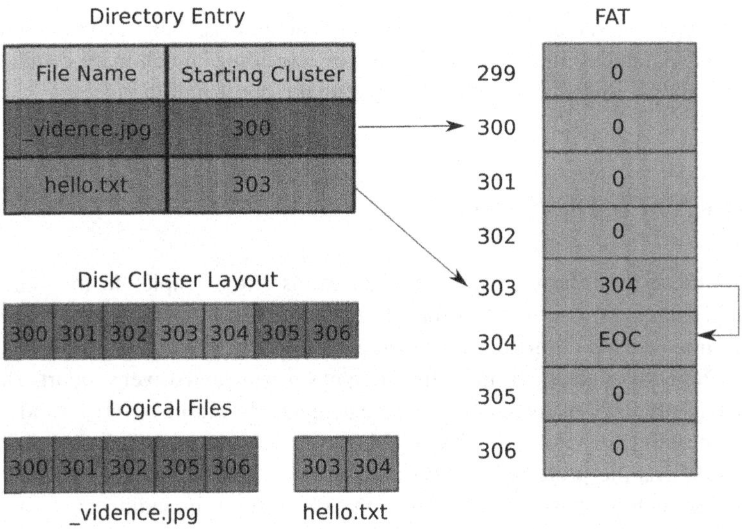


**Fig. 2** After deletion of `evidence.jpg`, the data are still there but the chain of clusters is lost ('_' represents the byte $0xE5$)

For example, a recovery program may search among all the directory entries and look for entries starting with $0xE5$. Once such an entry is found, the program can look further in the entry to find its starting cluster number. After that, the recovery program can check the file allocation table to see if the first file cluster has been reused by other files. If not, the program can extract the clusters sequentially in attempt to recover the original file.

Such a recovery strategy works when (1) the file entry in the directory entry has not been used by other files, (2) the file clusters have not been used by other files, and (3) the file was not fragmented (i.e., all clusters belonging to the file are contiguous and in sequence).

However, when the file is fragmented as in the case of `evidence.jpg`, from the file system metadata alone there is no way for the recovery program to tell which clusters had belonged to the original file. There are some useful tricks that can be used by the recovery program but the result is not guaranteed.

In the example using `evidence.jpg`, a traditional file recovery program would find the file entry starting with the byte $0xE5$ followed by `vidence.jpg` and recognize that it is a deleted file. It then finds out that the starting cluster number of the deleted file was 300. From this point, some recovery programs will only recover a partial file consisting clusters 300, 301, and 302 and stop when it seems that cluster 303 is in use by `hello.txt`. Smarter recovery programs would skip the clusters in use (303 and 304) and continue recovery from cluster 305, and would be able to recover the entire content of the file `evidence.jpg`. However, when the fragments of the deleted file happen to be far away from each other and separated by unallocated clusters, the recovery program would not be able to tell the difference between an original cluster with one that has not been allocated or had belonged to another file that was also deleted. In this case, file recovery becomes much more difficult, and would be almost impossible without analyzing the content of the file.

## 2.2 The NTFS File System

The FAT file systems have a number of limitations. For example, the maximum size of a single file is limited to 4 GB for all FAT variants, and the sizes of clusters and disk volumes are also limited. Some of these limits are due to the design of the file systems, and others are the result of insufficient software and driver support. The FAT file systems also lack many features that are demanded by more advanced and sophisticated operating systems. As a result, Microsoft introduced NTFS with Windows NT in 1993, as the preferred file system for its modern operating systems. Since then, many features have been added to the file system, including quota tracking, encryption, journaling, logging, and so on.

### 2.2.1 File Allocation

There are many features in NTFS that make data access more efficient. For example, the directory entries in NTFS are sorted and stored in a B-Tree [3], so that it is much more efficient to handle than the unordered linear structure in FAT.

The clusters used by files are stored in NTFS as cluster chains just like FAT. However, in NTFS there is an additional data structure called a bitmap. The bitmap of a volume is an array of bits where each bit maps to a cluster in the volume. The value of the corresponding bit of a cluster represents its allocation status, where a zero means it is free space available for file allocation, and a one means it has already been allocated to some file. By parsing the bitmap, NTFS is able to quickly determine the free space and the best way to allocate the clusters for a new file.

### 2.2.2 Deletion

Similar to FAT file systems, when a file is deleted in NTFS, the actual data clusters are not erased. Instead, the bits corresponding to the file clusters in the bitmap is reset to zero. Unlike the FAT file systems, the cluster chain that belongs to the file is left unchanged after the file has been deleted. As a result, if the entry of the deleted file is present in the directory entry, the entire cluster chain of the deleted file can be located, provided that those clusters have not been allocated to other files. This makes file recovery much easier with NTFS.

### 2.2.3 Recovery

As we have mentioned, file recovery is quite straightforward in NTFS when the file entry is still present, since the cluster chain of the file should also be present, and all that needs to be done is to verify that the cluster chain has not been overwritten, or are not allocated to another file. In the case where the file system is damaged or the entries of the deleted files have been removed, file system metadata alone would be insufficient for recovery.

## *2.3 Fragmentation*

In a file system, when the clusters belonging to a file are not contiguous and in sequence, we say that the file is fragmented. In other words, when a file is fragmented, if we extract the data from the first cluster to the last cluster of the file, the extracted data would not be the original file.

For example, the `evidence.jpg` in Fig. 1 is fragmented because it starts from cluster 300 and ends at cluster 306, but the two clusters 303 and 304 are not part of the file. In this case, `evidence.jpg` consists of two fragments, one from cluster 300 to cluster 302 and the other from 305 to cluster 306. Such files with two fragments are often referred to as *bifragmented*. Garfinkel showed that bifragmentation is the most common type of file fragmentation [11]. Nevertheless, files fragmented into three or more pieces are not uncommon.

Garfinkel's fragmentation statistics [11] come from over 350 disks containing various file systems including FAT, NTFS, and Unix file system (UFS). It is shown that while fragmentation in a typical disk is low, the fragmentation rate of forensically important files (user files) like that of e-mails, JPEG images, and Microsoft Word documents are high. The fragmentation rate of JPEG images was found to be 16, 17% of the Microsoft Word documents are fragmented, audio video interleave (AVI) video files have a fragmentation rate of 22%, and personal information store (PST) files generated by Microsoft Outlook are fragmented at a whopping rate of 58%. Fragmentation can be caused by many factors, which typically include low disk space, appending/editing of files, wear-leveling algorithms in next generation devices, and the file system itself.

Low Disk Space

If the free space of a disk volume is running low, the available clusters in the volume may not be contiguous but instead form small groups. In this case, when a new file is created such that the file size is larger than any of the group of free clusters, fragmentation of the new file will likely occur.

For example, if the largest group of consecutive clusters available for allocation contains 10 clusters, and a file needs 14 clusters for allocation, most file systems would fill the first 10 clusters with the beginning part of the file, and store the remaining four clusters in another area on the disk that has unallocated clusters available, which results in file fragmentation.

### Appending/Editing of Files

When editing an existing file in a file system, fragmentation may occur. For example, when data are appended to or inserted into a file such that an additional cluster is required to store the file, but the disk cluster following the last cluster of the file has already been occupied by another file, the file system may choose another location for the new cluster, which results in fragmentation.

There are some techniques that can be used to avoid fragmentation due to editing or appending. Some file systems, like the Amiga Smart File System [12], may attempt to move the whole file to a different location when fragmentation is going to occur otherwise. Some file systems, such as the UNIX File Systems (UFS), may attempt to pre-allocate contiguous chunks called *extents* in anticipation of the increase of the file size [21]. Another technique used in file systems such as XFS [28] and ZFS, called *delayed allocation*, reserves file system clusters but attempts to delay the physical allocation of the clusters until the operating system forces a flushing of the contents.

While these techniques are able to reduce fragmentation to some extent, they are unable to eliminate fragmentation completely. Constant editing, deletion, and additions of e-mails are the primary reasons why PST files (Microsoft Outlook files) are highly fragmented (58% of PST files analyzed in the wild [11]).

### Wear-Leveling Algorithms in Next Generation Devices

Solid-State Drives (SSDs) and flash memories often employ proprietary wear-leveling algorithms [10]. These devices, unlike magnetic disks, have storage units that can only be erased for a limited number of times before they become unreliable, and once a storage unit is not usable, the entire device will be unusable. As a result, wear leveling is necessary to spread out the erasure operations evenly to all storage units to extend the life expectancy of the devices.

Some wear leveling algorithms accomplish their goals by remapping the logical addresses of disk sectors to physical locations on the devices, so that when a cluster is modified, its content will be written to a new location instead of overwriting the original location, which would cause an erasure. Some wear leveling algorithms exploit the idea of journaling file systems to an extreme, and the entire file system becomes a log, so that new data are always written to the end of the log, and free space is reclaimed according to certain policies.

In these wear leveling algorithms, since erasure operations are avoided as much as possible, it becomes very likely that any modification to a file would make it fragmented. For example, if a single bit of a file cluster is changed, instead of modifying it in-place, as most magnetic disk controllers would do, the cluster is copied to a

**Table 1** Definition of Terms

| Term | Definition |
|---|---|
| Cluster | Smallest data unit that can be written to a disk. We use $b_y$ to denote the cluster numbered $y$ in the access order. |
| Header | A cluster that contains the beginning of a file. |
| Footer | A cluster that contains the end of a file. |
| Fragment | One or more consecutive clusters of a file that are not connected to other clusters of the same file. Fragmented files are those with two or more fragments. Fragments of a file are considered to be separated from each other by an unknown number of clusters. |
| Base-Fragment | The starting fragment of a file that contains the header as its first cluster. |
| Fragmentation Point | The last cluster of a fragment except for the last fragment of a fragmented file. A file may have multiple fragmentation points. |
| Fragmentation Area | A set of consecutive clusters $b_y, b_{y+1}, b_{y+2}, \ldots$, containing the fragmentation point. |

new location with the modified bit, hence creating a fragmentation. Therefore, if the controller of an SSD or a flash memory gets corrupted or damaged, it would be very difficult to determine the correct sequence of clusters for the files.

#### File System Forced Fragmentation

In some rare cases, fragmentation occurs even when there is sufficient contiguous free space, which is a result of the file allocation algorithm of the file system. For example, the UNIX File System will fragment files that are long or have bytes at the end of the file that will not fit into an even number of sectors [3].

In Table 1, we give the basic definitions concerning fragmentation, which are illustrated in Fig. 3.

### *2.4 File Carving*

When a file is fragmented, and the corresponding cluster chain recorded in the file system metadata is lost or damaged, the file would not be recoverable using traditional recovery techniques. Furthermore, for newer and more esoteric file systems, recovery may not be possible based solely on file system metadata, unless such support is directly provided by the file systems.

To recover files when the file system metadata is absent, unreliable or insufficient, we have to employ techniques that can analyze the syntactic or even the semantic structures of the file. These techniques are called *file carving*.

We note that a file carver may not recognize the file system being used, and it may or may not trust the information in the file system to be accurate. As a result,

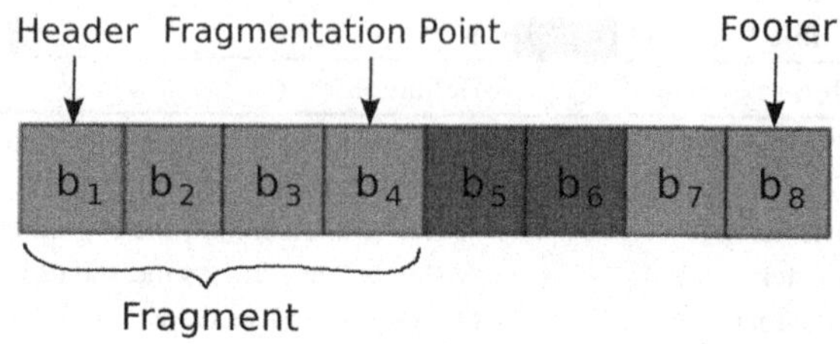


**Fig. 3** A fragmented file with two fragments

the carver has to determine which clusters actually contain useful data to carve from. This may involve all clusters in the disk. To reduce the amount of data to analyze, a number of forensic applications are able to identify a large number of files that are common to operating systems and applications based on their MD5 Hash and keywords. Both Encase and Forensic Toolkit (FTK), the two leading commercial disk forensic software providers, provide this option to quickly eliminate common and well-known files.

Nevertheless, the amount of remaining clusters to analyze can still be huge. This requires file carvers to be very efficient and accurate, which has stimulated many recent studies. In fact, file carvers have been evolving from straightforward carving of non-fragmented files to those with the capabilities of performing complex syntactic and semantic analysis. Next, we detail the process required to recover both fragmented and non-fragmented files using different forms of file carving.

## 3 Carving Non-fragmented Files

If a file is not fragmented, what is required to recover the file is to know where the file starts and ends, even when the file system metadata is unreliable. In other words, if we can find the header and the footer of a file, we will be able to recover it if it is not fragmented.

Fortunately, many files contain in their headers special structures that are specific to their types, which can be exploited by file carvers. More specifically, a file may contain a *magic number* at its beginning, which is special byte sequences at a prescribed offset, typically used by the operating system to identify the type of the file and to determine how the file should be opened. Hence, the header of such a file can be located by searching for the magic number. Similarly, some files may also contain special byte sequences at the end, so as to indicate the end of the data, which can be used to identify the footer. These special byte sequences are used by the first-generation file carvers to find headers and footers, so that a file can be recovered by grouping all the clusters from a header to a footer, provided that it has not been fragmented.

For example, for JPEG images, the data stream must begin with the byte sequence 0xFFD8 and ends with the byte sequence 0xFFD9. When a file carver locates a cluster that begins with 0xFFD8, it knows that this could be the beginning of a JPEG image. If it sees a byte sequence 0xFFD9 in another cluster after the header, it knows that this could be a JPEG footer. The file carver can then attempt to recover the JPEG image from the header to the footer.

It is worth to note that, there are many file types for which we may not be able to find special sequences at the end. In other words, they may not contain clearly identifiable footers. Instead, these files may contain other information that can be exploited by a file carver to recover the data, such as the file size.

An example of this is the Windows BMP image files. A BMP image file contains information about its size in bytes near the beginning of the file, which we can learn from the header cluster. In this case, traditional file carvers would recover files by identifying the header and then merging as many sequential clusters following the header as required, so that the total size matches the information in the header.

Foremost [15], originally developed by the United States Air Force Office of Special Investigations and The Center for Information Systems Security Studies and Research, was one of the first file carvers that implemented sequential header to footer carving and also implemented carving based on a header and the size of the file. Scalpel [26] is another file carver, which was built by rewriting the Foremost engine, and greatly improved its performance and memory usage.

The main limitation of the first-generation file carvers is that they simply extract data between a known header and footer (or ending point determined by the file size), with the assumption that the file is not fragmented and there is no missing information between the header and the footer. As a result, these file carvers frequently provide results that have "garbage" in the middle.

In Fig. 4, we can see an example of a JPEG file incorrectly carved based solely on the header-to-footer technique. This JPEG image is from the DFRWS 2006 Forensics Challenge [6], which does not contain any file system metadata. The DFRWS 2006 and 2007 challenge test sets [6, 7] are specifically created to test and stimulate research in the area of fragmented file carving.

It can be seen from the figure that a large number of clusters are decoded but are not part of the image. Therefore, even when we use a JPEG decoder to validate the extracted data, we may be able to find that the image is corrupted, but we would not be able to find the fragmentation point precisely. We will come back to this subtle issue later.

## 4 Carving Bifragmented Files

A file carver that carves files with two fragments using *fast object validation* is given by Garfinkel [11], based on the observation that bifragmentation is the most common among all types of fragmentation.

Object validation refers to the process of verifying the syntactic consistency of structured data, given that its type or format is known. This can be done in many

**Fig. 4** Incorrect recovery of a fragmented image from DFRWS 2006

ways, and often depends on methods that are specific to the actual data type. For example, an image in Portable network Graphics (PNG) format is composed from a 8-byte PNG signature followed by a series of *chunks*, and each chunk begins with a 4-byte length field and ends with a 4-byte Cyclic Redundancy Check (CRC) code. In this case, a validator can determine its type from the signature, and process the chunks after it one-by-one until it reaches a special IEND chunk that indicates the end of the file. For each chunk, the validator computes the CRC based on the chunk length, and compares the result with the CRC stored in the data stream. A mismatch would indicate that either the data is corrupted or fragmentation has occurred.

Unfortunately, not many file formats provide error-checking capabilities as the CRC code used in PNG images. For those files, a validator would have to go deeper to analyze the structure of the data, as if the data is used by its intended application. For example, for many media file formats, such as Joint Photographic Experts Group (JPEG) images and Moving Picture Experts Group (MPEG) videos, a *decoding* process takes place when they are viewed or played. During decoding, such a media file would be transformed from its compressed form to another form that can be used

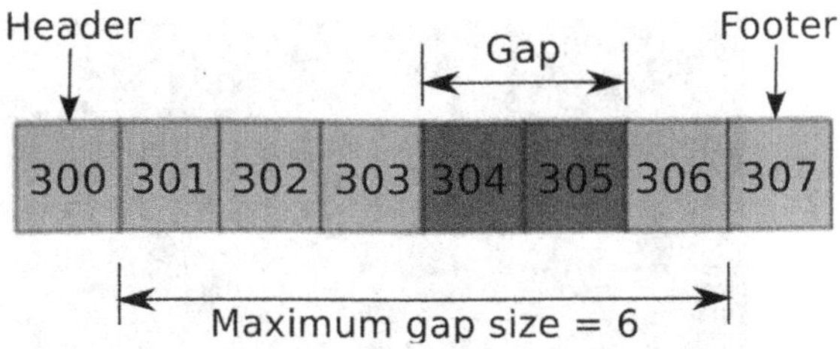


**Fig. 5** An example of bifragment gap carving, where a gap size $g = 2$ yields correct recovery of clusters $300 - 303$ and clusters $306 - 307$

for immediate display or playback. If an error is found during the decoding process, we would know that the data are corrupted or fragmented. Similarly, for compressed archives such as ZIP files, a validator would try to decompress the archives and would detect fragmentation based on whether errors occur during the process.

When files do not have structures or do not require any decoding before they can be used by their intended applications, it would be very difficult to design efficient and effective validators for them. For example, if a validator is given pieces of plain text, there would be very little useful structure (if any at all) to exploit in order to determine if fragmentation has occurred. The Windows BMP images are another example of such files. Although each BMP image has a small header that itself is structured, the actual pixel data can be anything, hence providing no useful information for fragmentation detection.

When the header and the footer of a file is known, and assume that the file is bifragmented, a validator for the file type can be used to find the fragmentation point by exhaustively searching all possible scenarios of bifragmentation until a successful decoding/validation is found. This technique is often referred to as Bifragment Gap Carving (BGC), and is the basis of Garfinkel's carving method [11].

In particular, given a bifragmented file, let $b_h$ be the header, $b_f$ be the last cluster of the first fragment (i.e., the fragmentation point), $b_s$ be the starting cluster of the second fragment, and $b_z$ be the footer. The job of the file carver is then to determine the locations of $b_f$ and $b_s$ given $b_h$, $b_z$ and all clusters in between. In this case, the Garfinkel's carver first chooses a gap size $g$, where $1 \leq g \leq z - h - 1$, and try all possible locations of $b_f$ and $b_s$ such that $s - f - 1 = g$. For each combination of $b_f$ and $b_s$, a validator is then run on the file formed by clusters $b_h$ to $b_f$ followed by clusters from $b_s$ to $b_z$. If the validation is successful, those clusters will be deemed to be the recovered file, otherwise the next combination is tried until all possible values of $g$'s and corresponding combinations of $b_f$ and $b_s$ are exhausted.

Figure 5 shows a file that is split into two fragments. The first fragment starts from cluster 300 and the last cluster for the fragment is 303. The second fragment starts with cluster 306 and ends with cluster 307. A bifragment gap carving algorithm would try all possible gaps with size from 0 to 6 and validate the resulting file with fast object validation. In this case, the correct recovery would be found when the gap size is $g = 2$, and the cluster indices $f = 303$ and $s = 306$.

Bifragment gap carving performs satisfactorily when the two fragments are close to each other (i.e., when gap size is small). However, it has the following limitations

**Fig. 6** Fully validated but incorrect JPEG image from DFRWS 2007 test set

for more general cases:

1. The technique does not scale for files fragmented with large gaps. If $n$ is the number of clusters between the header $b_h$ and footer $b_z$, in the worst case, $n^2$ object validations may be required before a successful recovery.
2. It was not designed for files with more than two fragments.
3. Successful validation does not always imply that a file was reconstructed correctly. Figure 6 from DFRWS 2007 is an example of a successfully decoded but incorrectly recovered JPEG image.

4. It only works with files that have a structure that can be validated or decoded.
5. Missing or corrupted clusters will often result in the worst case.

## 5 Graph Theoretic Carving

The problem of recovering fragmented files without reliable information from the file system has also stimulated graph theoretic research. Since data fragments can be considered as nodes in a graph and the relationship between consecutive fragments of the same file can be considered as edges, it is natural to formulate the problem of file carving as a graph theoretic problem where the goal is to reassemble the nodes by finding the correct edges. In this section, we are going to introduce some of these graph theoretic approaches to the problem of file recovery.

### *5.1 Reassembly as Hamiltonian Path Problem*

Shanmugasundaram and Memon [27] were some of the first to tackle recovery of fragmented files. They showed that the problem of finding the optimal ordering of file fragments is equivalent to that of finding a maximum weight Hamiltonian path in a complete graph, and providing an alpha–beta heuristic approach from game theory to solve this problem.

In particular, given a set of clusters $b_0, b_1, \ldots, b_n$ belonging to a document $A$, the goal of a file carver is to compute a permutation $P$ of the set that represents the original structure of the document. In other words, the carver needs to identify fragment pairs that are adjacent in the original document. To achieve that, a candidate weight $W_{x,y}$ is assigned to every pair of clusters $b_x$ and $b_y$, which represents the likelihood that cluster $b_y$ follows $b_x$. It is then conjectured that the correct permutation of the clusters, among all possible permutations, would maximize the total candidate weights of adjacent clusters. Therefore, to find the reassembly sequence is equivalent to finding the permutation such that the total candidate weights in adjacent clusters is maximized over all permutations of degree $n$.

We can take the set of all candidate weights to form an adjacency matrix of a complete graph of $n$ vertices, where vertex $x$ represents cluster $b_x$, and the edge weight $W_{x,y}$ represents the likelihood of cluster $b_y$ following cluster $b_x$. The proper reassembly sequence $P$ is a path in this graph that traverses all the vertices and maximizes the sum of candidate weights along the path, which is a Hamiltonian path in the graph with maximum weight. In this way, the problem is abstracted to a graph theoretic one.

As you might have noticed, the success of such a formulation depends on a way to assign candidate weights between all cluster pairs such that the conjecture holds. In other words, given two clusters, we have to have a way to determine the likelihood that one follows the other.

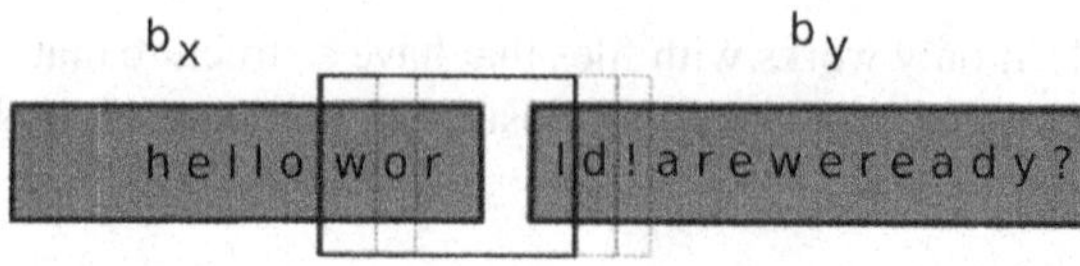


**Fig. 7** Prediction by Partial Matching with a windows size of four

Shanmugasundaram and Memon [27] proposed a method to assign candidate weights for a pair clusters called *prediction by partial matching* (PPM). PPM is a finite-order context modeling technique first introduced by Cleary and Witten [4] and has become a benchmark for lossless data compression. PPM employs a suite of fixed-order context models, from zero up to some predetermined maximum $k$, to predict the upcoming characters.

Using PPM, each cluster is individually processed and the resulting statistics are combined to form a single model for a document. The weight $W_{x,y}$ is then calculated by sliding a window across the ending bytes of cluster $b_x$ and the starting bytes of cluster $b_y$.

Figure 7 shows the weight being calculated using a sliding window of size four. The window is initially put on the edge of the cluster $b_x$ and the model is looked up to see the probability of `wor` being followed by the letter `l`. The window then slides one byte and now the probability of the `orl` followed by `d` is found and this is multiplied to the previous probability. This process is repeated until no part of the window is in cluster $b_x$. This provides us with the candidate weight $W_{x,y}$.

While PPM is shown to be good for structured data like text, its performance is not as good for images and compressed documents, where the data stream appears to be more random.

Pal et al. [24] extended the technique proposed by Shanmugasundaram and Memon [27] by introducing a system for fragmented file recovery for images, and tested the technique using fragmented 24-bit Windows BMP images with no file system meta-data. In their technique, to determine the weight of an edge between two clusters, the image boundaries between each cluster pair is compared and a weighting algorithm is developed, which analyzes the pixel differences between two clusters as shown in Fig. 8. The sum of differences of the pixels across the boundary as shown in the figure is calculated and then stored as the weight. If $w$ is the width of an image then the last $w$ pixels of $b_x$ are compared against the first $w$ pixels of cluster $b_y$ when computing $W_{x,y}$. The weight assignment is asymmetric in the sense that the weight $W_{x,y}$ for cluster pair $(b_x, b_y)$ is typically different from weight $W_{y,x}$ for the cluster pair $(b_y, b_x)$.

## 5.2 Reassembly as k-Vertex Disjoint Path Problem

The primary problem with the Hamiltonian path formulation is that it does not take into account that in real systems multiple files are fragmented together and that recovery can be improved if the statistics of multiple files are taken into account.

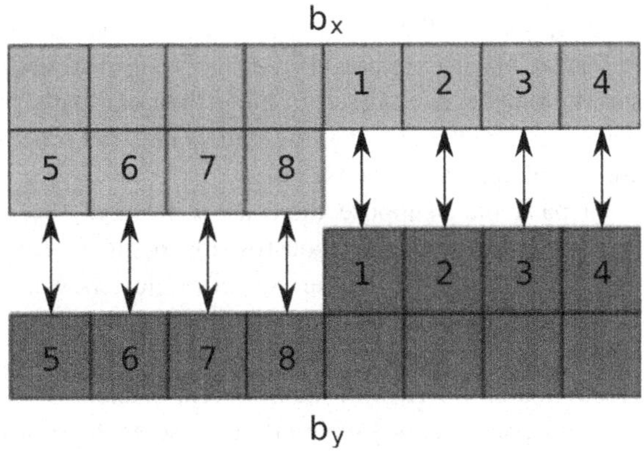


**Fig. 8** Weight assignment for bit map image cluster pairs

To address this problem, Pal and Memon [23] subsequently refined the formulation by considering the recovery of multiple files at the same time. Similar to the Hamiltonian path formulation, clusters are modeled as vertices in a complete graph, and weights are computed for all pairs of clusters. From these clusters, headers and footers are identified. Suppose there are $k$ headers in the given clusters, the problem of recovering the files becomes that of finding $k$ paths in the graph that start with the $k$ headers, respectively and end with the footers, such that they do not share any vertices (clusters) in common, and the total weight along the paths is maximized. This is referred to as a $k$-vertex disjoint path problem, which is known known to be non-deterministic polynomial-time (NP)-hard. In this section, we are going to describe a few algorithms proposed by Pal and Memon [23] in attempts to solve the problem efficiently.

### 5.2.1 Parallel Unique Path

Pal and Memon [23] give eight algorithms to solve the $k$-vertex disjoint path problem, among which are unique path (UP) algorithms. In a unique path algorithm, as we build the $k$ paths, each vertex can only be assigned to one and only one path, since the final paths should be disjoint.

As mentioned earlier, each cluster in a file system normally belongs to no more than one file. Therefore, UP algorithms are intuitive. However, an UP algorithm has the disadvantage that, during the building of the paths, if a vertex that belongs to one path is incorrectly assigned to another, both paths would be built wrongly. In other words, errors can cascade. Of the UP algorithms given by Pal and Memon [23], we will discuss the two that provide the best results.

The parallel unique path (PUP) algorithm, which is a variation of Dijkstra's single source shortest path algorithm [8], is used to recover all the files simultaneously. First, the clusters are processed to find all the headers. From each header $b_h$, the algorithm picks the cluster $b_x$ from the remaining clusters that yields the best match, which is determined by the edge weights discussed earlier. After the cluster $b_x$ is assigned to header $b_h$, it is removed from the pool of available clusters so that we can be sure no cluster is assigned to more than one path. Next, the best match for $b_x$ is determined and assigned to the same path. This process continues until all the paths are completed with footers.

In particular, suppose there are $k$ headers $(b_{h_1}, b_{h_2}, \ldots, b_{h_k})$, which are the starting vertices in the reconstruction paths $P_i$ for each $1 \leq i \leq k$. Let $S = \{b_{s_1}, b_{s_2}, \ldots, b_{s_k}\}$ be the current set of clusters waiting to be processed, where $b_{s_i}$ is the current cluster for the $i$th path. Initially, all the $k$ starting headers are stored as the current clusters for each path (i.e., $b_{s_i} = b_{h_i}$ for $1 \leq i \leq k$). The best greedy match for each of the $k$ current clusters is then found and stored in the set $T = \{b_{t_1}, b_{t_2}, \ldots, b_{t_k}\}$ where $b_{t_i}$ represents the best match for $b_{s_i}$. From the set $T$ the cluster with the overall best matching metric is chosen.

Assuming that this best cluster is $b_{t_i}$, we follow the steps below to build all the $k$ paths:

1. Append $b_{t_i}$ to reconstruction path $P_i$, (i.e., $P_i = P_i || b_{t_i}$).
2. Replace the current cluster in $S$ for $P_i$ (i.e., $b_{s_i} = b_{h_i}$).
3. Evaluate the new set $T$ of best matches for $S$.
4. Again find best cluster $b_{t_i}$ in $T$.
5. Repeat from Step 1 until all paths are built (i.e., footers are reached).

Figure 9 illustrates an example of the PUP algorithm where there are three paths $P_1$, $P_2$, and $P_3$ being reconstructed. Figure 9a shows the headers $H_1$, $H_2$, and $H_3$ and their best matches. The best of all the matches is the $H_2 - 6$ pair as shown with a dotted line. Figure 9b shows the new set of best matches after cluster 6 has been appended to the path $P_2$. Now cluster 4 is chosen once each for clusters $H_1$ and 6. However, the pair $H_1 - 4$ is the best and therefore cluster 4 is appended to the path $P_1$ and the next best match for cluster 6 is determined (Fig. 9c). This process continues until all paths are reconstructed.

### 5.2.2 Shortest Path First

Shortest path first (SPF) is a graph theoretic algorithm that is widely used in many applications [16]. In the context of data fragment reassembly, we assume that the best reassembly of fragments is equivalent to having the lowest average path cost. The average path cost is defined as the sum of the weights between the clusters of a path (which is equivalent to a recovered file) divided by the number of clusters in the path. Therefore, we can use a shortest path first algorithm to find the paths that would yield the lowest average path cost to achieve the best reassembly results.

This algorithm reconstructs one path at a time. Starting from a header, we find the path to a footer by repeatedly joining the vertex with the lowest weight to the existing path. After that we choose another header and repeat the process until all paths are built. Unlike the unique path algorithms, after we put a vertex (cluster) into a path, we do not remove it from the pool of available vertices for other paths.

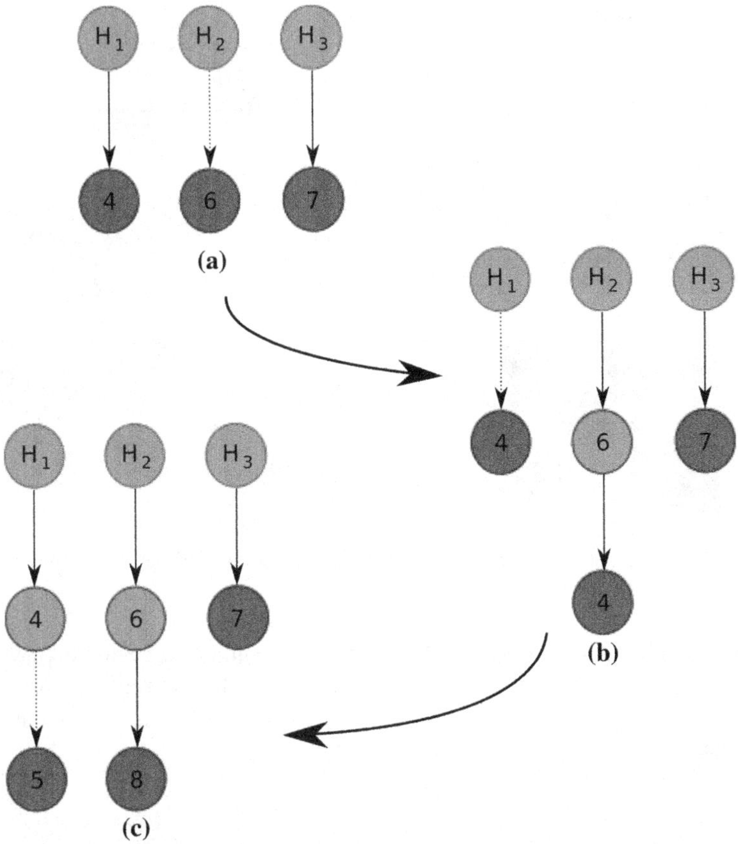


**Fig. 9** Simplified PUP example

In other words, although vertices must be unique in one path, some vertices may appear in more than one path. After we have built all the paths in this way, we choose the one with the lowest average path cost and consider it as a final path. We then check if any vertex in this path has appeared in other paths. If such a vertex is found, it is removed from other paths as well as the pool of available vertices, and all affected paths will be rebuilt using the same algorithm as before, and their average path cost recalculated. This process is repeated until all paths are final.

The performance of the carver is measured by the reassembly accuracy, which is defined as the rate of successfully reconstructed files to the total number of files. As a comparison, the SPF algorithm achieves a high accuracy of 88%, whereas the PUP algorithm has an accuracy of 83% [23]. However, the PUP algorithm is substantially faster and scales better than the SPF algorithm.

It is shown that the SPF technique could be used for image recovery under highly fragmented scenarios, but the primary issue with the techniques is that they do not scale very well. Since edge weights are precomputed and sorted so as to find the best

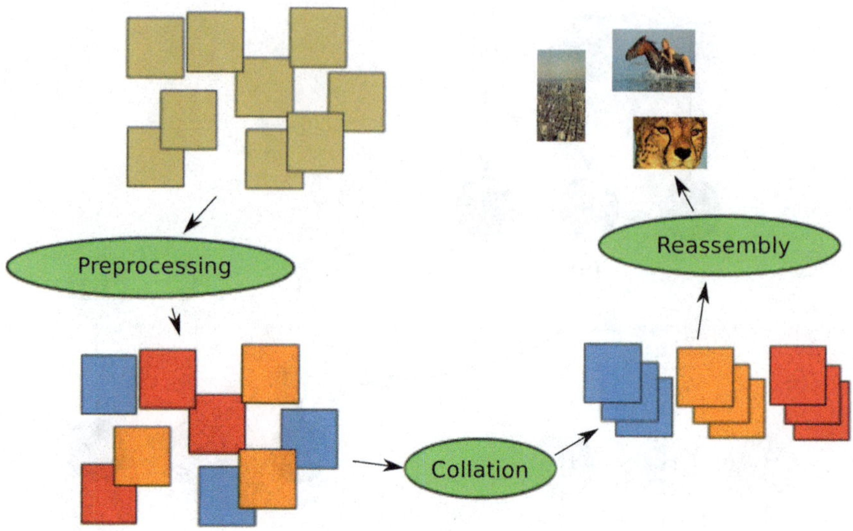


**Fig. 10** The three phases of SmartCarver: **a** preprocessing, **b** collation, and **c** reassembly

matches quickly, the complexity of the algorithm is dominated by this step, which is $O(n^2 \log n)$. Typical disk volumes today may contain millions of clusters. As a result, precomputing weights between all potential clusters can become computationally infeasible.

## 6 SmartCarver

To handle fragmented disk volumes in real-world applications, a file carver needs to be capable of processing files with arbitrary number of fragment, and efficient enough to handle a huge number of clusters. Pal et al. [25] recognize the complexity of the problem and give a design of a *SmartCarver* that uses both many useful heuristics as well as carefully designed algorithms to recover fragmented files in three phases as shown in Fig. 10.

The first phase is *preprocessing*, where file system data clusters are extracted, decrypted ,or decompressed as needed. The next phase is *collation*, where data clusters are categorized and classified as belonging to certain file types. Optionally, an additional step is made to determine if consecutive clusters belong to the same file, and if they do, the clusters are merged to reduce the amount of computation in the last phase. The final phase is *reassembly*, where clusters identified and merged during collation are pieced together to reconstruct the files.

## *6.1 Preprocessing*

The *preprocessing phase* of the SmartCarver prepares the given data for further processing. This may include decryption, decompression, and elimination of known files based on file system metadata available on the volume.

Some disk volumes are compressed or encrypted. For example, some Microsoft Windows platforms, including the Ultimate and Enterprise editions of Windows Vista and Windows 7, Windows Server 2008 and Windows Server 2008 R2, come with a feature called *BitLocker Drive Encryption*, which provides full data encryption on entire disk volumes.

To recover files from such encrypted disk volumes, the data must be decrypted before further processing. This is possible, for example, by exploiting recent development of memory-based attacks that can determine the encryption key for encrypted volumes on a computer by analyzing ghost images of the memory.

Similarly, if the volume utilizes compression technologies, the data clusters need to be decompressed as well. Regardless of the encryption/compression used, the data clusters must be decrypted/decompressed to continue the recovery process.

The preprocessing phase also analyzes the file system metadata and removes all the known (allocated) clusters from the volume since they can be carved directly without further processing. By eliminating the known clusters, we reduce the number of clusters to analyze, which makes further processing more efficient and more accurate as well. However, if it is suspected that the file system metadata has been tampered with or is corrupt, this step may be skipped and a full scale recovery will be performed. In this case, all clusters are considered to be "unallocated". Once the preprocessing phase is complete, the collation phase is started.

## *6.2 Collation*

In the collation phase, the SmartCarver works on the unallocated clusters and put them into different categories by a *cluster classification* process, where each cluster is classified as belonging to one or more file types. A file type represents a specific file format, like that of JPEG image or the Microsoft Office document format.

Cluster classification is important because it allows us to reduce the number of clusters to consider for each file being recovered. For example, a cluster containing pure ASCII English text would probably not belong to a Windows BMP image, hence should be excluded when reconstructing those images. As a result, the cluster classification not only improves the speed of processing, but also increases the accuracy of the results.

Research has been conducted in classifying byte streams, whether they are network traffic, disk clusters, or complete files. Some well-known techniques used for file type identification include keyword/pattern matching, fingerprinting, statistical analysis, and machine learning.

*Keyword/Pattern Matching* involves searching for special byte sequences (sometimes at specific byte offsets) to determine the type of the file a cluster belongs to. This technique is most useful in identifying the headers and their types. Many files begin with special byte sequences called *magic numbers*, which can be used by the operating system and applications to identify their types. For example, JPEG images always begins with the byte sequence `0xFFD8`, which is referred to as the *Start of Image* (SOI) marker. Files in Portable Document Format (PDF) always begin with the ASCII text `%PDF` followed by a version number. An HyperText Markup Language (HTML) file typically begins with the character sequences `<HTML>` or `<html>`. When we find those patterns at the beginning of a cluster, it would be very likely that the cluster is a header of the corresponding type. Similar principles can be applied to clusters that are not headers. For example, a cluster that has a lot of `href=` strings may be considered to be an HTML cluster.

While keyword/pattern matching is beneficial in identifying potential cluster types, it is not always reliable since the same pattern may appear in multiple types of files. For example, an HTML tutorial written in Microsoft Word or Portable Document Format may contain the string `href=` describing how Web links are defined, and the cluster containing this part of the tutorial may be incorrectly identified as belonging to an HTML document.

Therefore, additional tests are necessary to verify that a cluster indeed belongs to a particular type of file. Typically, these tests either employ heuristic rules or some statistical analysis. For example, given a cluster of data, we can examine the frequency at which American Standard Code for Information Interchange (ASCII) characters appear. If the frequency is very high, it would suggest that it is very unlikely that the cluster belongs to a video, audio, or image file, since such files often contain mostly compressed binary data, despite a small portion that may contain ASCII text metadata. Such a simple heuristic is, however, not reliable in identifying certain file types. For example, it will not distinguish between an HTML document, a plain text file, or a Microsoft Word document containing uncompressed text.

A very efficient and useful analysis that can be used to classify data fragments is the entropy analysis. In information theory, entropy is a measure of the uncertainty of a random variable. In the context of forensics data analysis, the (empirical) entropy of a piece of data is usually computed by considering the data as a set of 8-bit samples drawn from a discrete distribution. We can think of entropy as the level of redundancy of the given data, which in turn indicates whether the data can be easily compressed. It is known that the entropy of English plain text is slightly more than 3 bits, while that of JPEG images is typically no less than 7 bits. From these facts, we can easily distinguish between low entropy file types and high entropy ones. However, such analysis would be ineffective in distinguishing file types with similar entropy signatures. For example both ASCII plain text documents and HTML documents have low entropy values. As a result it would be very difficult to tell them apart based only on the entropy analysis.

A classification method based on file fingerprints is given by McDaniel and Heydari [20], and it is one of the first robust techniques to determine the type of files. Their main idea is to generate different *fingerprints* for different file types, which are then compared with the given data to determine its type. In their work, three fingerprinting methods were given.

One way of generating a fingerprint for a file type is to compute the *byte frequency distribution* (BFD), which is the byte histogram of the data. A byte histogram is a vector with 256 values, which indicate the frequencies of the occurrences of the 256 possible values represented by a byte. A BFD fingerprint is created by choosing multiple different files of a file type (for example choosing a few hundreds of different JPEG images), creating the BFD for each file and then averaging the BFD.

In addition, among different files of the same type, certain byte frequencies are more consistent than others. For example, in an HTML document, the bytes representing the forward slashes (/) and angle brackets (< and >) occur frequently. The byte values that have consistent frequencies are considered to be strong indicators of the file type, and this information is taken into consideration by calculating a *correlation strength* for each byte value. A complete BFD fingerprint for a particular file type then consists of the byte frequency distribution and the correlation strengths.

The second type of fingerprint is the *byte frequency cross-correlation* (BFC), which is computed from the correlation strengths between all pairs of byte frequencies in the byte frequency distribution. Such a fingerprint is based on the observation that in certain file types, the byte frequencies may be highly correlated. For example, in an HTML document, the bytes representing angle brackets < and > would have roughly the same frequency.

Unfortunately, the accuracy of the BFD and BFC techniques were found to be about 30% and 45%, respectively; hence, a third technique was introduced. Their third algorithm uses header and footer identification techniques (such as keyword/pattern matching techniques as described earlier) to help determine the file type. By doing so, the identification accuracy was reported to be 95%. However, for the identification of data clusters, this technique will not work very well since the majority of the clusters are neither header nor footer, which gives little useful information for header/footer analysis.

Another data fragment classification method was developed by Wang and Stolfo [31], which is used to detect anomalous payloads in network traffic. Their method is based on building a traffic profile using byte frequency distribution and the standard deviations, and later computing a statistical distance between the profile and the given network traffic data. This method was later enhanced in a later study by Li et al. [19] where *fileprints* based on $n$-gram analysis are created for different file types, and each fileprint may contain multiple $n$-gram models so that different statistical features can be captured for each file type. The accuracy of their fileprints-based algorithms were much improved compared to the previous algorithm based only on BFD. For some file types they were able to achieve a 100% accuracy and the lowest accuracy occurred for JPEG images at 77%. However, their results clearly indicate that the accuracy decreases as the number of bytes available for analysis is reduced, since their $n$-gram models require a sufficiently large number of samples to be accurate. In the case of volume clusters, since their sizes are typically no more than $4,096$ bytes, it becomes questionable whether such methods can be applied with high accuracy.

Karresand and Shahmehri [13, 14] were the first to look at classifying individual data clusters without headers or footers. They proposed a method called *Oscar*, which is a combination of a number of techniques. The main idea of Oscar is to build statistical models for different file types, which are then combined with keyword and pattern matching techniques to achieve high classification accuracy. The first statistical model Oscar uses is a centroid model created by computing the mean and standard deviation of the byte frequencies. When this model is combined with JPEG marker matching, Oscar achieves an accuracy of 97% when identifying JPEG images.

Besides byte frequency distributions, other statistical analysis were also introduced to improve the classification accuracy. In particular, Oscar employs another model that is also a centroid model based on the statistics of the differences between adjacent bytes, which was referred to as the *rate of change* (RoC). The rational is that although byte frequency distributions are useful, they do not capture the relationship among neighboring bytes. By using the RoC model, the accuracy of identifying JPEG images can be improved to 99%.

Another similar cluster classification method was given by Veenman [29], where the statistical model is built from a combination of byte frequency distribution, entropy, and the Kolmogorov complexity. The Kolmogorov complexity was added as it measures byte order. Interestingly, this study shows that certain file types are frequently confused with one another (e.g., JPEG images and ZIP archives), and further research is required to differentiate them.

It may seem that as we add more statistical features and build more complex statistical models, it would be easier to differentiate different file types. However, the study conducted by Erbacher and Mulholland [9] showed that, among 13 different statistical features they considered, only 5 of them were significant. In a later study due to Moody and Erbacher [22], these 5 types of statistical features were carefully tested and evaluated on 200 files of 8 different types. Their results showed that, by using these statistics alone, it is still very difficult to differentiate among textual file types (e.g., HTML, plain text, and CSV files), and the performance in differentiating Windows BMP images, JPEG images, and XLS spreadsheets was poor.

Another line of research tries to systematically build statistical models from these statistical features using machine learning techniques. Calhoun and Coles [2] developed a method based on Fisher Linear Discriminant using many kinds of statistical features and different combinations of them. Although their method achieves reasonably high accuracy, it can only be applied on pairs of file types. For example, given a cluster in either JPEG or PDF, their method gives the answer accurately 98% of the time, but it does not work well when there are many different file types to consider.

Support Vector Machines (SVMs) are machine learning tools that are very useful in classification [1, 5]. An SVM classifies given data samples (in the form of vectors) by mapping them to high dimensional spaces and constructs hyperplanes that divide the data into partitions such that data belonging to the same class will be put into the same partition with high probability and those with different classes will likely end up in different partitions.

A recent study by Li et al. [17] shows that, by using support vector machines, the classification accuracy can be quite high among high entropy file types (including JPEG images, MP3 audio files and PDF documents), which were shown to be difficult to differentiate in previous studies. Furthermore, their method only uses the byte frequency distribution alone to build the statistical model. It can be expected that the accuracy can be further improved when this technique is combined with others such as keyword and pattern matching.

## 6.3 Reassembly

Reassembly is the process of linking a (probably huge) set of data clusters so that files are reconstructed from them. During the first two stages, we already know the headers and footers, and we know, with a certain level of confidence, to which types of files the clusters belong. In a way, during the reassembly process we can concentrate on only one file type at a time since the clusters are put into different categories.

To recover a file, essentially we start from the header of the file, and determine if the next cluster actually follows the header in the original file. If it does, we then merge the first two clusters and check the next cluster. Otherwise, we will search for the next cluster among a set of clusters to find the cluster that should follow. This process is repeated until all the clusters of the file are found.

A central challenge in the reassembly stage is the problem of *fragmentation point detection*. Essentially, given a series of clusters from a header $b_h$ to a footer $b_f$ ($h < f$), we need to determine if fragmentation ever occurs. If not, the file is already reconstructed. If fragmentation occurs, we need to determine exactly after which cluster $b_x$ this happens, so that we can merge the clusters from $b_h$ to $b_x$ into a fragment $F$ and continue searching for the next cluster that should follow $b_x$ among the pool of available clusters. During the search for the next cluster, what happens is that we check a new cluster $b_z$, put it after $b_x$, and determine if fragmentation happens at $b_z$. If it does happen, we drop $b_z$ and continue to try other clusters, until a $b_z$ is found such that it does not cause a fragmentation. We then merge $b_z$ into $F$ and continue searching until a footer is found. In practice, we cannot always be sure if fragmentation happens at $b_z$, in which case we assign weights to all the candidates and use other algorithms (such as those graph theoretic algorithms discussed earlier) to determine which cluster should come next. In this case, a highly accurate and efficient fragmentation point detection algorithm is necessary for practical file carvers, especially when the total number of clusters is huge.

### 6.3.1 Keyword/Dictionary-Based Techniques

For textual document clusters, it is often possible to exploit the semantics of the text as a way to determine if fragmentation occurs. A straightforward technique is to use a list of keywords or a standard word dictionary to determine if a given cluster should be merged with an existing fragment.

Dictionary-based merging techniques can be applied when when a word is formed between the boundary of two clusters. For example, if a cluster $b_x$ ends with the character sequence `he` and the cluster $b_{x+1}$ begins with `llo world`, the word `hello` would be formed if the clusters were merged. Therefore, we can conclude that these two clusters are likely to be in sequence and should be merged.

In addition to plain text documents, dictionaries can also be built for textual documents where keywords (i.e., special byte sequences) would occur that imposes structures on the document. For example, the cluster $b_x$ has been identified as belonging to an HTML document and it ends with the character sequence `</cen` and the cluster $b_{x+1}$ starts with `ter>`, the HTML keyword `</center>` would be formed if the clusters were merged.

These techniques can be repeatedly applied until two consecutive clusters are found such that they do not share a keyword or dictionary word, thereby identifying a fragmentation point. The most obvious limitation of these techniques is that it is not always easy (if possible at all) to find keywords or dictionaries that can be used to reliably detect fragmentation points.

Even for binary files, knowledge of the syntactical structures can be exploited to detect fragmentation points. This is very similar to the bifragmented carving as we described earlier. For example, since we know that each of the chunks in PNG images begins with a length field and ends with a CRC code, we can repeatedly try verifying the CRC code based on the length of the chunk, and a failed CRC code would indicate that a fragmentation has occurred.

### 6.3.2 Sequential Fragmentation Point Detection

A major limitation of keyword/dictionary-based techniques is that, when there is no useful keyword or dictionary words within the given cluster in question, the fragmentation point detection would fail.

For example, a cluster $b_x$ contains the start of a PNG image chunk, from which we know the size $s$ of the chunk. However, the value of $s$ can be larger than the size of a cluster, from which we can only know that the CRC lies in the cluster $b_{x+2}$. If the CRC in $b_{x+2}$ fails, we would have no idea if the fragmentation occurs from cluster $b_{x+1}$ or $b_{x+2}$.

*Sequential hypothesis testing* (SHT) developed by Wald [30] is a statistical analysis where a given hypothesis is tested using increasing number of samples. The hypothesis is accepted or rejected when sufficiently significant results are observed, otherwise the testing continues with more samples.

It is recognized by Pal et al. that sequential hypothesis testing can be used for reliable detection of fragmentation points [25]. The primary idea being that for every cluster added to an existing path, there is a high likelihood that the subsequent clusters belong to the path as well. Therefore, sequential hypothesis testing can be applied to determine if consecutive clusters should be merged together, not only based on the file type structure but also on the content of individual files. This modified reassembly algorithm is referred to as sequential hypothesis-parallel unique path (SHT-PUP) algorithm, which is a modified version of the parallel unique path algorithm developed earlier.

The reassembly algorithm works as follows. Assume that there are $k$ files to recover, starting from $k$ headers $h_1, h_2, \ldots, h_k$, respectively. Let $P_1, \ldots, P_k$ be the $k$ paths for the $k$ files, respectively. Initially each path $P_i$ only contains the header $h_i$, and we add more clusters to the paths as we process them. Let $S = \{s_1, s_2, \ldots, s_k\}$ be the current set of the last clusters in the $k$ paths, respectively. We refer to this set of clusters as the *working set*. Initially the working set contains only the headers (i.e., $s_i = h_i$ for all $1 \leq i \leq k$). As we join a new cluster $b$ to the $i$th path, the $i$th element in the working set is updated to $b$. This process continues until all the clusters in the working sets are footers, which indicates that all paths are built.

For the current working set, the SHT-PUP reassembly algorithm updates the working set as follows. First, the best match for each of the $k$ clusters in the working set is found. These clusters are recorded in a candidate set $T = \{t_1, \ldots, t_k\}$, where $t_i$ is the best match for $s_i$ $(1 \leq i \leq k)$.

From the candidate set $T$, we first choose the cluster $t_j \in T$ with the best matching among all the clusters in $T$, and perform the following steps.

1. Add $t_j$ to reconstruction path of $j$-th path, (i.e., $P_i = P_i || t_i$).
2. Replace the $j$-th cluster in the working set $S$ by $t_j$ (i.e., $s_i = t_i$).
3. Sequentially analyze the clusters immediately after $t_j$ until fragmentation is detected or the path is built, and update the working set $S$ accordingly.
4. Evaluate new candidate set $T$ of best matches for $S$.
5. Again find best cluster $t_j$ in $T$.
6. Repeat from the first step until all paths are built.

Compared with the parallel unique path algorithm described earlier, the main enhancement here lies in Step 6.3.2, where the weight is not precomputed, but dynamically determined using sequential analysis. This technique is referred to as the *sequential fragmentation point detection* [25], which is done as follows.

Let us assume that for the path $P_i$, we have already built part of it, and the last cluster of the path (i.e., the $i$-th cluster in the working set) is $s_i$. Now we consider subsequent data clusters $b_1, \ldots, b_n$, and append them in that order to the path $P_i$. We compute a weight $W_j$ for each of the cluster $b_j$ that is appended $(1 \leq j \leq n)$.

We further define the hypotheses $H_0$ and $H_1$ as the following:

$H_0$: Clusters $b_1, b_2, \ldots, b_n$ belong to the current path in that order.
$H_1$: Clusters $b_1, b_2, \ldots, b_n$ do not belong in sequence to the current path.

For any $j$ such that $1 \leq j \leq n - 1$, if the evaluated data clusters $b_1, \ldots, b_j$ do not yield a conclusive decision, the test continues with the inclusion of cluster $b_{j+1}$ until one of the hypotheses is confirmed.

When hypothesis $H_0$ is true, the evaluated clusters are merged to the path $P_i$ and a new test is started. Each time the test starts with a new data cluster in sequence, the weight is computed with respect to the part of the path that has already been built.

The test procedure finalizes after one of the following conditions is satisfied:

1. $H_1$ is achieved (i.e., clusters do not belong to the path).
2. The file is completely recovered.
3. No more data cluster of the same file type is available.

Let $\mathbf{W}$ sequence of weights $W_1, \ldots, W_n$, where the weight $W_i$ is the pair-wise weight between the cluster $b_i$ and the cluster before it. During a sequential hypothesis test, a test statistic $\Lambda$ is computed as the likelihood ratio of the sequence $W$ under the two hypotheses. $\Lambda$ is expressed as the ratio of the conditional distributions of observed weights under $H_0$ and $H_1$ as below.

$$\Lambda(\mathbf{W}) = \frac{\Pr[\mathbf{W} \mid H_1]}{\Pr[\mathbf{W} \mid H_0]}. \tag{1}$$

In this equation, the two conditional probabilities can be obtained experimentally in advance using large data sets. For example, for JPEG images, we can simply split a large number of JPEG images into clusters and compute the pair-wise weights among consecutive clusters to obtain the empirical probability $\Pr[\mathbf{W} \mid H_0]$. To obtain the probability $\Pr[\mathbf{W} \mid H_1]$, we can randomly generate a large number of fragmented JPEG images, and compute the pair-wise weights right at the fragmentation points.

Finally, with these probabilities, a decision is made by comparing $\Lambda$ to appropriately selected thresholds $\tau^+$ and $\tau^-$ as below.

$$\mathbf{O} = \begin{cases} H_1, \ \Lambda(\mathbf{W}) > \tau^+ \\ H_0, \ \Lambda(\mathbf{W}) < \tau^- \\ \bot, \ \tau^- \leq \Lambda(\mathbf{W}) \leq \tau^+ \end{cases} \tag{2}$$

where $\mathbf{O}$ represents the outcome of the sequential fragmentation point detection, and $\bot$ represents a decision that is inconclusive. In other words, if the test statistic (likelihood ratio) is larger than the threshold $\tau^+$, we assume that hypothesis $H_1$ is true and we have found the fragmentation region. If, on the other hand, the test statistic is smaller than $\tau^-$, hypothesis $H_0$ is true and all the clusters are merged and the test continues from the next sequential cluster. Finally, if neither case is true, testing continues until one of the thresholds is exceeded or a footer is reached, which indicates that the file has been recovered.

Ultimately, the success of the sequential fragment point detection method depends on two factors. The first factor is the choice of the weight whose design has to take into consideration different file types and to capture semantic or syntactic characteristics of the files. The second factor is the accurate determination of the conditional probability mass functions under the two hypotheses.

In the study by Pal et al. [25], the reconstruction results of JPEG images were provided on well-known test data sets from the 2006 and 2007 DFRWS challenges [6, 7]. It was shown that their technique based on sequential analysis gave the best recovery rate among other known file carving tools, and was able to recover files fragmented into more than four pieces with the presence of clusters belonging to many other file types.

The major challenge with this technique is the difficulties involved in building accurate models to determine the thresholds and defining good weight computation algorithms. In their study, the image clusters were weighted by examining pixel differences at the boundary, as described in earlier sections on graph theoretic carving techniques. For text they used PPM across a sliding window between two clusters.

In a recent study [18], more sophisticated yet efficient weight computation algorithms for JPEG images were proposed and analyzed. It was shown that the anomaly in the DCT coefficients in JPEG images can be a good indicator of fragmentation. Another edge detection algorithm is also studied and showed to achieve reasonable results. This study, however, focused on pair-wise cluster weight computation and did not consider sequential hypothesis. Nevertheless, it can be expected that these techniques can be easily employed in a sequential analysis to achieve good accuracy.

## 7 Summary

Extracting digital image evidence from storage volumes is a challenging task, especially when the underlying file system is damaged or purposely modified, or the file system met-data is not reliable.

The recovery of digital image evidence involves searching for image data clusters, locating the starting and ending positions of the images, detecting fragmentation points, and finally reassembling the fragments to reconstruct the original images.

The first generation of forensic tools that aim at file recovery often make simplistic assumptions, such that the files are not fragmented or bifragmented, or that the total size of the data to be considered is reasonably small. As a result, such tools would fail to recover many files in real application scenarios when files can be arbitrarily fragmented and the size of data can be huge.

Graphic theoretic approaches have been studied as a means to handle huge amount of data systematically given that multiple fragmentation points can exist for a file. Due to the complexity of the problem, the best known techniques are greedy in nature and exploit heuristics together with possibly sophisticated statistical analysis and machine learning.

Many studies attempt to solve this problem by exploiting the structures of the files, as well as the analysis of the semantics of the file content. For images, it is often useful to check the presence (or the lack of) particular structural keywords at certain locations, and it is often desirable to examine the image content and analyze its consistency across different clusters of data.

Future research on the searching and extracting images is likely to see a combination of keyword/dictionary matching, heuristic searching, graph theoretic methods, statistical analysis, image processing, and machine learning.

## References


1. Boser BE, Guyon I, Vapnik V (1992) A training algorithm for optimal margin classifiers. In: Proceedings of the Fifth Annual Workshop on Computational Learning Theory, pp 144–152. ACM Press
2. Calhoun WC, Coles D (2008) Predicting the types of file fragments. In: Proceedings of the 8th Digital Forensics Research Conference (DFRWS), vol 5 supp. 1 of Digital Investigation, pp S14–S20, Sep 2008
3. Carrier B (2005) File System Forensics Analysis. Addison Wesley, New Jersey
4. Cleary JG, Witten IH (1984) Data compression using adaptive coding and partial string matching. IEEE Trans Commun 32(4):396–402
5. Cortes C, Vapnik V (1995) Support-vector networks. Mach Learn 20(3):273–297
6. DFRWS forensics challenge, 2006
7. DFRWS 2007 forensics challenge, 2007
8. Dijkstra EW (1959) A note on two problems in connexion with graphs. Numerische Mathematik 1:269–271
9. Erbacher RF, Mulholland J (2007) Identification and localization of data types within large-scale file systems. In: Proceedings of the 2nd International Workshop on Systematic Approaches to Digital Forensic, Engineering, pp 55–70, Apr 2007
10. Gal E, Toledo S (2005) Algorithms and data structures for flash memories. ACM Comput Surv, 37(2)
11. Garfinkel S (2007) Carving contiguous and fragmented files with fast object validation. In: Digital Forensics Research Workshop, vol 4S of Digital Investigation, pp S2–S12
12. Hendrikx J (1998) Amiga smart filesystem
13. Karresand M, Shahmehri N (2006) File type identification of data fragments by their binary structure. In: Proceedings of the 7th IEEE Information Assurance Workshop, pp 140–147, June 2006
14. Karresand M, Shahmehri N (2006) Oscar-file type identification of binary data in disk clusters and RAM pages. In: Security and Privacy in Dynamic Environments, vol 201 of IFIP International Federation for Information Processing, pp 413–424. Springer, Boston, 2006
15. Kendall K, Kornblum J (2001) Foremost
16. Kolliopoulos SG, Stein C (1998) Approximating disjoint-path problems using greedy algorithms and packing integer programs. In: Proceedings of the 6th Integer Programming and Combinatorial Optimization Conference (IPCO VI), number 1412 in LNCS, pp 152–168
17. Li Q, Ong AY, Suganthan PN, Thing VLL (2010) A novel support vector machine approach to high entropy data fragment classification. In: 5th International Annual Workshop on Digital Forensics & Incident Analysis
18. Li Q, Sahin B, Chang E-C, Thing VLL (2011) Content based JPEG fragmentation point detection. In: IEEE International Conference on Multimedia & Expo
19. Li W-J, Wang K, Stolfo SJ, Herzog B (2005) Fileprints: Identifying file types by n-gram analysis. In: IEEE Workshop on Information Assurance, pp 64–67
20. McDaniel M, Heydari MH (2003) Content based file type detection algorithms. In: Proceedings of the 36th Hawaii International Conference on System Sciences (HICSS'03), p 332. IEEE Computer Society
21. McVoy LW, Kleiman SR (1991) Extent-like performance from a UNIX file system. In: USENIX, pp 33–43
22. Moody SJ, Erbacher RF (2008) Sadi—statistical analysis for data type identification. In: Proceedings of the 3rd IEEE International Workshop on Systematic Approaches to Digital Forensic, Engineering, pp 41–54, May 2008
23. Pal A, Memon N (2006) Automated reassembly of file fragmented images using greedy algorithms. IEEE Trans. Image Process 5(2):385–393
24. Pal A, Sencar HT, Memon N (2003) Reassembling image fragments. In: ICASSP, pp IV-732-5
25. Pal A, Sencar HT, Memon N (2008) Detecting file fragmentation point using sequential hypothesis testing. In: Digital Forensics Research Workshop, vol 5 of Digital Investigation, pp S2–S13

26. Richard III GG, Roussev V (2005) Scalpel: A frugal, high performance file carver. In: Digital Forensics Research Workshop, Digital Investigation
27. Shanmugasundaram K, Memon N (2003) Automatic reassembly of document fragments via context based statistical models. In: Computer Security Applications Conference, pp 152–159
28. Sweeney A, Doucette D, Hu W, Anderson C, Nishimoto M, Peck G (1996) Scalability in the xfs file system. In: USENIX, pp 1–14
29. Veenman CJ (2007) Statistical disk cluster classification for file carving. In: Proceedings of the IEEE 3rd International Symposium on Information Assurance and Security, pp 393–398
30. Wald A (1945) Sequential tests of statistical hypotheses. The Ann Math Stat 16(2):117–186
31. Wang K, Stolfo SJ (2004) Anomalous payload-based network intrusion detection. In: Recent Advances in Intrusion Detection, vol 3224 of LNCS, pp 203–222

[illegible]

# Part II
# Techniques Attributing an Image to Its Source

# Image and Video Source Class Identification

**Alex C. Kot and Hong Cao**

**Abstract** Advances in digital technology have brought us numerous cheap yet good-quality imaging devices to capture the visionary signals into discrete form, i.e., digital images and videos. With their fast proliferation and growing popularity, a security concern arises since electronic alteration on digital multimedia data for deceiving purposes becomes incredibly easy. As an effort to restore the traditional trust on the acquired media, multimedia forensics has emerged to address mainly the challenges concerning the origin and integrity of multimedia data. As one important branch, source class identification designs the methodologies to identify the source classes and software tools based on its content. Through investigating and detecting the source features of various forms, a large number of identification methodologies have been developed in recent years and some achieved very good results. In this chapter, we review the history and major development in image and video source class identification. By first establishing an overall picture of multimedia forensics, we discuss the role of source identification in relation with other forensic fields and elaborate the existing methodologies for identification of different source types.

## 1 Introduction

Images and videos are traditionally labeled as "trustworthy" as they are known as being captured through analog acquisition devices to depict the real-world happenings [1]. This traditional trustworthiness is largely established upon the remarkable difficulties in modifying the multimedia content. For example,

A. C. Kot (✉)
School of Electrical and Electronic Engineering, Nanyang Technological University, Singapore
e-mail: eackot@ntu.edu.sg

H. Cao
Institute for Infocomm Research, Agency for Science, Technology and Research (A*STAR), Singapore
e-mail: hcao@i2r.a-star.edu.sg

H. T. Sencar and N. Memon (eds.), *Digital Image Forensics*,
DOI: 10.1007/978-1-4614-0757-7_5,

a film-based photo could only be altered by specialists through darkroom tricks, which were time-consuming, costly, and often came with a limited alteration extent. In contrast, digital revolution has brought us today countless powerful devices and handy electronic tools, including high-resolution multimedia acquisition devices, high-performance computers, high-speed Internet connections, high-fidelity displays, highly advanced editing tools, etc. These tools largely replace their analog counterparts to enable easy digital content creation, processing and communication at affordable cost and in mass scale. While one enjoys the conveniences of using these ubiquitous tools, a serious question shall be asked as "Can we still trust the images and videos acquired?" The popular Photoshop software has millions of registered users and fans worldwide. It can facilitate a computer-literate user to modify almost anything on a digital image in several simple steps. Similar user-friendly tools are also available for modifying video content. Most of the time, the human eyes can hardly differentiate a genuine scene from a deliberately altered scene. As such, increasing fraudulent cases involving multimedia forgeries have appeared and we mention several recent impactful cases. In the first case [2], an image of the most wanted terrorist, Osama Bin Laden, was found to be a syntactic creation instead of a real photo. In the second case [3], the inaugural photo of the Israel cabinet was digitally tampered with two female cabinet members being replaced. One can refer to [2, 3] for the stories of these forgeries. More famous forgery examples can be found in [4] and these forged media were mostly published in the world-renowned newspapers and magazines. Note that publishing forgery media content is an unethical behavior. Once exposed, the forgeries would generally lead to degraded credibility of the parties who publish them. In view of the potential damages caused by electronic alterations on the digital media in the publishing industry, the US National Press Photographer Association (NPPA) has clearly stated in their guidelines for news photographs that "altering the editorial content of photograph, in any degree, is breach of the ethical standards recognized by the NPPA." [1]. Figure 1 also demonstrates several created forgery examples in comparison with their original versions. These forgeries have tested to be hard to identify by the human eyes when singly presented.

Besides the visual content, another fragile yet important piece of information is the multimedia source, i.e., information about the acquisition device used, its model, type, its inclusive components, and processing modules. When multimedia content is presented as an evidence in important scenarios, e.g., in a courtroom, its source information often represents useful forensic clues that need to be assured in advance. For instance, knowing the source device that captures a multimedia evidence can facilitate verification or tracing of the device owner as well as the camera taker. Knowing the device model or brand information can help a forensic analyst to tap into rich online resources to know more about its acquisition device. This potentially improves the chance in detecting the underlying forgeries on the content. Although the source model information of a media may be read out directly from the metadata headers, e.g., from the exchangeable image file format (EXIF) of a photograph, these headers could be easily removed or counterfeited by a forger using ubiquitous editing tools. And such header alterations are believed to be difficult to recognize and trace.

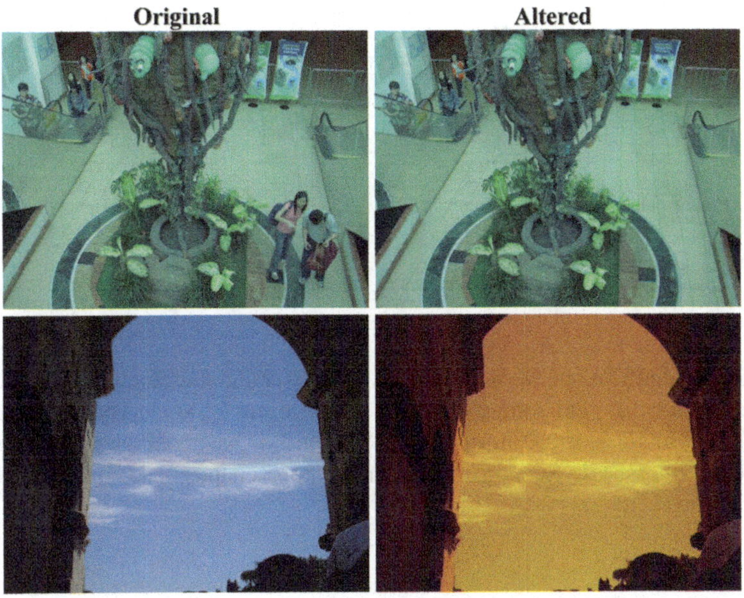


**Fig. 1** Image forgery examples. On the *top*, two people are delicately removed. At the *bottom*, color hue is modified to show as if the picture was captured in the evening time

While images and videos are still popularly used as supporting evidences and historical records in a growing number of applications from police investigation, law enforcement, journalist reporting, surveillance, insurance claims, medical and dental examinations, military, museum and consumer photography, scientific means in media source identification and forgery detection are in urgent need. This urgency is stimulated by several new trends: (1) Proliferation of digital cameras and other multimedia acquisition devices is fast due to the improving photographic technology and the reducing cost. Digital cameras attached on mobile handsets have now become a necessary feature [5] for mobile phone users; (2) Media sharing on the Internet has become increasingly easy and popular. Media sharing and social network giants such as *YouTube* and *Facebook* gain huge success through promoting users to share their own multimedia contents online. At the same time, they also provide common platforms for the forgery medias to spread quickly to make great social and political impacts; (3) A significant portion of multimedia from the public have good news value and are sourced by various news agencies for making journalist reports [6, 7]. However, one big challenge [6] is how to authenticate the contents from an unreliable public source; (4) Following the human stem-cell fake [8], which involves obvious image forgeries, scientific journal editors [9] have been actively seeking efficient forensic tools, which detect the deliberate image tweaks.

This chapter focuses on the topic of image and video source class identification. The remainder of this chapter is organized as follows: In Sect. 2, we give a brief overview of multimedia forensics and introduce the forensic role

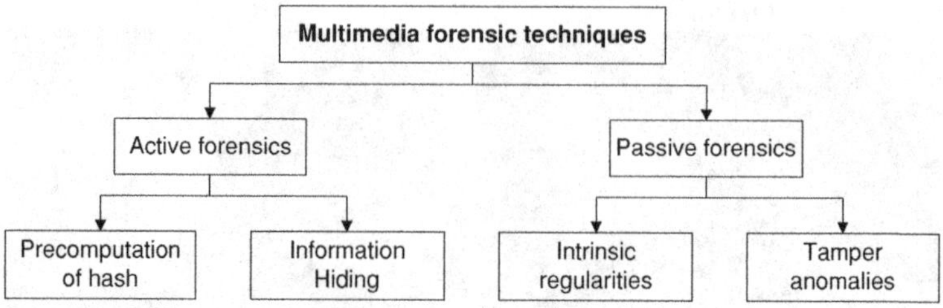


**Fig. 2** Classification of multimedia forensics techniques

of source identification. Section 3 reviews the existing source class identification techniques for different source types. Section 4 discusses some common machine learning techniques used in forensics. Section 5 talks about anti-forensics and the countermeasures. Section 6 summarizes this chapter and indicates future opportunities.

## 2 Overview of Multimedia Forensics

### *Active and Passive Forensics*

With an ultimate goal to restore the "trust" toward digital multimedia and to alert deliberate tampering, multimedia forensics has attracted considerable attention in the past decade. As shown in Fig. 2, the proposed techniques generally fall into two categories, the active and the passive categories. The active approaches commonly require pre-computation of some fragile property, e.g., a cryptographic hash of the media, or prior insertion of some secret protection data through information hiding before the media which are sent through an unreliable public channel. Upon receiving these media, forensic conclusions can be reached by comparing the recomputed media property or the extracted hidden data with their original versions. For instance, two early works actively modify the camera structure to build the "trustworthy" or the "secure" digital camera to protect the integrity of pictures. In [10], an encrypted digital signature file is separately generated for each photo in a "trustworthy camera" for content authentication purposes. In [11], a watermark inclusive of iris biometric data is embedded in a "secure camera" during image creation and the extracted watermark can be used for both image integrity verification and camera taker identification. More active approaches can be found in [12] for forensic applications such as content authentication and digital rights management for various multimedia types. A recent approach in [13] also proposes to protect the integrity of the emerging electronic ink data through a lossless data embedding technique.

Although the active approaches can be highly effective in securing multimedia data from malicious operations and for ownership verification, the strict requirement on cooperative end-to-end protocols is hardly met in many practical forensic scenarios. Moreover, the protective data embedded through hiding techniques are considered as artificial noises, which potentially degrade the media's quality. On the other hand, the passive approaches are non-intrusive in nature and do not impose the above constraints. Hence, they are readily applicable to a wide range of forensic applications. Two recent surveys can be found in [14, 15]. The common philosophy behind is that multimedia acquired through common acquisition devices are natural, containing rich source-induced fingerprints due to the imperfection of processing pipelines. These media occupy only a small and highly regularized space in the entire high-dimensional space. If accurately modeled and detected, the fingerprints or source features can be used to trace back the multimedia source from their content. Tampering likely disturbs some intrinsic regularity, introduces new tampering anomalies, and leads to many forms of inconsistencies. If detected, they can be used as tell-tale signs to expose the forgery. As shown in Fig. 2, passive forensics can be further divided into two subcategories, through detecting some intrinsic regularity or through detecting some tampering anomalies. The formal focuses on detecting some unique intrinsic footprints for source identification or to assess their disturbances for tamper forensics. The latter detects specific anomalies left over by some common tampering operations, which rarely exist in an untampered media.

## *Passive Forensics and Source Identification*

Although content forgery detection is the primary goal in multimedia forensics, the current research has encompassed a breadth of topics listed below:

- *Source identification related* [16–72]: Which individual device, which source class, e.g., sensor type, processor type, model and brand, is the device used in capturing a given media? What is the media processing software used? Is a media acquired with a source as claimed? Given a set of media, can we cluster them according to their sources?
- *Tampering discovery* [24, 50, 53]: Has a given media been tampered and where is the altered region?
- *Steganalysis*: Has a media been concealed with a secret message? What is the message length, content, and the embedding method used?
- *Recovery of processing history*: What processes have a media gone through? What are the parameters used?
- *Recapturing identification*: Is a given media captured on real scenery or on artificially created scenery using the modern display technology?
- *Computer graphics identification*: Is a given media acquired using a camera device or artificially created using photorealistic computer graphic (PRCG) software?
- *Device temporal forensics*: When is a given media captured? How can we order a set of media according to their acquisition time?

These topics address different security aspects of multimedia data, including their origin, content integrity, innocence against secret data embedding, status of being acquired by a device versus synthetic creation, processing history, and their acquisition time. Out of these topics, source identification stands out as one important topic by itself. Although source identification addresses a separate issue and other forensic topics, e.g., forgery detection, are not necessarily related with source identification, we find in many existing solutions that they are closely linked in the following ways:

- Source identification methodologies can be used directly to address one type of forgery, where a forger modifies the metadata tags and falsely claims a media from one source as acquired from another source;
- Multimedia forgery often involves mixing signals from multiple sources, which can lead to many source-dependant inconsistencies. These inconsistencies can be characterized either through using the source features or based on the outcomes of the source identification techniques at localized pieces of the media;
- Many statistical source features are found to be good to address the tamper forensics issue due to their fragile nature toward tampering and manipulations. Similarly, features originally proposed for other forensic topics, e.g., image quality metrics for steganalysis and high-order wavelet statistics for detection of computer graphic images, are later found to work well in source device model identification;
- Source identification can serve as an initial step for other forensic analyses, e.g., tampering discovery. For instance, a forensic analyst can identify the source of a multimedia first and then apply the corresponding source-specific template to discover the local tampering.

Source identification is commonly formulated as a machine learning task. In the training phase, the labeled representative data from $N$ sources are provided. A set of source features are then extracted from the training data to form the corresponding set of feature vectors. As an optional step, we can learn the knowledge from the training data and perform feature refinement such as dimensionality reduction, feature selection, re-weighting, or filtering in order to retain a compact set of highly discriminative features. The learned knowledge $T$ is also applied in the test phase to perform similar refinement. Based on the refined features and their labels, a template for each source class or a classifier $C$ is constructed through using a supervised machine learning technique. In the test phase, similar feature extraction and refinement are performed to have the refined test feature vector. The predicted source label is then obtained as the classification output of $C$ on this vector. The source identification performance can be largely affected by several factors such as representativeness of the learning data, goodness of the forensic features, and the learning algorithm used. The data representativeness issue can be alleviated if sufficient training data with good content variety are provided. The good forensic features are those that are stable against media content variations and comprehensively capture the unique characteristics of the source. The desired learning algorithm not only forms a good decision boundary to classify the training data, but also generalizes well for the test data.

# 3 Multimedia Source Class Identification

As a subtopic of source identification and different from individual source device identification, source class identification is the science to identify certain attributes of the acquisition device for a given media, e.g., the device model, software tool, which are often shared by a class of devices. Depending on the outcomes, we categorize source class identification into source category identification and software tools identification. In the following section, we review the techniques for each category.

## *Source Category Identification*

Since media acquisition devices of the same class, e.g., the same model or brand, often share similar software processing modules and hardware components; the different source models can be identified by extracting features that are intrinsic and unique to each device class. The commonly used features are the processing regularities associated with some digital processing modules, e.g., demosaicing regularity and color features, or the regularities, e.g., noise statistics, image quality metrics, lens distortions, which are associated with some attributes of the hardware, e.g., quality of the sensor and distortion characteristics of the lens system.

### Lens Distortion Features

Lens radial distortion (LRD) is purported as the most severe among the different types of lens distortions such as spherical aberration, astigmatism, field curvature, chromatic aberration, etc. LRD is easily visible for the inexpensive wide-angle lens and manifests itself as rendering straight lines into curved lines in the acquired visual signal. The direction of LRD is along the radial axis and its magnitude is a function of the distance between the object and the lens central axis. Based on a second-order LRD model, Choi et al. [16] proposed to compute the distortion parameters through measuring the distortions on the extracted straight-line edge segments. By using two estimated parameters as features together with the Kharrazi's features in [61], an accuracy of 89% is achieved in identifying five camera models based on a fixed optical zooming. However, the accuracy of using the distortion parameters alone can be severely affected if optical zooming of the lens is varied.

Lateral chromatic aberration (LCA) is formed due to the varying refractive indexes of lens materials for different light wavelengths so that light of different colors cannot be perfectly focused. This causes misalignment in different color channels, which are small near the optical center, but becomes larger at the locations that are farther away from the optical center. Based on a linear LCA model, Van et al. [17] estimate the global model parameters, e.g, coordinates of the optical center and a scalar ratio, using an image registration technique that maximizes the mutual entropy between two different color channels. By using these parameters as features for support vector machine (SVM) classification, an identification accuracy of 92% was achieved for three different mobile phone models. This work also provided some results indicating that the LCA features cannot be used to distinguish two individual mobile cameras of the same model.

### Demosaicing Regularity

Demosaicing regularity represents the most significant difference in processing among various DSC models as the choice of CFA and demosaicing algorithm is usually fixed for a given model but it may differ for different models. Moreover, 2/3 of the color samples of a photo are interpolated with a demosaicing algorithm consisting of only a few formulas. As a key process that determines image fidelity [73], demosaicing has been extensively researched in the past to reconstruct full-color images from the raw sensor data. Based on a similar assumption that Bayer CFA is adopted, the existing demosaicing solutions differ largely in terms of their edge adaptiveness, their grouping strategy for the adaptive methods, the reconstruction domain, the low-pass filter type, size, parameters, and the follow-up enhancement, refinement procedures. Table 1 summarizes some of these differences for 14 representative demosaicing algorithms [74] published in the past two decades either as US patents or in top journals. These large differences enable each demosaicing algorithm to introduce some unique and persistent correlation throughout the output image, whose detection is desirable for multimedia forensics.

The early detection works use some simple pixel-level interpolation models within each color channel. Generally, a demosaiced sample at location $(x, y)$ of a color channel $c$ can be expressed as

$$I(x, y, c) = \sum_{\forall (i,j) \in \Omega} w_{ij} I(x+i, y+j, c) + \varepsilon_{xyc} \tag{1}$$

where $\{w_{ij}\}$ are the weights representing the correlation characteristics induced by the demosaicing process and $\varepsilon_{xy}$ denotes an interpolation error. $\Omega$ represents a predefined supporting neighborhood, which can be chosen accordingly. Figure 3 shows three interpolation models with different supporting neighborhoods. Based on the model in Fig. 3a, Popescu and Farid [51] proposed an expectation maximization (EM) algorithm for estimating a set of 24 interpolation weights by iteratively optimizing a posterior probability map and computing the variance of the interpolation errors. The probability map here contains the estimated probability for each sample being a demosaiced sample and the weights' solution is derived to minimize the expected interpolation errors. The iterative EM cycles stop when a stable solution of weights is reached. By using these weights as features for a linear discriminant classifier, an average accuracy of 97% was achieved in distinguishing 28 pairs of demosaicing algorithms, where the minimal pairwise accuracy is 88%. Bayram et al. [51, 52, 55] extended this EM algorithm to identify commercial DSCs by assuming a larger 5×5

**Table 1** Comparison of 14 representative demosaicing algorithms

| Algorithm | Edge adaptiveness | Domain (refinement) | Reconstructive filter (size) |
|---|---|---|---|
| Bilinear [51] | Non-adaptive | Intra-channel | Fixed (3×3) |
| Bicubic [51] | Non-adaptive | Intra-channel | Fixed (7×7) |
| Cok 87 | Non-adaptive | Color hue | Fixed (3×3) |
| Freeman 88 | Non-adaptive | Color difference | Median (3×3) |
| Laroche 94 | Adaptive | Mixed | Fixed (3×3) |
| Hamilton 97 | Adaptive | Color difference | Fixed (3×3) |
| Kimmel 99 | Adaptive | Color hue (inverse diffusion) | Adaptive (3×3) |
| Li 01 | Non-adaptive | Color difference | Adaptive |
| Gunturk 02 | Non-adaptive | Color difference (iterative) | Fixed (5×5) |
| Pei 03 | Adaptive | Color difference (iterative) | Adaptive (3×3) |
| Wu 04 | Adaptive | Color difference (iterative) | Fixed (3×3) |
| Chang 04 | Adaptive | Color difference (iterative) | Adaptive (5×5) |
| Hirokawa 05 | Adaptive | Color difference (iterative) | Median (5×5) |
| Alleysson 05 | Non-adaptive | Chrominance | Fixed (11×11) |

interpolation window, and analyzed patterns of periodicity in row-averaged second-order derivatives in the non-smooth and smooth image regions, respectively. The filter weights and the periodicity property are employed as features in constructing SVM classifier. An average accuracy of 96% was achieved in identifying three camera models. Long et al. [53] proposed to compute the normalized $13\times13$ pixel correlation coefficients based on the interpolation model in Fig. 3b, for each of the three color channels. These coefficients are used as features to train a neural network classifier to identify different source camera models. Using majority-vote fusion for decision outcomes from three color channels, a close to 100% accuracy was reported for identifying four commercial cameras of different models in the presence of 0–5% rejected or undetermined cases. This work also shows that the post-processes such as median filtering could severely affect the detected correlation as well as the identification performance.

For non-intrusive component analysis, Swaminathan et al. [54] introduced several new procedures in the demosaicing detection: (a) estimation of the underlying CFA; (b) division of the image into three regions including horizontal gradients, vertical gradients, and the smooth region; (c) separate estimation of the underlying demosaicing formulas for each region. Note that in the green-channel interpolation model used in Fig. 3c, each symbol "o" denotes one sensor sample in the $7\times7$ neighborhood of a demosaiced sample "$\Diamond$". These sensor samples are identified from the demosaiced samples after the underlying CFA is estimated. The interpolation models for the red and blue channels share a similar concept, except that the sensor samples are located at different sites. With an assumption that similar interpolation formula is applied in each region within each color channel, a total least square solution is employed to compute their corresponding weights as representation of the underlying demosaicing formulas. Experimentally, a total of 441 such parameters

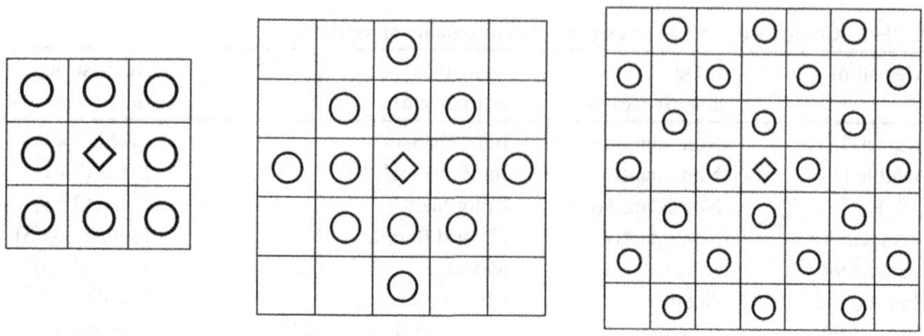


**Fig. 3** Intra-channel correlation models adopted for detection of demosaicing regularities where "◇" is a demosaiced sample and "o" indicates a neighboring support sample. **a** Popescu and Farid model [51]; **b** Long and Huang's model [54] and **c** Swaminathan, Wu and Liu's model [55] for green channel with Bayer CFA assumed

were computed as features in identifying 19 cameras of different models. With 200 images per camera, these features achieved an average identification accuracy of 86% using SVM.

Although some reasonably good results are reported in the above methods, the intra-channel interpolation models are limited to characterize the correlation within one color channel only. On the other hand, demosaicing algorithms, e.g., those in Table 1, often employ cross-color domains, e.g., the color difference plane, for reconstructing the missing samples. This inevitably introduces the cross-correlation between the red, green, and blue channels. Moreover, the criterion of dividing the demosaicing samples into different edge groups and the corresponding reconstruction formulas can vary greatly from one algorithm to another. To comprehensively capture these demosaicing differences, Cao and Kot [56–60] proposed partial derivative correlation models for demosaicing detection. With an assumption of Bayer CFA being the underlying CFA, a demosaiced image **I** and its sensor data **S** can be written as

$$\mathbf{I} = \{I_{xyc}\} = \begin{pmatrix} \{r, G, B\}_{11} & \{R, g, B\}_{12} & \cdots \\ \{R, g, B\}_{21} & \{R, G, b\}_{xy} & \\ \vdots & & \ddots \end{pmatrix}, \quad \mathbf{S} = \{S_{xy}\} = \begin{pmatrix} r_{11} & g_{12} & \cdots \\ g_{21} & b_{xy} & \\ \vdots & & \ddots \end{pmatrix}$$

where $R$, $G$ and $B$ denote demosaiced samples in red, green, and blue channels, respectively, and $r$, $g$, $b$ denote the sensor samples, which are known before the demosaicing process. $I_{xyc} = I(x, y, c)$ and $S_{xy} = S(x, y)$. For each demosaiced sample $I_{ijc}$ along $t$ axis, a derivative-based demosaicing model is written as

$$I_{xyc}^{(t)} = \sum_{\forall (i,j) \in \Omega} w_{ij}^{(t)} S_{x+i,y+j}^{(t)} + \varepsilon_{xyc} \tag{2}$$

where $I_{xyc}^{(t)}$ is the partial second-order derivative of $I_{xyc}$ along $t$ axis, $S_{xy}^{(t)}$ is the derivative for $S_{xy}$ and $\left\{w_{ij}^{(t)}\right\}$ are the derivative weights. Suppose that $t$ represents the $x$ axis, $I_{xyc}^{(t)}$ and $S_{xy}^{(t)}$ can be written as

$$\begin{aligned} I_{xyc}^{(t)} &= I_{x+1,yc} + I_{x-1,yc} - 2I_{xyc} \\ S_{xy}^{(t)} &= S_{x+2,y} + S_{x-2,y} - 2S_{xy} \end{aligned}$$

The above formulation makes use of the 2-by-2 periodicity property of the CFA lattice so that reconstruction of a missing sample along $t$ axis can be deduced equivalent to estimating its partial second-order derivative. The main advantage for using derivatives is that such interpolation model largely suppresses the detection variations caused by different contents as the scenery-dependent local DC components are removed in the derivatives. Also since the support neighborhood $\Omega$ contains sensor derivatives from all three color channels, it allows extending the detection across the boundaries of three color channels naturally. Based on the derivative models, the detection algorithm uses an expectation maximization reverse classification to iteratively divide all demosaiced samples in 16 categories by minimizing the model fittings errors. This takes care of the different demosaicing grouping strategies that are commonly used. The derivative weights for each category are then estimated using a least square solution. It is demonstrated in a comparison that the demosaicing formulas in terms of derivative weights can be better used to re-interpolate the demosaiced samples from the sensor samples with less error than the intensity weights. Correspondingly, demosaicing features including derivative weights, error cumulants, and sample distributions are suggested for source identification. The result shows that the selected 250 features using sequential floating forward selection (SFFS) achieved a remarkable accuracy of 98% in identification of 14 commercial camera models based on small cropped blocks of 512×512. Not only for high-quality DSC images, Cao and Kot [58] also extended these features to identify the low-end images acquired by mobile devices. Using eigenfeature regularization and extraction (ERE) [75], the reduced 20 features achieved 99% accuracy in identifying nine mobile cameras of dissimilar models, which outperformed several other statistical forensic features recommended for mobile camera identification. Some limitations are that the identification performance can be affected by lossy JPEG compression with low quality factors. Similar camera models may implement identical software modules so that they are hard to distinguish using processing regularities.

Although it is expectedly difficult to identify very similar camera models using demosaicing regularity, Cao and Kot [60] demonstrated that the reduced demosaicing features can still identify five similar Canon DIGIC processors with an average accuracy of 96% based on compact digital single lens reflex (DSLR) camera model templates and one-nearest neighbor classifier. The work also showed that identification accuracy increases significantly if more than one DSLR camera per model is used to train each model template.

### Other Statistical Features

Other statistical features for source class identification include color features, high-order wavelet statistics, image quality metrics, binary similarity metrics, noise statistics, statistic moments, etc.

For camera model identification, Kharrazi et al. [61] proposed 34 features including color features, image quality metrics, and wavelet coefficient statistics. With an SVM classifier, these features yield an accuracy of 88% in identifying five camera models, where three models are from the brand Canon. Celiktutan et al. [63] combined three sets of features including binary similarity measures, image quality metrics, and wavelet coefficients statistics for identification of cell-phone camera models. With feature-level fusion and SFFS feature selection, it reported an average identification accuracy of 95% for 16 cell phone models using SVM. This work also tested score-level fusion strategies with different rules, where a maximal accuracy of 98% is reported based on the product rule. Xu et al. [64] proposed a set of 390 statistical moments to characterize the image spatial representation, JPEG representation, and pixel co-occurrences. Using SVM, these features achieved 98% accuracy to identify six camera models from four brands when one model is represented by one camera. But in the case that each model is represented by over 150,000 *Flickr* images from 30 to 40 cameras, the reported accuracy drops to 86% for identifying eight models from four brands. The corresponding brand identification accuracy is 96%. Gloe et al. [65] combined color features [61], image quality metrics [63], and high-order wavelet statistics [63] to study several camera model identification related issues, such as the intra-model camera similarity, the influence of the number of devices per model and images per device in learning, and the influence of image size. Based on 44 cameras from 12 models and using SVM classification, the result shows that the overall model identification accuracy can improve significantly when a larger percentage of images are used for training. In contrast, increasing the number of devices per model only contributes slightly toward better performance. The result also shows that cropping all images to smaller size of $2560\times1920$ would considerably degrade the overall model identification accuracy from 96 to 90%.

For scanner model identification, Gou et al. [43, 44] proposed a set of 60 noise statistics features. These features characterize the noise residual statistics for different denoising algorithms, the distribution of wavelet coefficients, and the statistics of neighborhood prediction errors. Based on 14 scanners from 11 models, about 94% model identification accuracy was achieved using an SVM classifier. Khanna and Delp [66] proposed to compute a set of gray-level co-occurrences and difference histograms at character level and at block level to identify the source scanner models from scanned text documents. Based on five scanner models, these features achieve more than 95% identification accuracy for cropped image size of $512\times512$. In [67], the same authors also demonstrated that the block-level features show good robustness toward different font sizes and font shapes.

For device class identification, McKay et al. [47] combined the noise statistics features in [43] together with color interpolation coefficients [54] to identify different source classes including computer graphics, DSCs, mobile cameras and scanners, and their source models. An average accuracy of 94% was achieved for classification of four types of devices, 98% for five cell-phone camera models, 96% for four scanner models and 94% for four camera models. Fang et al. [68] suggested combining 72 wavelet coefficients features and 12 noise moments to distinguish DSLR cameras from compact cameras. With eight DSLR cameras and 12 compact cameras, a classification accuracy of 95% is achieved based on image size of $1024 \times 1024$. The results also show that the accuracy can drop to about 84% for a small image size of $256 \times 256$. Khanna et al. [45] proposed to compute statistics of the row correlation scores and the column correlation scores from noise residuals as features to distinguish scanned images and camera captured images. These features are designed to capture the differences between 1D sensor noise pattern used in scanners and the 2D pattern for DSCs. Experimentally, 17 features achieved 99% accuracy in distinguishing images from three scanner models and three cameras using SVM. In a more practical test that images from two cameras and two scanners are used for training and images from the remaining one camera and one scanner are for testing, the identification accuracy drops to 94%.

## *Software Tools Identification*

There are also interests to identify the commercial software tools such as RAW converters and source encoders. DSLR cameras are relatively high-quality DSCs that allow users saving the sensor data into proprietary RAW formats. RAW converters, also known as electronic darkroom, can then be used to develop these RAWs into full-color photos on a personal computer (PC). These RAW converters perform a similar function as the camera's internal software processing unit. However, since they are designed for PCs, sophisticated demosaicing and post-processing techniques that require substantial computing resources can be implemented. Moreover, they allow the users to experiment on the darkroom settings to achieve the desired effects. Cao and Kot [57] proposed to use the similar demosaicing features in [58, 59] for classification of ten commercial RAW converters such as ACDSee, Photoshop CS, PaintShop Pro, etc, based on cropped $512 \times 512$ blocks. For each converter, four types of cropped blocks were differentiated depending on the cropping locations in $2 \times 2$ periodical lattice. For a total of forty RAW-converter classes, their reduced 36 demosaicing features achieved a remarkable low identification error rate of 1%.

Images and videos are often lossily encoded to reduce the size before they are transmitted or stored. Image encoders can adopt different types of coding techniques such as transform coding, sub-band coding, and differential image coding. These coding techniques can be further divided based on the types of transformation used. Lin et al. [69] developed a set of techniques that determine the source encoder based on the intrinsic fingerprints in the output images. The algorithm starts by testing whether an image has gone through block-based processing, followed by estimation of the block size. The proposed test makes uses of the fact that block-based processing often introduces periodical discontinuities at the block boundaries. After this test, the work then developed separate intrinsic fingerprint analyses cum similarity measures for testing on transform coding, subband coding, and differential image coding, respectively. Experimentally, it achieved over 90% detection rate for identifying five types of lossy encoders with the peak signal to noise ratio as (PSNR) <36 dB. However, the performance would degrade if the compression quality factor is increased. For instance, the detection rate drops to 80% when the PSNR increases to 40 dB. For transform-based image encoders, Luo et al. [70] also proposed a feature-based technique that analyzes the histogram artifact of a quantized coefficient and transforms it into two features per coefficient. These features are designed to capture the distinctive histogram differences for quantized and unquantized coefficients. Experimentally, these features show good efficacy in detecting the presence of JPEG compression, in identification of different transform encoders, different block sizes and different wavelet decomposition levels. For instance, a 14-dimensional feature vector achieves a remarkable 99% detection rate in identifying seven different DCT and DWT transformations when the lossy compression PSNR is greater than 37 dB.

For commercial video encoder identification, Su et al. [71] demonstrated that different MPEG encoders can introduce different statistical distribution of their motion vectors. This is because different encoders could adopt different parameters, e.g., the threshold for static macro blocks and the maximal search window size, in their practical implementations. The authors proposed a set of statistical features to characterize the motion vectors of an MPEG encoder. Using $k$-nearest neighbor ($k$-NN) classifier, these features achieved an accuracy of 75% to identify eight commercial encoder tools based on the constant bit-rate video clips.

## 4 Forensics Learning Techniques

Since most multimedia forensic problems can be readily formulated as pattern classification tasks, employment of suitable pattern classification and related techniques plays an important role for achieving good forensic performances. Below, we discuss the feature reduction, classification, and fusion techniques used in previous forensics works.

### *Feature Reduction Techniques*

High feature dimensionality can easily cause increased computational complexity and incurs long classifier training time and slow responses. The relatively small size of training data as compared with feature dimension could also lead to insufficient classifier training with degraded test performance. It is often desirable to substantially reduce the feature dimensionality before the reduced features are used in forensic analysis. The feature reduction can be achieved using either subspace transformation or feature selection techniques.

The subspace approaches typically search for a compact set of projection axes or a subspace from the entire high-dimensional image or feature space. Through projecting the high-dimensional feature data onto each axis, a new feature is generated as a linear combination of the old features. Principal component analysis (PCA) and linear discriminant analysis (LDA) are two commonly used subspace methods in many pattern classification tasks. With an aim of minimizing the least square errors in the reconstructed signal, PCA computes the total covariance structure from a given set of training data. The compact PCA subspace is found through eigen decomposition on the total covariance matrix. By projecting the feature data onto the PCA subspace, a number of uncorrelated features will be obtained, which can be best used for applications such as signal reconstruction and data compression. LDA, on the other hand, is designed for discrimination purposes. With a set of labeled training feature data from different classes, LDA computes the within-class covariance matrix and the between-class covariance matrix. A set of linear discriminant axes are found with the goal of maximizing the ratio between the between-class projected variance and the within-class projected variance. Practically, LDA suffers several problems: (a) The subspace found by LDA are prone to estimation errors on the covariance matrices, especially when numerous close-to-zeros eigenvalues of the within-class covariance matrix are present; (b) The number of discriminant features is limited by the number of classes. For an $L$-class classification problem, only $L$-1 discriminant features can be extracted by LDA; (c) the discriminant features found by LDA are statistically correlated. A large number of works are available in pattern classification literature to improve the subspace-based discriminant feature extraction through addressing the above issues of the LDA. One recent work [75], called ERE, regularizes the unreliable portion of within-class eigenspectrum for the extraction of a compact set of discriminant forensic features. The features show good performance on the unseen data especially when the size of training data is small as compared with the feature dimensionality.

Unlike the subspace methods, feature selection techniques aim to select the most competent feature subset by removing the features identified to be unproductive or less productive toward a specific classification goal, e.g., maximizing the accuracy of a given classifier. Since the exhaustive searching of the best feature subset usually requires highly-intensive computation, the main issue becomes how to search for a best feature subset with manageable searching complexity. Pudil et al. [76] proposed a popular sequential forward floating search (SFFS) method together with nonmonotonic criterion function for feature selection. Starting from an empty feature set, the SFFS algorithm performs stepwise feature inclusion and conditional feature removal to select a feature subset. Compared with simple sequential searching algorithms, the backtracking mechanism, i.e., the conditional feature removal, employed by SFFS allows removing the previous selected but currently useless features due to inclusion of new features, which more likely results in an optimal feature subset. The SFFS algorithm has been applied to several prior forensic works [59, 62, 63].

In terms of feature compression ratio and their discriminative power, we find that subspace methods usually provide better classification performance with fewer features. For instance, our work [59] reported identification accuracy of 97.5% for 14 DSCs using 250 SFFS selected features and probabilistic SVM. Based on the same data set and classification method, a better accuracy of 98% can be achieved using only 20 ERE features. Moreover, the time taken for SFFS can be impractically long if both the original feature dimension and the number of features to be selected are high. Despite these disadvantages, feature selection is fundamentally different from the subspace methods and it has the advantage that the unselected features need not be computed in the test scenario. In practice, feature selection can be used in conjunction with subspace feature reduction to improve the tradeoff between feature extraction efficiency and the accuracy of the forensic analyzer.

## *Classification Techniques*

Among the different pattern classification techniques, SVM is so far the most popularly used in the existing source identification works. SVM is a powerful nonlinear pattern classification tool. By minimizing the structural risk [77] in classification, SVM optimizes the separation margin between different classes so that it typically exhibits good generalization performance on the unseen data. The kernel trick commonly employed in SVM also maps the current low-dimensional feature space into a high-dimensional space before a better separation plane can be found in the high-dimensional space. Several works, e.g. [78], further extended the traditional SVM into probabilistic SVM (or PSVM) by studying distribution of the SVM outputs and maps them into probabilistic scores by optimizing a sigmoid function.

Boosting is another promising domain of classification methods that can be applied to multimedia forensics. Boosting methods typically involve training a number of weak classifiers and incrementally combining them into a strong ensemble classifier. In the AdaBoost method [79], Freund and Schapire developed a theoretical framework to construct a strong ensemble classifier by incrementally minimizing the upper bound of an exponential loss function. Therefore, the separation margin is systematically improved to result in good generalization performances of an AdaBoost-learned classifier. In [80], Schapire et al. further elaborated several extensions of AdaBoost including the extension of the discrete AdaBoost into real-valued RealBoost algorithm and the extension of AdaBoost to multiclass pattern classification. On top of the RealBoost, Li et al. [81] proposed a FloatBoost algorithm to incorporate a floating search-based backtracking mechanism to exclude the unfavorable classifiers from the classifier ensemble. The boosting classifiers are suitable in real-time classification due to the simple ensemble structure. Moreover, boosting algorithms can be easily applied to asymmetric learning tasks, where huge differences on the sizes of different classes are present.

Some other classification methods applied in prior forensics works also include neural network [53], linear discriminant classifier [50], and probabilistic classifiers based on correlation similarity measure [23, 24].

### *Fusion Techniques*

Classifier fusion refers to the techniques of combining multiple classifiers or features for a given pattern classification task. In multimedia forensics, such techniques are useful for combining diversified forensic experts or features for better performances. Classifier fusion can be implemented at different levels, such as feature level, kernel level, score level, and decision level. At feature level, different features are simply combined before they are used in classification. Kernel trick is a common technique in the state-of-the-art pattern classification techniques, e.g., SVM and neural network, to enable nonlinear classification with improved performances. The commonly used kernels include linear kernel, polynomial kernel, RBF kernel, and sigmoid kernel. Fusion at kernel level can be implemented to compute the kernel values associated with different feature sets separately and combine these kernel values together to form a new hybrid kernel function. Score-level fusion is appropriate for combining classifiers, which provide confidence scores, e.g., probabilistic scores. The scores are combined and the classification decision is made based on the combined score. For decision-level fusion, each individual classifier makes its own classification decision and the decisions are combined to give the final classification decision. For fusion at the score level and the decision level, fusion rules are commonly applied and the frequently used rules include the Mean, Summation, Product, Max, Min, Median and Majority Vote, etc. In a previous forensic work, Celiktutan et al. [63] compare score-level fusion based on different rules with feature-level fusion in combining three diversified feature sets for identification of cellular phone camera models. Their comparison results show that the score-level fusion based on product rule gives the best result.

## 5 Anti-Forensics and Countermeasures

Similar to other security technologies, multimedia source identification is also subject to various attacks. The recent works in [82, 83] have proposed several attacks to forensic techniques, each targeted for a well-known methodology. These techniques demonstrated that source identification through detection of the photo-response non-uniformity (PRNU) noise patterns in [23] and the color-filter-array interpolation traces in [50] can all be defeated through faking the desired source fingerprints into a given forgery image. Although these attacks are specific to the selected forensic methodologies and each attack likely introduces some new artifacts detectable by other forensic methodologies, the works [82, 83] have shown that performing image forensics with a single tool can be easily attacked by a sophisticated attacker. It has also been pointed out in several early works that comprehensive forensic analysis shall be based on a suite of forensic tools which examine different image properties. In general, covering up all tampering artifacts and restoring all intrinsic regularities simultaneously into a media is believed to be more difficult than attacking on a single forensic tool.

The benefit of studying anti-forensics is that researchers can position them as sophisticated attackers to evaluate how trustworthy a forensic methodology is. At the same time, new opportunities arise since anti-forensics techniques would also likely leave different kinds of new traces, which are potentially employed to lodge countering anti-forensics claims. For instance, Goljian et al. [36] have developed a triangular test to determine whether a fake camera PRNU fingerprint has been estimated and artificially injected into a forgery image to frame an innocent camera owner. The assumption is that the attacker and camera owner would have access to different provisions of the image data from the camera. Also, the fake PRNU fingerprint estimated from one or only a few images would contain the remnant of these images that the camera owner also has access to. If the number of images for learning the fake PRNU fingerprint is not large, the triangle test can help the camera owner identify the images that have been used by the attacker. The triangular test is also applicable to the case that the camera owner has only two forgery images being injected with the same fake fingerprint.

In the future, there is no doubt that the technology to doctor photos, including those attacks targeted to disable forensic technology, will continuously improve and so will be the forensic techniques. It is hard to assure that the forensic side will eventually be the sole winner. However, by making electronic forgeries on multimedia a difficult task, the research effort in passive forensics will still pay off to restore the traditional trustworthiness on multimedia especially in the occasions where the security is important.

# 6 Conclusions and Future Opportunities

In this chapter, we introduce the forensic role of source identification in multimedia forensics and review the major developments in different types of source identification tasks. For source class identification, the media acquisition parameters and statistical regularities are commonly detected as forensic features. Among them, demosaicing regularity represents one most successful type of statistical features and its detection based on derivative-based correlation models show good results. Methods for identification of software tools, e.g., source RAW converters and multimedia encoders, are also developed through detection of demosaicing regularity and intrinsic compression fingerprints, respectively. As machine learning plays an important role in source identification tasks, we also briefly discuss the commonly used techniques, such as feature reduction, classification, and fusion.

Looking forward, source identification will still remain as an ongoing and challenging research field. As better photographic technology and new imaging device models from different manufacturers roll out fast, it is important to validate the assumptions for current forensic techniques and keep the source identification technology up to date. Also, the majority of the current works have been tested in relatively small scale. To improve their practicality, large-scale tests and comparison need to be conducted on common benchmarking data sets and learning platforms. The security aspect on current identification technology has started receiving more attention recently. Rigorous and thorough investigation is needed to evaluate how difficult it is for the current source identification technique to be counterfeited and the efficient countermeasures to be explored. There would also be a need to improve the sensitivity of the existing technology, i.e., the identification performance on only a small localized piece of the media.

## References


1. Brugioni DA (1999) Photo fakery: the history and techniques of photographic deception and manipulation. Brassey's, Dulles
2. Spanish MP's photo used for Osama Bin Laden poster. In: BBC News (16 Jan, 2009)
3. Israel: women photoshopped from cabinet picture to cater to the ultra-orthodox. In: The Huffington Post (18 Jul, 2009)
4. Farid H, Photo tampering throughout the history. In: http://www.cs.dartmouth.edu/farid/research/digitaltampering/
5. Mosleh F (Kodak) (2008) Cameras in handsets evolving from novelty to DSC performance, despite constraints. In: Embedded.com
6. Douglas T (2005) Shaping the media with mobiles. In: BBC News
7. Lewis J (2007) Don't just stand and stare, shoot it, too. In: The Singapore Straits, Times, (28 April, 2007)
8. Cyranoski D (2006) Verdict: Hwang's human stem cells were all fakes. Nature 439:122–123
9. Pearson H (2006) Forensic software traces tweaks to images. Nature 439:520–521
10. Friedman GL (1993) The trustworthy digital camera: restoring credibility to the photographic image. IEEE Trans Consumer Electron 39:905–910
11. Blythe P, Fridrich J (2004) Secure digital camera. In: Proceedings of the digital forensic research workshop (DFRWS)
12. Cox J, Miller ML, Bloom JA (2001) Digital watermarking. Morgan Kaufmann, San Francisco
13. Cao H, Kot AC (2010) Lossless data embedding in electronic Inks. IEEE Trans Inf Forensics Secur 5(2):314–323
14. Sencar HT, Memon N (2008) Overview of state-of-the-art in digital image forensics. In: Part of indian statistical institute platinum jubilee monograph series titled 'Statistical Science and Interdisciplinary Research'. World Scientific Press, Singapore
15. Mahdian B, Saic S (2010) A bibliography on blind methods for identifying image forgery. Signal Process Image Commun 25(6):389–399
16. Choi S, Lam EY, Wong KKY (2006) Automatic source camera identification using the intrinsic lens radial distortion. Opt Express 14(24):11551–11565
17. Van LT, Emmanuel S, Kankanhalli MS (2007) Identifying source cell phone using chromatic aberration. In: Proceedings of the ICME, pp 883–886
18. Kurosawa K, Kuroki K, Saitoh N (1999) CCD fingerprint method-identification of a video camera from videotaped images. In: Proceedings of the IEEE international conference on image processing, pp 537–540
19. Geradts ZJ, Bijhold J, Kieft M, Kurosawa K, Kuroki K, Saitoh N (2001) Methods for identification of images acquired with digital cameras. In: Proceedings of the SPIE, vol 4232, p 505
20. Kurosawa K, Kuroki K, Saitoh N (2002) An approach to individual video camera identification. J Forensic Sci 47(1):97–102

21. Saitoh N, Kurosawa K, Kuroki K, Akiba N, Geradts ZJ, Bijhold J (2002) CCD fingerprint method for digital still cameras. In: Proceedings of the SPIE, vol 4709, pp 37–48
22. Kurosawa K, Saitoh N (2003) Fundamental study on identification of CMOS cameras. In: Proceedings of the SPIE, vol 5108, p 202
23. Lukas J, Fridrich J, Goljan M (2006) Digital camera identification from sensor pattern noise. IEEE Trans Inf Forensics Secur 1(2):205–214
24. Chen M, Fridrich J, Goljan M, Lukas J (2008) Determining image origin and integrity using sensor noise. IEEE Trans Inf Forensics Secur 3(1):74–89
25. Khanna N, Mikkilineni AK (2007) Scanner identification using sensor pattern noise. In: Proceedings of the SPIE, vol 6505, p 65051K
26. Sutcu Y, Bayram S, Sencar HT, Memon N (2007) Improvements on sensor noise based source camera identification. In: Proceedings of the international conference on multimedia Expo, pp 24–27
27. Goljan M, Fridrich J (2008) Camera identification from cropped and scaled images. In: Proceedings of the SPIE, vol 6819, p 68190E
28. Zhang C, Zhang H (2008) Digital camera identification based on canonical correlation analysis. In: Proceedings of the IEEE workshop on MSP, pp 769–773
29. Zhang C, Zhang H (2010) Identifying color image origin using curvelet transform. In: Proceedings of the IEEE international conference on image processing, pp 2125–2128
30. Hu Y, Yu B, Jian C (2009) Source camera identification using large components of sensor pattern noise. In: Proceedings of the international conference on computer science and its applications, pp 1–5
31. Hu Y, Jian C, Li C-T (2010) Using improved imaging sensor pattern noise for source camera identification. In: Proceedings of the IEEE international conference on multimedia and Expo, pp 1481–1486
32. Li C-T (2010) Source camera identification using enhanced sensor pattern noise. IEEE Trans Inf Forensics Secur 5(2):280–287
33. Liu BB, Lee H-K, HuY, Choi C-H (2010) On classification of source cameras: a graph based approach. In: Proceedings of the IEEE international workshop on information forensics and security, pp 1–5
34. Liu B-B, Hu Y, LeeH-K (2010) Source camera identification from significant noise residual regions. In: Proceedings of the IEEE international conference on image processing, pp 1749–1752
35. Goljan M, Fridrich J (2009) Large scale test of sensor fingerprint camera identification. In: Proceedings of the SPIE electronic imaging, forensics, security, steganography, and watermarking of multimedia contents X, vol 7254, pp 0I–0J
36. Goljan M, Fridrich J, Chen M (2010) Sensor noise camera identification: countering counter forensics. In: Proceedings of the SPIE, electronic imaging, media forensics and security XII, pp 0S-01–0S-12
37. Goljan M, Chen M, Fridrich J (2007) Identifying common source digital camera from image pairs. In: Proceedings of the international conference on image processing, pp 125–128
38. Bloy J (2008) Blind camera fingerprinting and image clustering. IEEE Trans Pattern Anal Mach Intell 30(3):532–534
39. Alles EJ, Geradts ZJMH, Veenman CJ (2008) Source camera identification for low resolution heavily compressed images. In: Proceedings of the ICCSA, pp 557–567
40. Chen M, Fridrich J, Goljan M, Lukas J (2007) Source digital camcorder identification using sensor photo response non-uniformity. In: Proceedings of the SPIE, vol 6505
41. Choi C-H, Lee M-J, Lee H-K (2010) Scanner identification using spectral noise in the frequency domain. In: Proceedings of the IEEE international conference on image processing (ICIP'10), pp 2121–2124
42. Houten WV, Geradts Z (2009) Source video camera identification for multiply compressed videos originating from YouTube. Digit Investig 6(1–2):48–60
43. Gou H, Swaminathan A, Wu M (2007) Robust scanner identification based on noise features. In: Proceedings of the SPIE, vol 6505, p 65050S

44. Gou H, Swaminathan A, Wu M (2009) Intrinsic sensor noise features for forensic analysis on scanners and scanned images. IEEE Trans Inf Forensics Secur 4(3):476–491
45. Khanna N, Mikkilineni AK, Chiu GTC, Allebach JP, Delp EJ (2007) Forensic classification of imaging sensor types. In: Proceedings of the SPIE, vol 6505, p 65050U
46. Filler T, Fridrich J, Goljan M (2008) Using sensor pattern noise for camera model identification. In: Proceedings of the international conference on image processing, pp 1296–1299
47. McKay C, Swaminathan A, Gou H, Wu M (2008) Image acquisition forensics: forensic analysis to identify imaging source. In: Proceedings of the ICASSP, pp 1657–1660
48. Dirik AE, Sencar HT, Memon N (2008) Digital single lens reflex camera identification from traces of sensor dust. IEEE Trans Inf Forensics Secur 3:539–552
49. Dirik AE, Sencar HT, Memon N (2009) Flatbed scanner identification based on dust and scratches over scanner platen. In: Proceedings of the international conference on acoustics, speech and signal processing (ICASSP), pp 1385–1388
50. Popescu C, Farid H (2005) Exposing digital forgeries in color filter array interpolated images. IEEE Trans Signal Process 53(10):3948–3959
51. Bayram S, Sencar HT, Memon N, Avcibas I (2005) Source camera identification based on CFA interpolation. In: Proceedings of the international conference on image processing, vol 3, pp 69–72
52. Bayram S, Sencar HT, Memon N (2006) Improvements on source camera-model identification based on CFA interpolation. In: Proceedings of the WG 11.9 international conference on digital forensics
53. Long Y, Huang Y (2006) Image based source camera identification using demosaicking. In: Proceedings of the IEEE 8th workshop on multimedia, signal processing, pp 419–424
54. Swaminathan A, Wu M, Liu KJR (2007) Nonintrusive component forensics of visual sensors using output images. IEEE Trans Inf Forensics Secur 2(1):91–106
55. Bayram S, Sencar HT, Memon N (2008) Classification of digital camera-models based on demosaicing artifacts. Digit Investig 5(1–2):49–59
56. Cao H, Kot AC (2008) A generalized model for detection of demosaicing characteristics. In: Proceedings of the international conference on multimedia expo, pp 1513–1516
57. Cao H, Kot AC (2009) RAW-tool identification through detected demosaicing regularity. In: Proceedings of the international conference on image processing, pp 2885–2888
58. Cao H, Kot AC (2010) Mobile camera identification using demosaicing features. In: Proceedings of the ISCAS, pp 1683–1686
59. Cao H, Kot AC (2009) Accurate detection of demosaicing regularity for digital image forensics. IEEE Trans Inf Forensics Secur 4(4):899–910
60. Cao H, Kot AC (2011) Similar DSLR processor identification using compact model template. In: Asia-Pacific signal and information processing association annual summit and conference (APSIPA ASC'11)
61. Kharrazi M, Sencar HT, Memon N (2004) Blind source camera identification. In: Proceedings of the international conference on image processing, vol 1, pp 709–712
62. Tsai M-J, Wu G-H (2006) Using image features to identify camera sources. In: Proceedings of the ICASSP, vol 2, pp 297–300
63. Celiktutan O, Sankur B, Avcibas I (2008) Blind identification of source cell-phone model. IEEE Trans Inf Forensics Secur 3(3):553–566
64. Xu G, Shi YQ, Xu W (2009) Camera brand and model identification using moments of 1-D and 2-D characteristic functions. In: Proceedings of the international conference image processing, pp 2917–2920
65. Gloe T, Borowka K, Winkler A (2009) Feature-based camera model identification works in practice. Lecture Notes in Computer Science, vol 5806, pp 262–276
66. Khanna N, Delp EJ (2009) Source scanner identification for scanned documents. In: Proceedings of the IEEE international workshop on information forensics and security (WIFS'09), pp 166–170
67. Khanna N, Delp EJ (2010) Intrinsic signatures for scanned documents forensics: effect of font shape and size. In: Proceedings of the IEEE international symposium on circuits and systems (ISCAS'10), pp 3060–3063

68. Fang Y, Dirik AE, Sun X, Memon N (2009) Source class identification for DSLR and compact cameras. In: Proceedings of the multimedia signal processing (MMSP), pp 1–5
69. Lin WS, Tjoa SK, Zhao HV, Liu K (2009) Digital image source coder forensics via intrinsic fingerprints. IEEE Trans Inf Forensics Secur 4(2):460–475
70. Luo W, Wang Y, Huang J (2010) Detection of quantization artifacts and its applications to transform encoder identification. IEEE Trans Inf Forensics Secur 5(4):810–815
71. Su Y, Xu J, Dong B (2009) A source video identification algorithm based on motion vectors. In: Proceedings of the international workshop on computer science and engineering (WCSE'09), vol 2, pp 312–316
72. Chen M, Fridrich J, Goljan M, Lukas J (2007) Source digital camcorder identification using sensor photo response non-uniformity. In: Proceedings of the SPIE, vol 6505, pp 65051G-1–65051G-12
73. Ramanath R, Snyder WE, Yoo Y, Drew MS (Jan 2005) Color image processing pipeline. IEEE Signal Process Mag 22(1):34–43
74. Li X, Gunturk B, Zhang L (2008) Image demosaicing: a systematic survey. In: Proceedings of the SPIE, vol 6822
75. Jiang X, Mandal B, Kot AC (2008) Eigenfeature regularization and extraction in face recognition. IEEE Trans Pattern Anal Mach Intell 30(3):383–394
76. Pudil P, Ferri FJ, Novovicova J, Kittler J (1994) Floating search methods for feature selection with nonmonotonic criterion functions. In: Proceedings of the international conference on pattern recognition, vol 2, pp 279–283
77. Vapnik VN (1999) The nature of statistical learning theory, 2nd edn. Springer, New York
78. Scholkopf B, Platt JC, Talor JS, Smola AJ, Williamson RC (2001) Estimating the support of a high-dimensional distribution. Neural Comput 13(7):1443–1471
79. Freund Y, Schapire R (1997) A decision-theoretic generalization of on-line learning and an application to boosting. J Comput Syst Sci 55(1):119–139
80. Schapire RE, Singer Y (1999) Improved boosting algorithms using confidence-rated predictions. Mach Learn 37(3):297–336
81. Li SZ, Zhang Z (2004) FloatBoost learning and statistical face detection. IEEE Trans Pattern Anal Mach Intell 26(9):1112–1123
82. Gloe T, Kirchner M, Winkler A, Bohme R (2007) Can we trust digital image forensics? In: Proceedings of the international conference on multimedia, pp 78–86
83. Kirchner M, Bohme R (2009) Synthesis of color filter array pattern in digital images. In: Proceedings of the SPIE, vol 7254, p 72540K

# Sensor Defects in Digital Image Forensic

**Jessica Fridrich**

**Abstract** Just as human fingerprints or skin blemishes can be used for forensic purposes, imperfections of digital imaging sensors can serve as unique identifiers in numerous forensic applications, such as matching an image to a specific camera, revealing malicious image manipulation and processing, and determining an approximate age of a digital photograph. There exist several different types of defects that are of interest to the forensic analysts caused by imperfections in manufacturing, physical processes occurring inside the camera, and by environmental factors. This chapter begins with analyzing the pixel defects, while pointing out their forensic potential. Then, specific problems are formulated as tasks involving detection or matching of defects and noise patterns. Practical algorithms for these tasks are developed within the framework of parameter estimation and signal detection theory. The performance of the algorithms is demonstrated in real world examples.

## 1 Introduction

In the heart of every electronic device capable of taking digital pictures is an imaging sensor. There exist two types of sensors—Charge-Coupled Device (CCD) and Complementary Metal-Oxide Semiconductor (CMOS). Both sensors consist of a large number of photo detectors commonly called pixels. Pixels are made of silicon and capture light by converting photons into electrons using the photoelectric effect [27, 30]. The charge accumulated at every pixel is transferred out of the sensor, amplified, and then run through an AD converter that converts it to a digital signal. The digitized signal is further processed before the data is stored as an electronic

J. Fridrich (✉)
Department of Electrical and Computer Engineering,
T. J. Watson School of Applied Science and Engineering,
SUNY Binghamton, Binghamton, NY13902-6000, USA
e-mail: fridrich@binghamton.edu

H. T. Sencar and N. Memon (eds.), *Digital Image Forensics*,
DOI: 10.1007/978-1-4614-0757-7_6,

file (JPEG, TIFF, etc.) on the camera storage device. The pixels are several microns across and have a rectangular shape. In theory, the amount of electrons (charge) outputted by a pixel should depend solely on the intensity of the incident light. In reality, however, there are many factors that introduce both systematic and random deviations. It is exactly these fluctuations that find important applications in forensic analysis.

We will be interested primarily in *systematic* variations in pixel response that manifest themselves in a consistent manner in all images because random fluctuations that change independently from scene-to-scene would not be particularly useful. In the next section, we describe several types of such systematic sensor defects and how they affect the output of a sensor. By doing so, we arrive at a model of sensor output that will be useful later for deriving estimators of parameters that describe the defects and for constructing defect detectors. In Sect. 3, we introduce the concept of sensor fingerprint and derive a procedure using which the fingerprint can be estimated from images taken by the camera. The tasks of camera identification, device linking, and forgery detection can be approached by testing the presence of sensor fingerprint in images. This is the subject of Sects. 4, 4.1, and 4.4. The methods are tested on real images in Sect. 5. In Sect. 6, we develop methods that can address counter-forensic activities of the type when an adversary estimates a camera fingerprint and then adds it to an image from a different camera to frame an innocent victim. The presence or absence of defects can also provide temporal information in estimating an approximate age of digital photographs. This so-called temporal forensics is the subject of Sect. 7. The chapter is summarized in Sect. 8.

## *1.1 Notation*

Everywhere in this chapter, boldface font will denote vectors (or matrices) of length specified in the text, e.g., $\mathbf{X}$ and $\mathbf{Y}$ are vectors of length $m \times n$ and $\mathbf{X}(i)$ denotes the $i$th component of $\mathbf{X}$. Sometimes, we will index the pixels in an image using a 2-D index formed by the row and column index. Unless mentioned otherwise, all operations among vectors or matrices, such as product, ratio, raising to a power, etc., are *elementwise*. The Euclidean dot product of vectors is denoted as $\mathbf{X} \cdot \mathbf{Y}$ with $\|\mathbf{X}\| = \sqrt{\mathbf{X} \cdot \mathbf{X}}$ being the $L_2$ (Euclidean) norm of $\mathbf{X}$. Denoting the sample mean with a bar, the normalized correlation is

$$\mathrm{corr}(\mathbf{X}, \mathbf{Y}) = \frac{(\mathbf{X} - \overline{\mathbf{X}}) \cdot (\mathbf{Y} - \overline{\mathbf{Y}})}{\|\mathbf{X} - \overline{\mathbf{X}}\| \, \|\mathbf{Y} - \overline{\mathbf{Y}}\|}. \tag{1}$$

For a logical statement $P$, we also make use of the Iverson bracket defined as $[P] = 1$ when $P$ is true and $[P] = 0$ when $P$ is false.

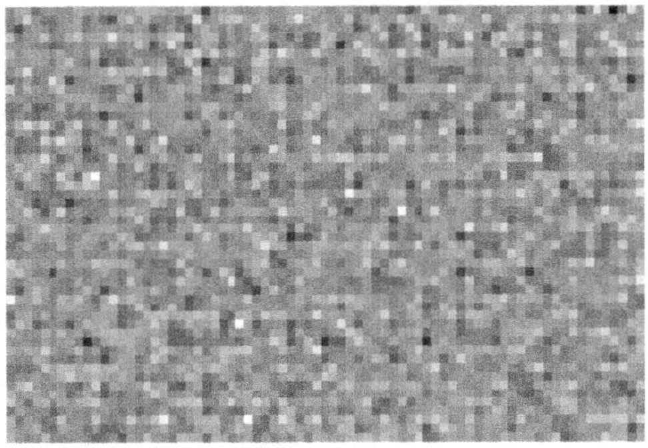

**Fig. 1** Close-up of the PRNU factor **K** enhanced for visualization

# 2 Imaging Sensors and Their Defects

In this section, we explain two types of systematic defects—the photo-response nonuniformity and the dark current—that are useful for several important forensic tasks, including camera identification and forgery detection. Then, we formulate a model of pixel output that will be used in the rest of this chapter to build all necessary mathematical tools for the forensic analyst.

## *2.1 Photo-Response Non-Uniformity*

The charge generated in a pixel depends on the physical dimensions of the pixel photosensitive area and on the homogeneity of silicon. The pixels' physical dimensions slightly vary due to imperfections in the manufacturing process. Also, the inhomogeneity naturally present in silicon contributes to variations in quantum efficiency among pixels (the ability to convert photons to electrons). The variations in quantum efficiency among pixels can be captured with a matrix $\mathbf{K} \in \mathbb{R}^{m \times n}$ of the same dimensions as the sensor. When an imaging sensor is illuminated with light intensity $\mathbf{I} \in \mathbb{R}^{m \times n}$, in the absence of other noise sources or imperfections, the sensor would register a noisy scene $\mathbf{I} + \mathbf{IK}$ instead (We remind that the product $\mathbf{IK}$ is an elementwise product of matrices). The term $\mathbf{IK}$ is usually referred to as the photo-response nonuniformity or PRNU. A large value of $\mathbf{K}$ leads to a point defect called "pixel with abnormal sensitivity."

One can say that the scene $\mathbf{I}$ is overlaid with a "noise pattern" $\mathbf{IK}$, which is essentially the matrix $\mathbf{K}$ modulated by the scene $\mathbf{I}$. Note that the energy of this noise pattern depends on the light intensity—it is larger for bright images and smaller in pictures of mostly dark scenes.

Figure 1 shows a magnified portion of the PRNU factor $\mathbf{K}$ from a 4-megapixel camera Canon G2. Bright dots correspond to pixels that consistently generate more electrons, while dark dots mark pixels whose response is consistently lower. To give the reader a sense of how weak the PRNU signal **IK** typically is, for a picture of a uniform background **I** with average grayscale in the middle of the dynamic range (grayscale $\mathbf{I} = 128$ for an 8-bit image), the average energy of $\mathbf{I}(i)\mathbf{K}(i)$ over all pixels $i$ is 0.5, which can also be formulated as SNR of 51 dB. The energy of the PRNU strongly varies among camera models.

In Sect. 3, it is shown how the PRNU factor **K** can be used as a sensor "fingerprint" for a variety of forensic tasks.

## 2.2 Dark Current

Even when a pixel is not exposed to light during picture taking, it contains a small number of free electrons due to thermal effects. Their number increases with temperature and exposure [26, 28, 30]. It is also affected by ISO setting. In the absence of all other defects, the pixel's output is $\mathbf{I} + \tau\mathbf{D} + \mathbf{c}$, where $\tau\mathbf{D}$ is called the dark current and $\mathbf{c}$ the offset. Here, $\tau \geq 0$ is a multiplicative factor whose value is determined by the temperature, exposure, and ISO (higher ISO leads to a larger value of $\tau$) and $\mathbf{D}, \mathbf{c} \in \mathbb{R}^{m\times n}$ are matrices. In images taken with a short exposure (e.g., 1/60th of a second or shorter), the dark current is usually very weak. However, it may start dominating the sensor output for dark scenes ($\mathbf{I} \approx 0$) when $\tau$ becomes large (large ISO and/or temperature and/or long exposure). An extremely high value of **D** produces the most common point defect called the hot pixel. A high value of the offset **c** leads to another defect type commonly recognized as the stuck pixel. Both defects were proposed for forensic tasks in 1999 by Kurosawa [35], who demonstrated that as long as a video clip contained some dark frames, hot/stuck pixels can be used to uniquely identify digital video cameras.

Hot/stuck pixels occur randomly and uniformly on the sensor independently of each other, which makes them useful for determining an approximate age of digital photographs (see Sect. 7.2).

## 2.3 Pixel Output Model

Considering the impact of all the defects discussed so far, we arrive at the following model for the raw output of a sensor:

$$\mathbf{Y} = \mathbf{I} + \mathbf{IK} + \tau\mathbf{D} + \mathbf{c} + \boldsymbol{\Theta}. \tag{2}$$

We remind that $\tau$ is a scalar multiplicative factor whose value is determined by exposure, temperature, and ISO settings. The matrix **c** is the matrix of offsets and dark current factor **D**, is a noise-like signal due to the leakage of electrons into pixels' electron wells (Fig. 2 bottom). Finally, the reader recognizes **K** as the PRNU factor. The modeling noise $\boldsymbol{\Theta}$ is the collection of all other noise sources, which are mostly random in nature and thus difficult to use for forensic purposes (readout noise, shot noise, also known as the photonic noise, quantization noise, etc.).

**Fig. 2** *Top*: A stuck pixel in an image and its close-up. The pixel happens to be red because it has a red color filter in front of it. *Bottom*: An example of a dark frame. Notice that there appears to be a degree of "hotness" among the pixels

It should be stressed that the defects represented by matrices $\mathbf{K}$, $\mathbf{D}$, and $\mathbf{c}$ usually represent quite small deviations with the exception of spike pixel defects, such as hot or stuck pixels. The three matrices could be estimated from multiple images taken by the camera.

To improve the signal-to-noise ratio between the signal of interest (defect) and the observable $\mathbf{Y}$, we usually work with the noise residual $\mathbf{W} = \mathbf{Y} - F(\mathbf{Y})$, obtained using a denoising filter $F$. In particular,

$$\begin{aligned} \mathbf{W} &= \mathbf{Y} - F(\mathbf{Y}) \\ &= \mathbf{IK} + \tau\mathbf{D} + \mathbf{c} + \mathbf{I} - F(\mathbf{Y}) + \boldsymbol{\Theta} \\ &= \mathbf{IK} + \tau\mathbf{D} + \mathbf{c} + \boldsymbol{\Xi}, \end{aligned} \tag{3}$$

where $\Xi$ stands for the sum of the modeling noise and the remnant of the content $\mathbf{I} - F(\mathbf{Y})$ present due to the inability of the denoising filter to separate content from noise. The term $\mathbf{I} - F(\mathbf{Y})$ is especially large in textured regions and around edges.

A variety of filters could be used in practice for this task. In this chapter, we will use the wavelet filter [43] designed to suppress a nonstationary Gaussian noise and a $3 \times 3$ median filter.

## 3 Sensor Fingerprint

The following five properties of PRNU represented with matrix **K** are the main reason why it was proposed to play the role of a sensor fingerprint.

1. *Dimensionality*. The matrix **K** appears random, which gives it a large information content and makes it unique to each sensor. The probability of two sensors having similar fingerprints is extremely low. Even two cameras of the same model have statistically independent fingerprints.
2. *Universality*. All imaging sensors exhibit PRNU.
3. *Generality*. The PRNU component **IK** is present in every picture independently of the camera optics, camera settings, or scene content, with the exception of completely dark images, where $\mathbf{I} \approx 0$.
4. *Stability*. The factor **K** is stable in time and under wide range of environmental conditions (temperature, humidity).
5. *Robustness*. The PRNU component **IK** survives lossy compression, filtering, gamma correction, and many other typical processing.

The elements of the matrix **K** are well modeled as independent and identically distributed (IId) realizations of a Gaussian random variable. By establishing the presence of the PRNU signal **IK** in an image, one can prove with a high level of certainty that the image was obtained by a *specific* camera whose fingerprint is *known*. This application is called sensor (camera) identification. Expanding this application further, by detecting the presence of the signal **IK** in individual image regions, one can reveal that certain regions were replaced or tampered with; this application is known as integrity verification or forgery detection. Alternatively, by proving that two images share a common signal, it is possible to establish that these two images came from the same device even when its fingerprint is not available. This is called device linking. Finally, the PRNU signal **IK** can be used as a template to recover geometrical processing the image has been subjected to, such as cropping, resizing, or rotation.

In the sections below, we explain a procedure for estimating the sensor fingerprint as well as methods for its detection in images to address the problem of camera identification, device linking, fingerprint matching, and forgery detection. The performance of these methods is evaluated on real imagery in Sect. 5.

## 3.1 Fingerprint Estimation

The problem of estimating the fingerprint $\mathbf{K}$ can be approached using standard techniques from parameter estimation. The estimator will depend on the model of sensor output. Ideally, the best estimator should be tailored to the specific camera make and model, while taking into account its processing pipeline. There is, however, substantial strength in approaching the estimation using a simplified model that is universally valid for virtually all camera makes and models. This avenue is taken in this chapter.

Starting with the model of the noise residual $\mathbf{W} = \mathbf{Y} - F(\mathbf{Y})$ (3), we simplify it by including the dark current and the offset into the noise term:

$$\mathbf{W} = \mathbf{IK} + \Xi. \tag{4}$$

It is easier to estimate the PRNU term from $\mathbf{W}$ than from $\mathbf{Y}$ because the image content is greatly suppressed in $\mathbf{W}$.

Let us assume that we have a database of $d \geq 1$ images, $\mathbf{Y}_1, \ldots, \mathbf{Y}_d$, obtained by the camera whose fingerprint we wish to estimate. For each pixel $i$, we model the sequence $\Xi_1(i), \Xi_2(i), \ldots, \Xi_d(i)$ as a white Gaussian noise (WGN) with variance $\sigma^2(i)$. The noise term is technically not independent of the PRNU signal $\mathbf{IK}$ due to the content leftover $\mathbf{I} - F(\mathbf{Y})$ in $\Xi$. However, because the energy of this term is small compared to $\mathbf{IK}$, the assumption that $\Xi$ is independent of $\mathbf{IK}$ is reasonable.

From (4), we can write for each $k = 1, \ldots, d$ in a matrix form:

$$\frac{\mathbf{W}_k}{\mathbf{I}_k} = \mathbf{K} + \frac{\Xi_k}{\mathbf{I}_k}. \tag{5}$$

Under our assumption about the noise term, the log-likelihood of observing a given $\mathbf{K}$ is

$$L(\mathbf{K}) = -\frac{d}{2}\sum_{k=1}^{d} \log(2\pi\sigma^2/\mathbf{I}_k^2) - \sum_{k=1}^{d}\left(\frac{\mathbf{W}_k/\mathbf{I}_k - \mathbf{K}}{2\sigma^2/\mathbf{I}_k^2}\right)^2. \tag{6}$$

By taking partial derivatives of $L$ with respect to individual elements of $\mathbf{K}$ and solving for $\mathbf{K}$, we obtain the maximum likelihood estimate:

$$\hat{\mathbf{K}} = \frac{\sum_{k=1}^{d} \mathbf{I}_k \mathbf{W}_k}{\sum_{k=1}^{d} \mathbf{I}_k^2}. \tag{7}$$

We remind that all operations in (7) are elementwise. To be able to use this estimator in practice, we can simply set $\mathbf{I}_k = \mathbf{Y}_k$ or $\mathbf{I}_k = F(\mathbf{Y}_k)$ because the PRNU term is weak.

We also define the quality of a fingerprint estimate as

$$q = \operatorname{corr}(\mathbf{K}, \hat{\mathbf{K}}). \tag{8}$$

The Cramer–Rao Lower Bound (CRLB) [31] gives us the bound on the variance of $\hat{\mathbf{K}}$

$$\frac{\partial^2 L(\mathbf{K})}{\partial \mathbf{K}^2} = -\frac{\sum_{k=1}^{d} \mathbf{I}_k^2}{\sigma^2}, \tag{9}$$

which implies

$$\mathrm{Var}(\hat{\mathbf{K}}) \geq \left(-E\left(\frac{\partial^2 L(\mathbf{K})}{\partial \mathbf{K}^2}\right)\right)^{-1} = \frac{\sigma^2}{\sum_{k=1}^{d} \mathbf{I}_k^2}. \tag{10}$$

Because the sensor model is linear, the CRLB tells us that the maximum likelihood estimator is minimum variance unbiased and its variance is proportional to $d$. Therefore, the best images for estimating the fingerprint are those with high luminance (but not saturated) and small $\sigma^2$ (images with a smooth content). If the camera under investigation is available to the analyst, unsaturated out-of-focus images of bright cloudy sky would be the best. In practice, good estimates of the fingerprint may be obtained from as few as 20 natural images depending on the camera. If sky images are used instead of natural images, only approximately half of them would suffice to obtain an estimate with a comparable accuracy.

### 3.1.1 Fingerprint Post-Processing

The fingerprint estimate $\hat{\mathbf{K}}$ contains all components that are systematically present in every image, including artifacts introduced by color interpolation, JPEG compression, on sensor signal transfer [15], and sensor design. While the PRNU is unique to the sensor, the above Non-Unique Artifacts (NUAs) are shared among cameras of the same model or sensor design. Consequently, PRNU factors estimated from two different cameras may be slightly correlated, which undesirably increases the false identification rate. Fortunately, since most of these artifacts are due to demosaicking algorithms that depend on the Color Filter Array (CFA) and are periodic in nature, they can be removed by zero-meaning the rows and columns of separately for each pixel type as defined by the CFA. We explain the procedure on the example of the Bayer CFA.

Assuming $\mathbf{I}$ has $m \times n$ pixels, for the Bayer CFA there are four types of pixels forming four interleaved submatrices $\hat{\mathbf{K}}^T$, $T \in \{R, G1, G2, B\}$. where $\hat{\mathbf{K}}^T$ is of dimension $(m/2) \times (n/2)$. The operation of zero-meaning is described using Algorithm 1.

Reassembling the four submatrices into one $m \times n$ matrix again, the magnitude of the final fingerprint estimate is further processed in the DFT domain using a Wiener filter with noise variance $\sigma^2$, $W(., \sigma^2)$, to further suppress any remaining NUAs, such as nonperiodic artifacts [8] (all operations are again elementwise):

$$\mathbf{F} = \mathcal{F}(\hat{\mathbf{K}}), \quad \hat{\mathbf{K}} \leftarrow \mathrm{Real}\left[\mathcal{F}^{-1}\left(\mathbf{F} \cdot \frac{|\mathbf{F}| - W(|\mathbf{F}|, \sigma^2)}{|\mathbf{F}|}\right)\right], \tag{11}$$

**Algorithm 1** The procedure of zero-meaning removes NUAs from the fingerprint estimated using (7).

$r_i = 1/n \sum_{j=1}^{n} \hat{\mathbf{K}}^T(i, j)$ \\ compute row averages
**for** $i = 1$ to $m$ \\ zero-mean rows
  $\hat{\mathbf{K}}^T(i, j) \leftarrow \hat{\mathbf{K}}^T(i, j) - r_i$ for $j = 1$ to $n$
**end**
$c_j = 1/m \sum_{i=1}^{m} \hat{\mathbf{K}}^T(i, j)$ \\ compute column averages
**for** $j = 1$ to $n$ \\ zero-mean rows
  $\hat{\mathbf{K}}^T(i, j) \leftarrow \hat{\mathbf{K}}^T(i, j) - c_j$ for $i = 1$ to $m$
**end**

**Fig. 3** The NUAs in the form of a linear pattern for a Canon S40 camera

where $\mathcal{F}$ is the orthonormal Fourier transform and $\sigma^2 = \frac{1}{mn} \sum_{i,j} \hat{\mathbf{K}}^2(i, j)$.

The difference between the original estimate (7) and the post-processed $\hat{\mathbf{K}}$ is called the linear pattern (see Fig. 3) and it is a useful forensic entity by itself—it can be used to classify a camera fingerprint to a camera model or brand. The reader is referred to [14] for more details.

For color images, the PRNU factor can be estimated for each color channel separately, obtaining thus three fingerprints of the same dimensions $\hat{\mathbf{K}}_R$, $\hat{\mathbf{K}}_G$, and $\hat{\mathbf{K}}_B$. As these three fingerprints are highly correlated due to in-camera processing, such as demosaicking or color interpolation, an analyst may choose to work with a single fingerprint obtained by converting the three color fingerprints using the usual conversion from RGB to grayscale:

$$\hat{\mathbf{K}} = 0.3\hat{\mathbf{K}}_R + 0.6\hat{\mathbf{K}}_G + 0.1\hat{\mathbf{K}}_B. \tag{12}$$

# 4 Digital Forensics Using Sensor Fingerprint

Historically the first application of sensor fingerprint was camera identification [40]. The goal is to determine whether an image under investigation was taken with a specific camera whose fingerprint is available. This does not necessarily mean that the camera needs to be physically available to the analyst because the fingerprint can be estimated from images that provably came from the camera. The identification is achieved by testing whether the noise residual of the image under investigation contains traces of the camera fingerprint.

We formulate the hypothesis testing problem for camera identification in a setting that is general enough to conveniently cover the remaining forensic tasks—device linking and fingerprint matching. In device linking, two images are tested if they came from the same camera (the camera itself is not available). The task of matching two estimated fingerprints occurs in matching two video clips because individual video frames from each clip can be used as a sequence of images from which an estimate of the camcorder fingerprint can be obtained (here, again, the cameras/camcorders may not be available to the analyst).

## 4.1 Device Identification

We consider a more general scenario in which the image under investigation has possibly undergone a geometrical transformation, such as scaling, rotation, or cropping. For simplicity of explanation, we will also assume that before applying any geometrical transformation, the image was in grayscale represented with an $m \times n$ matrix $\mathbf{I}(i, j)$. The geometrical transformation is a complication because the image and the sensor fingerprint are no longer synchronized.

Let us denote $\mathbf{u}$ as the (unknown) vector of parameters describing the geometrical transformation, which we denote $T_{\mathbf{u}}$. For example, $\mathbf{u}$ could be a scaling ratio or a 2-D vector consisting of the scaling parameter and an unknown angle of rotation. In device identification, we wish to determine whether or not the transformed image $\mathbf{Z}$ was taken with a camera with a known fingerprint estimate $\hat{\mathbf{K}}$. We will assume that the geometrical transformation is downgrading (such as downsampling) and thus it will be more advantageous to match the inverse transform with the fingerprint rather than matching $\mathbf{Z}$ with a transformed version of $\hat{\mathbf{K}}$.

We now formulate the detection problem in a slightly more general form to cover all three forensic tasks mentioned at the beginning of this section within one framework. The fingerprint detection is the following two-channel hypothesis testing problem

$$\begin{aligned} H_0 &: \quad \mathbf{K}_1 \neq \mathbf{K}_2, \\ H_1 &: \quad \mathbf{K}_1 = \mathbf{K}_2, \end{aligned} \tag{13}$$

where

$$\mathbf{W}_1 = \mathbf{I}_1\mathbf{K}_1 + \Xi_1,$$
$$T_{\mathbf{u}}^{-1}(\mathbf{W}_2) = T_{\mathbf{u}}^{-1}(\mathbf{Z})\mathbf{K}_2 + \Xi_2. \tag{14}$$

Here, all signals are observed with the exception of the noise terms, and the fingerprints $\mathbf{K}_1$ and $\mathbf{K}_2$. In particular, for the device identification problem, we have $\mathbf{I}_1 = 1$, $\mathbf{W}_1 = \hat{\mathbf{K}}$ estimated in the previous section, and $\Xi_1$ is the estimation error of the PRNU. $\mathbf{K}_2$ is the PRNU from the camera that took the image, $\mathbf{W}_2$ is the geometrically transformed noise residual, and $\Xi_2$ is a noise term. In general, $\mathbf{u}$ is an unknown nuisance parameter. Note that since $T_{\mathbf{u}}^{-1}(\mathbf{W}_2)$ and $\mathbf{W}_1$ may have different dimensions, the formulation (13)–(14) involves an unknown spatial shift between both signals, $\mathbf{s}$.

Modeling the noise terms and as white Gaussian noise with known variances $\sigma_1^2, \sigma_2^2$, the generalized likelihood ratio test [32] for this two-channel problem was derived in [29]. The test statistic $t$

$$t = \max_{\mathbf{u},\mathbf{s}}\{E_1(\mathbf{u},\mathbf{s}) + E_2(\mathbf{u},\mathbf{s}) + C(\mathbf{u},\mathbf{s})\}, \tag{15}$$

is a sum of three terms: two energy-like quantities and a cross-correlation term:

$$E_1(\mathbf{u},\mathbf{s}) = \sum_{i,j} \frac{\mathbf{I}_1^2(i,j)(\mathbf{W}_1(i+s_1,j+s_2))^2}{\sigma_1^2\mathbf{I}_1^2(i,j) + \sigma_1^4\sigma_2^{-2}\left(T_{\mathbf{u}}^{-1}(\mathbf{Z})(i+s_1,j+s_2)\right)^2}, \tag{16}$$

$$E_2(\mathbf{u},\mathbf{s}) = \sum_{i,j} \frac{\left(T_{\mathbf{u}}^{-1}(\mathbf{Z})(i+s_1,j+s_2)\right)^2\left(T_{\mathbf{u}}^{-1}(\mathbf{W}_2)(i+s_1,j+s_2)\right)^2}{\sigma_2^2\left(T_{\mathbf{u}}^{-1}(\mathbf{Z})(i+s_1,j+s_2)\right)^2 + \sigma_2^4\sigma_1^{-2}\mathbf{I}_1^2(i,j)}, \tag{17}$$

$$C(\mathbf{u},\mathbf{s}) = \sum_{i,j} \frac{\mathbf{I}_1\mathbf{W}_1(i,j)(T_{\mathbf{u}}^{-1}(\mathbf{Z})(i+s_1,j+s_2))(T_{\mathbf{u}}^{-1}(\mathbf{W}_2)(i+s_1,j+s_2))}{\sigma_2^2\mathbf{I}_1^2(i,j) + \sigma_1^2\left(T_{\mathbf{u}}^{-1}(\mathbf{Z})(i+s_1,j+s_2)\right)^2}. \tag{18}$$

The complexity of evaluating these three expressions is proportional to the square of the number of pixels, $(mn)^2$, which makes this detector unusable in practice. Thus, we simplify this detector and give it the form of a Normalized Cross-Correlation (NCC) that can be evaluated using the fast Fourier transform. Under $H_1$, the maximum in (15) is mainly due to the contribution of the cross-correlation term, $C(\mathbf{u},\mathbf{s})$, that exhibits a sharp peak for the proper values of the geometrical transformation. Thus, a much faster suboptimal detector is the NCC between $\mathbf{X}$ and $\mathbf{Y}$ maximized over all shifts $s_1$, $s_2$, and $\mathbf{u}$,

$$\mathrm{NCC}(s_1,s_2,\mathbf{u}) = \frac{\sum_{i,j=1}^{m,n}(\mathbf{X}(i,j)-\overline{\mathbf{X}})(\mathbf{Y}(i+s_1,j+s_2)-\overline{\mathbf{Y}})}{\|\mathbf{X}-\overline{\mathbf{X}}\|\,\|\mathbf{Y}-\overline{\mathbf{Y}}\|}, \tag{19}$$

which we view as an $m \times n$ matrix parametrized by $\mathbf{u}$, where

$$\mathbf{X} = \frac{\mathbf{I}_1\mathbf{W}_1}{\sqrt{\sigma_2^2\mathbf{I}_1^2 + \sigma_1^2\left(T_{\mathbf{u}}^{-1}(\mathbf{Z})\right)^2}}, \quad \mathbf{Y} = \frac{T_{\mathbf{u}}^{-1}(\mathbf{Z})T_{\mathbf{u}}^{-1}(\mathbf{W}_2)}{\sqrt{\sigma_2^2\mathbf{I}_1^2 + \sigma_1^2\left(T_{\mathbf{u}}^{-1}(\mathbf{Z})\right)^2}}. \tag{20}$$

A more stable detection statistics, whose meaning will become apparent from error analysis later in this section, that we strongly advocate to use for all camera identification tasks, is the Peak to Correlation Energy measure (PCE):

$$\mathrm{PCE}(\mathbf{u}) = \frac{NCC(\mathbf{s}_{\mathrm{peak}}, \mathbf{u})^2}{\frac{1}{mn-|\mathcal{N}|}\sum_{\mathbf{s}\notin\mathcal{N}} NCC(\mathbf{s}, \mathbf{u})^2}, \tag{21}$$

where for each fixed $\mathbf{u}$, $\mathcal{N}$ is a small region surrounding the peak value of NCC, $\mathbf{s}_{\mathrm{peak}}$, across all shifts $s_1, s_2$.

For device identification from a single image, the fingerprint estimation noise $\Xi_1$ is much weaker compared with $\Xi_2$—the noise residual of the image under investigation. Thus, $\sigma_1^2 = \mathrm{Var}(\Xi_1) = \mathrm{Var}(\Xi_2) = \sigma_2^2$ and (19) simplifies to a normalized cross-correlation between

$$\mathbf{X} = \mathbf{W}_1 = \hat{\mathbf{K}} \quad \text{and} \quad \mathbf{Y} = T_{\mathbf{u}}^{-1}(\mathbf{Z})T_{\mathbf{u}}^{-1}(\mathbf{W}_2). \tag{22}$$

Recall that $\mathbf{I}_1 = 1$ for device identification when its fingerprint is known.

In practice, the maximum PCE value can be found by a search on a grid obtained by discretizing the range of $\mathbf{u}$. Unfortunately, because the statistic is noise-like for incorrect values of $\mathbf{u}$ and only exhibits a sharp peak in a small neighborhood of the correct value of $\mathbf{u}$, gradient methods do not apply and we are left with a potentially expensive grid search. The grid has to be sufficiently dense in order not to miss the peak. As an illustrative example, we will provide next additional details about how one can carry out the search for the simpler case when the image is known to have been subjected only to a combination of scaling and cropping, in which case $\mathbf{u} = r$ is an unknown scaling ratio. The reader is advised to consult [21] for more details.

### 4.1.1 Identification From Scaled Images

Assuming the image under investigation $\mathbf{Z}$ has dimensions $M \times N$, we search for the scaling parameter at discrete values $r_k \leq 1$, $k = 0, 1, \ldots, R$, from $r_0 = 1$ (no scaling, just cropping) down to $r_R = \max\{M/m, N/n\} < 1$:

$$r_k = \frac{1}{1 + 0.005k}, \quad k = 0, 1, 2, \ldots. \tag{23}$$

This particular form of search grid is fine enough not to miss the correct scaling ratio but not as dense as that would slow down the search. After the ratio is found, it is possible to further refine the estimate by a secondary search around a small neighborhood of the ratio just found.

For a fixed scaling parameter $r_k$, the cross-correlation (19) does not have to be computed for all shifts **s** but only for those that move the upsampled image $T_{r_k}^{-1}(\mathbf{Z})$ within the dimensions of $\hat{\mathbf{K}}$ because only such shifts can be generated by cropping. Given that the dimensions of the upsampled image are $M/r_k \times N/r_k$, we have the following range for the spatial shift $\mathbf{s} = (s_1, s_2)$:

$$0 \le s_1 \le m - M/r_k \quad \text{and} \quad 0 \le s_2 \le n - N/r_k. \tag{24}$$

The peak of the 2-D NCC across all spatial shifts **s** is evaluated for each $r_k$ using PCE($r_k$). If $\max_k PCE(r_k) > \tau$, we decide $H_1$ (camera and image are positively matched). Moreover, the value of the scaling parameter at which the PCE attains its maximum determines the scaling ratio $r_{\text{peak}}$. The location of the peak, $\mathbf{s}_{\text{peak}}$, in the normalized cross-correlation determines the cropping parameters. Thus, as a by-product of this algorithm, we can determine the processing history of **Z** (see Fig. 4 bottom). The fingerprint is essentially playing the role of a synchronizing template. It can also be used for reverse engineering of in-camera processings, such as digital zoom [21] or blind estimation of focal length at which the image was taken for cameras that corrects lens distortion inside the camera [22].

In any forensic application, it is important to keep the false alarm rate low. For camera identification tasks, this means that the probability, $P_{FA}$, that a camera that did not take the image is falsely identified must be below a certain user-defined threshold (Neyman-Pearson setting). Thus, we need to obtain a relationship between $P_{FA}$ and the threshold on the PCE. Note that the threshold will depend on the size of the search space, which is in turn determined by the dimensions of the image under investigation (Fig. 5).

Under hypothesis $H_0$ for a fixed scaling ratio $r_k$, the values of the normalized cross-correlation NCC($\mathbf{s}$, $r_k$) as a function of **s** are well-modeled [21] as white Gaussian noise $\xi_k \sim N(0, \sigma_k^2)$ with variance that may depend on $k$. Estimating the variance of the Gaussian model using the sample variance of NCC($\mathbf{s}$, $r_k$) over **s** after excluding a small central region $\mathcal{N}$ surrounding the peak,

$$\hat{\sigma}_k^2 = \frac{1}{mn - |\mathcal{N}|} \sum_{\mathbf{s} \notin \mathcal{N}} NCC(\mathbf{s}, r_k)^2. \tag{25}$$

We now calculate the probability $p_k$ that the NCC would attain the peak value NCC($\mathbf{s}_{\text{peak}}$, $r_{\text{peak}}$) or larger by chance:

$$p_k = \int\limits_{\text{NCC}(\mathbf{s}_{\text{peak}}, r_{\text{peak}})}^{\infty} \frac{1}{\sqrt{2\pi}\hat{\sigma}_k} \exp(x^2/2\hat{\sigma}_k^2)\, \mathrm{d}x,$$

**Fig. 4** The original image and its cropped and scaled version. The scaling ratio was $r = 0.51$. The light gray rectangle shows the original image size, while the dark gray frame shows the size after cropping but before resizing

$$= \int_{\hat{\sigma}_{\text{peak}}\sqrt{PCE_{\text{peak}}}}^{\infty} \frac{1}{\sqrt{2\pi}\hat{\sigma}_k} \exp(x^2/2\hat{\sigma}_k^2)\,\mathrm{d}x,$$

$$= Q\left(\frac{\hat{\sigma}_{\text{peak}}}{\hat{\sigma}_k}\sqrt{PCE_{\text{peak}}}\right), \tag{26}$$

where $Q(x) = 1 - \Phi(x)$ with $\Phi(x)$ denoting the cumulative distribution function of a standard normal variable $N(0, 1)$ and $PCE_{\text{peak}} = PCE(r_{\text{peak}})$.

As explained above, during the search for the cropping vector **s**, we need to search only in the range (24), which means that we are taking maximum over $l_k = (m - M/r_k + 1) \times (n - N/r_k + 1)$ samples of $\xi_k$. Thus, the probability that the maximum value of $\xi_k$ would not exceed $\text{NCC}(\mathbf{s}_{\text{peak}}, r_{\text{peak}})$ is $(1 - p_k)^{l_k}$. After $R$ steps in the search, the probability of false alarm is

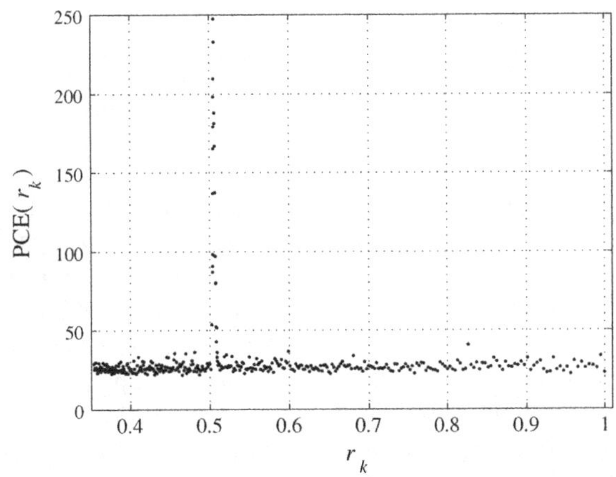


**Fig. 5** Search for the scaling ratio. The peak in $PCE(r_k)$ was detected around the correct ratio of 0.51

$$P_{FA} = 1 - \prod_{k=1}^{R} (1 - p_k)^{l_k}. \tag{27}$$

Since we can stop the search after the PCE reaches a certain threshold, we have $r_k \leq r_{\text{peak}}$. Because $\hat{\sigma}_k$ is non-decreasing in $k$, $\hat{\sigma}_{\text{peak}}/\hat{\sigma}_k \geq 1$. Because $Q(x)$ is decreasing, we have $p_k \leq Q(\sqrt{PCE_{\text{peak}}})$. Thus, because $l_k \leq mn$, we obtain an upper bound on $P_{FA}$

$$P_{FA} \leq 1 - (1 - p)^{l_{\max}}, \tag{28}$$

where $l_{\max} = \sum_{k=0}^{R-1} l_k$ is the maximal number of values of the parameters $r$ and $\mathbf{s}$ over which the maximum of (15) could be taken. Equation (28), together with $p = Q(\sqrt{\tau})$, determines the threshold for PCE, $\tau = \tau(P_{FA}, M, N, m, n)$.

This finishes the technical formulation and solution of the camera identification algorithm from a single image if the camera fingerprint is known. To provide the reader with some sense of how reliable this algorithm is, we include some experiments on real images in Sect. 5. This algorithm can also be used with small modifications for device linking and fingerprint matching. A large-scale test of this methodology appears in [19].

Another extension of the identification algorithm to work with cameras that correct images for lens barrel/pincushion distortion on the fly appears in [22]. Such cameras are becoming very ubiquitous as manufacturers strive to offer customers a powerful optical zoom in inexpensive and compact cameras. Since the lens distortion correction is a geometrical transformation that depends on the focal length (zoom), there can be a desynchronization between the noise residual and the fingerprint if both were estimated at different focal lengths. Here, a search for the distortion parameter is again needed to resynchronize the signals.

## 4.2 Device Linking

The detector derived in the previous section can be readily used with only a few changes for determining whether two images, $\mathbf{I}_1$ and $\mathbf{Z}$, were taken by the exact same camera [18], an application called device linking. Note that in this problem the camera or its fingerprint are not necessarily available.

The device linking problem corresponds exactly to the two-channel formulation (13) and (14) with the GLRT detector (15). Its faster, suboptimal version is the PCE (21) obtained from the maximum value of $\mathrm{NCC}(\mathbf{s}_{\mathrm{peak}}, r_{\mathrm{peak}})$ over all $s_1, s_2, \mathbf{u}$ (see (19) and (20)). In contrast to the camera identification problem, the power of both noise terms, $\Xi_1$ and $\Xi_2$, is now comparable and needs to be estimated from observations. Fortunately, because the PRNU term $\mathbf{IK}$ is much weaker than the modeling noise $\Xi$, reasonable estimates of the noise variances are simply $\hat{\sigma}_1^2 = \mathrm{Var}(\mathbf{W}_1)$, $\hat{\sigma}_2^2 = \mathrm{Var}(\mathbf{W}_2)$.

Unlike the camera identification problem, the search for unknown scaling must now be enlarged to scalings $r > 1$ (upsampling) because the combined effect of unknown cropping and scaling for both images prevents us from easily identifying which image has been downscaled with respect to the other one. The error analysis carries over from Sect. 4.1.1. Due to space limitations we do not include experimental verification of the device linking algorithm. Instead, the reader is referred to [18].

## 4.3 Fingerprint Matching

The last fingerprint matching scenario corresponds to the situation when we need to decide whether or not two estimates of potentially two different fingerprints are identical. This happens, for example, in video clip linking because the fingerprint can be estimated from all frames forming the clip [10].

The detector derived in Sect. 4.1 applies to this scenario, as well. It can be further simplified because for matching fingerprints we have $\mathbf{I}_1 = \mathbf{Z} = 1$ and (19) becomes the normalized cross-correlation between $\mathbf{X} = \hat{\mathbf{K}}_1$ and $\mathbf{Y} = T_{\mathbf{u}}^{-1}(\hat{\mathbf{K}}_2)$.

Video identification has its own challenges that do not necessarily manifest for digital still images. The individual frames in a video are typically harshly quantized (compressed) to keep the video bit rate low. Thus, one needs many more frames for a good quality fingerprint estimate (e.g., thousands of frames). The compression artifacts carry over to the fingerprint estimate in a specific form of NUAs called "blockiness." Simple zero-meaning combined with Wiener filtering as described in Sect. 3.1.1 needs to be supplemented with a more aggressive procedure. The original paper [10] describes a notch filter that removes spikes due to JPEG compression in the frequency domain. The reader can also consult this source for an experimental verification of the fingerprint matching algorithm when applied to video clips.

## 4.4 Forgery Detection

A different, but nevertheless important, use of the sensor fingerprint is verification of image integrity. Certain types of tampering can be identified by detecting the fingerprint presence in smaller regions. The assumption is that if a region was copied from another part of the image (or an entirely different image), it will not have the correct fingerprint on it. The reader should realize that some malicious changes in the image may preserve the PRNU and will not be detected using this approach. A good example is changing the color of a stain to a blood stain.

The forgery detection algorithm tests the presence of the fingerprint in each $B \times B$ sliding block $\mathcal{B}_b$ separately and then fuses all local decisions. For simplicity, we will assume that the image under investigation did not undergo any geometrical processing. For each block, $\mathcal{B}_b$, the detection problem is formulated as a binary hypothesis testing problem:

$$\begin{aligned} H_0 &: \ \mathbf{W}_b = \Xi_b, \\ H_1 &: \ \mathbf{W}_b = a_b \mathbf{I}_b \hat{\mathbf{K}}_b + \Xi_b. \end{aligned} \tag{29}$$

Here, $\mathbf{W}_b$ is the block noise residual, $\hat{\mathbf{K}}_b$ is the corresponding block of the fingerprint, $\mathbf{I}_b$ is the block intensity, $a_b$ is an unknown attenuation factor due to possible processing of the forged image, and $\Xi_b$ is the modeling noise assumed to be a white Gaussian noise with an unknown variance $\sigma^2_{\Xi,b}$. The likelihood ratio test for this problem is the normalized correlation

$$\rho_b = \mathrm{corr}(\mathbf{I}_b \hat{\mathbf{K}}_b, \mathbf{W}_b). \tag{30}$$

In forgery detection, we may desire to control both types of error—failing to identify a tampered block as tampered and falsely marking a region as tampered. To this end, we will need to estimate the distribution of the test statistic $\rho_b$ under both hypotheses. The probability density under $H_0$, $p(x|H_0)$, can be estimated by correlating the known signal $\mathbf{I}_b \hat{\mathbf{K}}_b$ with noise residuals from other cameras. The distribution of $\rho_b$ under $H_1$, $p(x|H_1)$, is much harder to obtain because it is heavily influenced by the block content. Dark blocks will have a lower value of the correlation due to the multiplicative character of the PRNU. The fingerprint may also be absent from flat areas due to strong JPEG compression or saturation. Finally, textured areas will have a lower value of the correlation due to stronger modeling noise. This problem can be resolved by building a predictor of the correlation that will tell us what the value of the test statistics $\rho_b$ and its distribution would be if the block $b$ was not tampered and indeed came from the camera.

The predictor is a mapping that needs to be constructed for each camera. The mapping assigns an estimate of the correlation $\hat{\rho}_b = \mathrm{Pred}(i_b, f_b, t_b)$ to each triple $(i_b, f_b, t_b)$, where the individual elements of the triple stand for a measure of intensity, saturation, and texture in block $b$. The mapping $\mathrm{Pred}(., ., .)$ can be constructed for example using regression [8, 11] or machine learning techniques by training on a database of image blocks coming from images taken by the camera. The block size cannot be too small (because then the correlation $\rho_b$ has too large a variance). On the other hand, large blocks would compromise the ability of the forgery detection algorithm to localize. For full-size camera images with several megapixels, blocks of $64 \times 64$ or $128 \times 128$ pixels seem to work well.

A reasonable measure of intensity is the average intensity in the block:

$$i_b = \frac{1}{|\mathcal{B}_b|} \sum_{i \in \mathcal{B}_b} \mathbf{I}(i). \tag{31}$$

We take as a measure of flatness the relative number of pixels, $i$, in the block whose sample intensity variance $\sigma_{\mathbf{I}}^2(i)$ estimated from the local $3 \times 3$ neighborhood of $i$ is below a certain threshold

$$f_b = \frac{1}{|\mathcal{B}_b|} \left| \{ i \in \mathcal{B}_b | \sigma_{\mathbf{I}}(i) | < c\mathbf{I}(i) \} \right|, \tag{32}$$

where $c \approx 0.03$ (for Canon G2 camera). The best value of $c$ varies with the camera model.

The texture measure evaluates the amount of edges in the block. Among many available options, we give the following example

$$t_b = \frac{1}{|\mathcal{B}_b|} \sum_{i \in \mathcal{B}_b} \frac{1}{1 + \mathrm{Var}_5(\mathbf{H}(i))}, \tag{33}$$

where $\mathrm{Var}_5(\mathbf{H}(i))$ is the sample variance computed from a local $5 \times 5$ neighborhood of pixel $i$ for a high-pass filtered version of the block, $\mathbf{H} = H(\mathcal{B}_b)$, such as one obtained using an edge detector $H$.

Since one can obtain potentially hundreds of blocks from a single image, only a small number of images (e.g., 10) are needed to train (construct) the predictor. Figure 6 shows the performance of the predictor for a Canon G2 camera. The form of the mapping Pred() was a second-order polynomial:

$$\mathrm{Pred}(x, y, z) = \sum_{k+l+m \le 2} \lambda_{klm} x^k y^l z^m, \tag{34}$$

where the unknown coefficients $\lambda_{klm}$ were determined using a least squares estimator.

The data used for constructing the predictor can also be used to estimate the distribution of the prediction error $\upsilon_b$:

$$\rho_b = \hat{\rho}_b + \upsilon_b, \tag{35}$$

where $\hat{\rho}_b$ is the predicted value of the correlation for block $b$. Say that for a given block under investigation, we apply the predictor and obtain the estimated value $\hat{\rho}_b$.

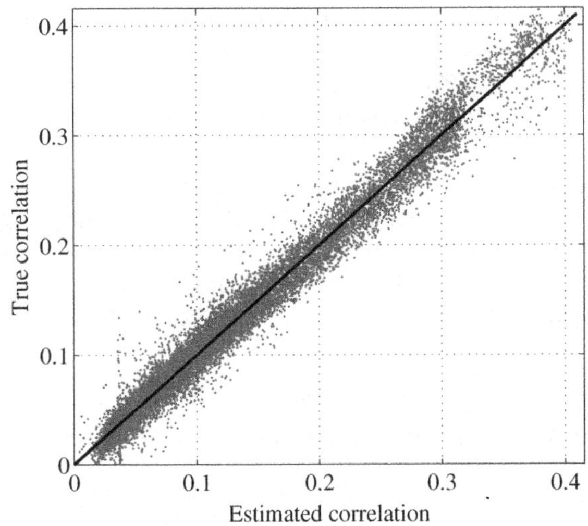


**Fig. 6** Scatter plot of the true correlation $\rho_b$ versus the estimate $\hat{\rho}_b$ for 30,000 $128 \times 128$ blocks from 300 TIFF images by Canon G2

The distribution $p(x|H_1)$ is obtained by fitting a parametric probability density function to all points in Fig. 6 whose estimated correlation is in a small neighborhood of $\hat{\rho}_b$, $(\hat{\rho}_b - \epsilon, \hat{\rho}_b + \epsilon)$ for some small $\epsilon$. A sufficiently flexible model that allows both thin and thick tails is the generalized Gaussian model with density $\alpha/(2\sigma\Gamma(1/\alpha)) \exp(-(|x-\mu|/\sigma)^\alpha)$ with variance $\sigma^2\Gamma(3/\alpha)/\Gamma(1/\alpha)$, mean $\mu$, and shape parameter $\alpha$.

We now continue with the description of the forgery detection algorithm using sensor fingerprint. The algorithm proceeds by sliding a block across the image and evaluates the test statistics $\rho_b$ for each block $b$. The decision threshold $\tau$ for the test statistics $\rho_b$ needs to be set to bound $\Pr(\rho_b > \tau|H_0)$—the probability of misidentifying a tampered block as un-tampered. (In our experiments shown in Sect. 5.2, we requested this probability to be less than 0.01.)

Block $b$ is marked as potentially tampered if $\rho_b < \tau$ but this decision is attributed only to the central pixel $i$ of the block. Through this process, for an $m \times n$ image we obtain an $(m - B + 1) \times (n - B + 1)$ binary array $\mathbf{Z}(i) = [\rho_b < \tau]$ ($[\cdot]$ is the Iverson bracket) indicating the potentially tampered pixels with $\mathbf{Z}(i) = 1$.

The above Neyman–Pearson criterion decides "tampered" whenever $\rho_b < \tau$ even though $\rho_b$ may be "more compatible" with $p(x|H_1)$. This is more likely to occur when $\rho_b$ is small, such as for highly textured blocks. To control the amount of pixels falsely identified as tampered, we compute for each pixel $i$ the probability of falsely labeling the pixel as tampered when it was not

$$p_{MD}(i) = \int_{-\infty}^{t} p(x|H_1)\,\mathrm{d}x. \tag{36}$$

Pixel $i$ is labeled as un-tampered (we reset $\mathbf{Z}(i) = 0$) if $p_{MD}(i) > \beta$, where $\beta$ is a user-defined threshold. (In experiments in the next section, $\beta = 0.01$.) The resulting binary map $\mathbf{Z}$ identifies the forged regions in their raw form.

The final map $\mathbf{Z}$ is obtained by post-processing $\mathbf{Z}$ using morphological filters. The block size imposes a lower bound on the size of tampered regions that the algorithm can identify. We thus remove from $\mathbf{Z}$ all simply connected tampered regions that contain fewer than $64 \times 64$ pixels. The final map of forged regions is obtained by dilating $\mathbf{Z}$ with a square $20 \times 20$ kernel. The purpose of this step is to compensate for the fact that the decision about the whole block is attributed only to its central pixel and we may miss portions of the tampered boundary region.

# 5 Real-World Examples

In this section, we demonstrate how the forensic methods proposed in the previous sections may be implemented in practice and also include some sample experimental results to give the reader an idea about how the methods work on real imagery. The reader is referred to [8, 9, 11, 19] for more extensive tests and to [18] and [10] for experimental verification of device linking and fingerprint matching for video clips. Camera identification from printed images appears in [20].

## *5.1 Camera Identification*

The experiment in this section was carried out on images from a Canon G2 camera with a 4-megapixel CCD sensor. The camera fingerprint was estimated for each color channel separately using the maximum likelihood estimator (7) from 30 blue sky images acquired in the TIFF format. The estimated fingerprints were preprocessed as described in Sect. 3.1 to remove any residual patterns (NUAs) not unique to the sensor. This step is very important because these artifacts would cause unwanted interference at certain spatial shifts, $\mathbf{s}$, and scaling factors $r$, and thus decrease the PCE and substantially increase the false alarm rate. The fingerprints estimated from all three color channels were combined into a single fingerprint using formula (12). All other images involved in this test were also converted to grayscale before applying the detectors described in Sect. 4.

The same camera was further used to acquire 720 images containing snapshots or various indoor and outdoor scenes under a wide range of light conditions and zoom settings spanning the period of 4 years. All images were taken at the full CCD resolution and with a high JPEG quality setting. Each image was first cropped by a random amount up to 50% in each dimension. The upper left corner of the cropped region was also chosen randomly with uniform distribution within the upper left quarter of the image. The cropped part was subsequently downsampled by a randomly

**Fig. 7** $PCE_{\text{peak}}$ as a function of the scaling ratio $r$ for 720 images matching the camera. The detection threshold $\tau$, which is outlined with a horizontal line, corresponds to $P_{FA} = 10^{-5}$

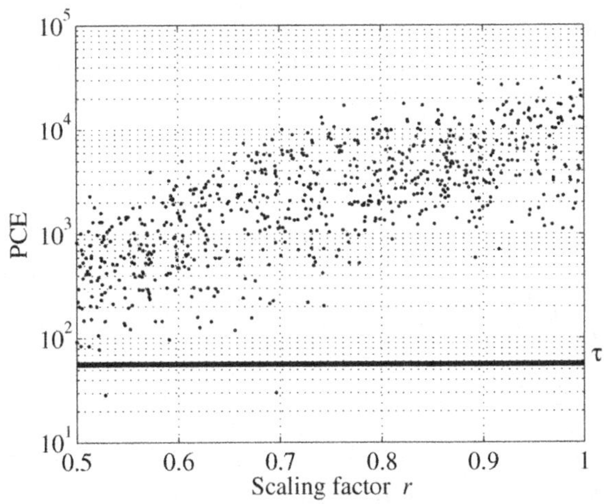


**Fig. 8** $PCE_{\text{peak}}$ for 915 images not matching the camera. The detection threshold $\tau$ is again outlined with a horizontal line and corresponds to $P_{FA} = 10^{-5}$

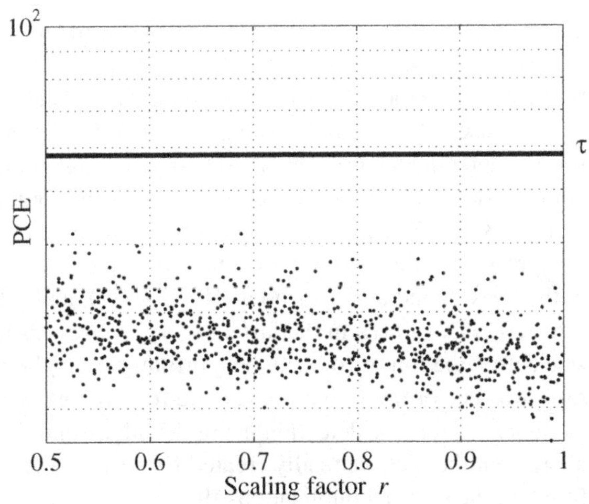


chosen scaling ratio $r \in [0.5, 1]$. Finally, the images were converted to grayscale and compressed with 85% quality JPEG.

The detection threshold $\tau$ was chosen to obtain the probability of false alarm (28) $P_{FA} = 10^{-5}$. The camera identification algorithm was run with $r_{\text{min}} = 0.5$ on all images. Only two missed detections were encountered (Fig. 7). In the figure, the PCE is displayed as a function of the randomly chosen scaling ratio. The missed detections occurred for two highly textured images. In all successful detections, the cropping and scaling parameters were detected with accuracy better than two pixels in either dimension.

To test the false identification rate, we used 915 images from more than 100 different cameras downloaded from the Internet in native resolution. The images were cropped to 4 megapixels (the size of Canon G2 images) and subjected to the same random cropping, scaling, and JPEG compression as the 720 images before. The threshold for the camera identification algorithm was set to the same value as in the previous experiment. All images were correctly classified as not coming from the tested camera (Fig. 8).

The camera identification algorithm (without considering scaling and cropping) was subjected to a large scale test in [19]. The authors carried out experiments on 1,024,050 images from 6,896 cameras of 150 different models downloaded from the public image sharing web site Flickr.com. With the decision threshold fixed to $\tau = 60$, the overall missed detection rate was $P_{MD} = 0.024$ with the false alarm rate $P_{FA} = 2.4 \times 10^{-5}$.

## 5.2 *Forgery Detection*

The forgery detection algorithm was tested on an image from a 4-megapixel Olympus C765 digital camera equipped with a CCD sensor. Figure 9a shows the original image taken in the raw format. Using Photoshop, the girl in the middle was covered by pieces of the house siding from the background (b). The letters (c)–(f) show the result of the forgery detection algorithm after the forged image was further processed using JPEG compression with quality factor 75 (c), after the forgery was subjected to denoising using a $3 \times 3$ Wiener filter with default value of $\sigma$ in Matlab followed by JPEG compression with quality factor 90 (d), after processing using gamma correction with $\gamma = 0.5$ and again saved as JPEG 90 (e), and after downscaling to 60% of its size and JPEG 90 (f). In the last case, the image was upsampled back to its original size before the forgery detection algorithm was applied (i.e., no search for the scaling ratio was performed). In all cases, the forged region was accurately detected. More examples of forgery detection using this algorithm, including the results of tests on a large number automatically created forgeries as well as non-forged images, can be found in the original publications [9, 11].

# 6 Fighting the Fingerprint-Copy Attack

Since the inception of camera identification methods based on sensor fingerprints in 2005 [41], researchers have realized that the sensor fingerprint can be copied onto an image that came from a different camera in an attempt to frame an innocent victim. In the most typical and quite plausible scenario, Alice, the victim, posts her images on the Internet. Eve, the attacker, estimates the fingerprint of Alice's camera and superimposes it onto another image. Indeed, as already shown in the original publication and in [17, 46], correlation detectors cannot distinguish between a genuine fingerprint and a fake one.

**Fig. 9** From upper left corner by rows: **a** the original image, **b** its tampered version, the result of the forgery detection algorithm after the forged image was processed using **c** JPEG with quality factor 75, **d** $3 \times 3$ Wiener-filtered plus JPEG 90, **e** gamma correction with $\gamma = 0.5$ plus JPEG 90, and **f** scaled down by factor of 0.6 and JPEG 90

In this section, we describe a countermeasure [12, 23] against this fingerprint-copy attack that enables Alice to prove that she has been framed (that the fingerprint is fake). We assume that Alice owns a digital camera $C$. Eve takes an image $\mathbf{J}$ from a different camera $C'$ with fingerprint $\mathbf{K}' \neq \mathbf{K}$ and makes it appear as if it was taken by $C$. She does so by first estimating the fingerprint of $C$ from some set of Alice's images and then properly adds it to $\mathbf{J}$.

In particular, let us assume that Eve has access to $N$ images, from $C$ and estimates its fingerprint $\hat{\mathbf{K}}_E$ using the estimator (7). We note that Eve certainly is free to use a different estimator. However, any estimation procedure will be some form of averaging of noise residuals $\mathbf{W}$. Having estimated the fingerprint, Eve may preprocess $\mathbf{J}$ to suppress the PRNU term $\mathbf{JK}'$ introduced by the sensor in $C'$ and/or to remove any artifacts in $\mathbf{J}$ that are incompatible with $C$. Because suppressing the PRNU term is not an easy task [45], quite likely the best option for Eve is to skip this step altogether.

This is because the PRNU component $\mathbf{JK}'$ in $\mathbf{J}$ is very weak to be detected per se and because it is unlikely that Alice will gain access to $C'$. In fact, Eve should avoid processing $\mathbf{J}$ too much as it may introduce artifacts of its own.

If Eve saves her forgery as a JPEG file, she needs to make sure that the quantization table is compatible with camera $C$, otherwise Alice will know that the image has been manipulated and did not come directly from her camera. If camera $C'$ uses different quantization tables than $C$, Eve will inevitably introduce double compression artifacts into $\mathbf{J}$, giving Alice again a starting point of her defense.

Unless $C$ and $C'$ are of the same model, the forged image may contain color-interpolation artifacts of $C'$ incompatible with those of $C$. Alice could leverage techniques developed for camera brand/model identification [7] and prove that there is a mismatch between the camera model and the color-interpolation artifacts. A knowledgeable attacker may, in turn, attempt to remove such artifacts of $C'$ and introduce interpolation artifacts of $C$, for example, using the method described in [4].

It should now be apparent that it is far from easy to create a "perfect" forgery. While it is certainly possible for Alice to utilize traces of previous compression or color-interpolation artifacts, no attempt is made in this section to exploit these discrepancies to reveal the forged fingerprint. Our goal is to develop techniques capable of identifying images forged by Eve even in the most difficult scenario for Alice when $C'$ is of exactly the same model as $C$ to avoid any incompatibility issues discussed above. Thus, Alice cannot take advantage of knowing any a priori information about $C'$.

The final step for Eve is to plant the estimated fingerprint in $\mathbf{J}$, creating thus the forged image $\mathbf{J}'$. In her attempt to mimic the acquisition process, and in accordance with (4), Eve superimposes the fake fingerprint multiplicatively, which is what would happen if $\mathbf{J}$ was indeed taken by $C$:

$$\mathbf{J}' = [\mathbf{J}(1 + \alpha\hat{\mathbf{K}}_E)], \tag{37}$$

where $\alpha > 0$ is a scalar fingerprint strength and $[x]$ is the operation of rounding $x$ to integers forming the dynamic range of $\mathbf{J}$. Finally, Eve saves $\mathbf{J}'$ as JPEG with the same or similar quantization table as that of the original image $\mathbf{J}$.

Formula (37) should be understood as three equations for each color channel of $\mathbf{J}$. This attack indeed succeeds in fooling the camera identification algorithm in the sense that the response of the fingerprint detector on $\mathbf{J}'$ (either the correlation (19), the generalized matched filter in [11], or the PCE (21)) will be high enough to indicate that $\mathbf{J}'$ was taken by camera $C$.

A very important issue for Eve is the choice of the strength $\alpha$. Here, we grant Eve the ability to create a "perfect" forgery in the sense that $\mathbf{J}'$ elicits the same response of the fingerprint detector implemented with the true fingerprint $\mathbf{K}$ when $\mathbf{J}'$ was indeed taken by $C$. A good estimate of the detector response can be obtained using a predictor, such as the one described in Sect. 4.4. This way, Eve makes sure that the fingerprint is not suspiciously weak or strong. While it certainly is true that Eve cannot easily construct the predictor because she does not have access to the true fingerprint $\mathbf{K}$, she may select the correct strength $\alpha$ by pure luck. To be more precise here, we grant Eve the ability to guess the right strength instead of giving her access to $\mathbf{K}$. We note that similar assumptions postulating a clairvoyant attacker are commonly made in many branches of information security.

## 6.1 Detecting Fake Fingerprints

Here, we describe a test using which Alice can decide whether an image came from her camera or whether it was forged by Eve as described above. For simplicity, we only discuss the case when Eve created one forged image $\mathbf{J}'$ and Alice has access to some of the $N$ images used by Eve to estimate $\hat{\mathbf{K}}_E$ but Alice does not know which they are. She has a set of $N_c \geq N$ candidate images that Eve may have possibly used. This is a very plausible scenario because, unless Eve gains access to Alice's camera and takes images of her own and then removes them from the camera before returning the camera to Alice, Eve will have little choice but to use images taken by Alice, such as images posted by Alice on the Internet. In this case, Alice can prove that the forged image did not originally come from her camera by identifying among her candidate images those used by Eve.

We now explain the key observation based on which Alice can construct her defense. Let $\mathbf{I}$ be one of the $N$ images available to Alice that Eve used to forge $\mathbf{J}'$. Since the noise residual of $\mathbf{I}$, $\mathbf{W}_\mathbf{I}$, participates in the computation of $\hat{\mathbf{K}}_E$ through formula (7), $\mathbf{J}'$ will contain a scaled version of the *entire* noise residual $\mathbf{W}_\mathbf{I} = \mathbf{IK} + \Xi_\mathbf{I}$. Thus, besides the PRNU term, $\mathbf{W}_\mathbf{I}$ and $\mathbf{W}_{\mathbf{J}'}$ will share another signal—the noise $\Xi_\mathbf{I}$. Consequently, the correlation $c_{\mathbf{I},\mathbf{J}'} = \mathrm{corr}(\mathbf{W}_\mathbf{I}, \mathbf{W}_{\mathbf{J}'})$ will be larger than what it would be if the only common signal between $\mathbf{I}$ and $\mathbf{J}'$ was the PRNU component (which would be the case if $\mathbf{J}'$ was not forged). As this increase may be quite small and the correlation itself may fluctuate significantly across images, the test that evaluates the statistical increase must be calibrated. We call this test the triangle test.

Alice starts her defense by computing an estimate of the fingerprint of her camera $\hat{\mathbf{K}}_A$ from images guaranteed to not have been used by Eve. For instance, she can take new images with her camera $C$. Then, for a candidate image $\mathbf{I}$, she computes $c_{\mathbf{I},\mathbf{J}'}$, $c_{\mathbf{I},\hat{\mathbf{K}}_A} = \mathrm{corr}(\mathbf{W}_\mathbf{I}, \hat{\mathbf{K}}_A)$, and $c_{\mathbf{J}',\hat{\mathbf{K}}_A} = \mathrm{corr}(\mathbf{W}_{\mathbf{J}'}, \hat{\mathbf{K}}_A)$ (follow Fig. 10). The test is based on the fact that for images $\mathbf{I}$ that were not used to forge $\mathbf{J}'$, the value of $c_{\mathbf{I},\mathbf{J}'}$ can be estimated from $c_{\mathbf{I},\hat{\mathbf{K}}_A}$ and $c_{\mathbf{J}',\hat{\mathbf{K}}_A}$ while when $\mathbf{I}$ was used in the forgery, the correlation $c_{\mathbf{I},\mathbf{J}'}$ will be higher than its estimate.

In order to obtain a more accurate relationship, similar to our approach to forgery detection in Sect. 4.4, we will work by blocks of pixels, denoting the signals constrained to block $b$ with subscript $b$. We adopt the model (4) for the noise residuals and a similar model for Alice's fingerprint:

$$\mathbf{W}_{\mathbf{I},b} = a_{\mathbf{I},b}\mathbf{I}_b\mathbf{K}_b + \Xi_{\mathbf{I},b}, \tag{38}$$

$$\mathbf{W}_{\mathbf{J}',b} = a_{\mathbf{J}',b}\mathbf{J}'_b\mathbf{K}_b + \Xi_{\mathbf{J}',b}, \tag{39}$$

$$\hat{\mathbf{K}}_A = \mathbf{K}_b + \xi_b. \tag{40}$$

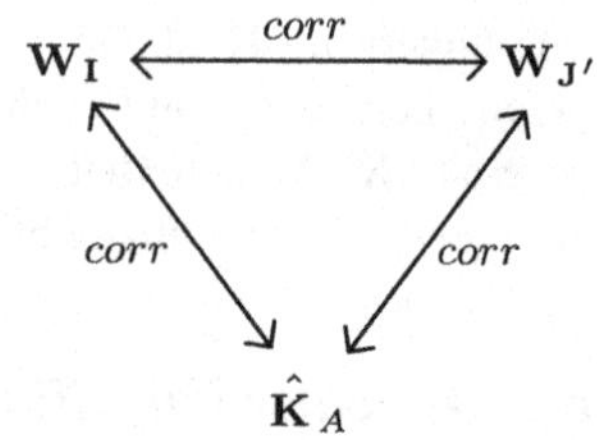


**Fig. 10** Diagram for the triangle test

The block-dependent attenuation factor $a$ in (39) has been introduced due to the fact that various image processing that might have been applied by Eve to $\mathbf{J}$ may affect the PRNU factor differently in each block.

When $\mathbf{I}$ was not used by Eve, under some fairly mild assumptions about the noise terms in (39), the following estimate of $c_{\mathbf{I},\mathbf{J}'}$ is derived in [12, 23]

$$\hat{c}_{\mathbf{I},\mathbf{J}'} = \mathrm{corr}(\mathbf{W_I}, \hat{\mathbf{K}}_A)\mathrm{corr}(\mathbf{W_{J'}}, \hat{\mathbf{K}}_A)\mu(\mathbf{I}, \mathbf{J}')q^{-2}, \tag{41}$$

where $\mu(\mathbf{I}, \mathbf{J}')$ is the "mutual-content factor,"

$$\mu(\mathbf{I}, \mathbf{J}') = \frac{\sum_b a_{\mathbf{I},b} a_{\mathbf{J}',b} \overline{\mathbf{I}_b \mathbf{J}'_b}}{\sum_b a_{\mathbf{I},b} \overline{\mathbf{I}_b} \cdot \sum_b a_{\mathbf{J}',b} \overline{\mathbf{J}'_b}} N_B, \tag{42}$$

and the bar denotes the sample mean as before. The integer $N_B$ is the number of blocks and $q \leq 1$ is the quality of $\hat{\mathbf{K}}_A$, $q^{-2} = 1 + (SNR_{\hat{\mathbf{K}}_A})^{-1}$, $SNR_{\hat{\mathbf{K}}_A} = \|\mathbf{K}\|^2 / \|\xi\|^2$.

The attenuation factors can be estimated by computing the following block-wise correlations:[1]

$$a_{\mathbf{I},b} = \frac{\|\mathbf{W}_{\mathbf{I},b}\|}{\sqrt{\overline{\mathbf{I}_b^2}} \, \|\hat{\mathbf{K}}_{A,b}\|} \mathrm{corr}(\mathbf{W}_{\mathbf{I},b}, \hat{\mathbf{K}}_{A,b})q^{-2}. \tag{43}$$

Continuing the analysis of the case when $\mathbf{I}$ was not used by Eve, we consider $c_{\mathbf{I},\mathbf{J}'}$ and $\hat{c}_{\mathbf{I},\mathbf{J}'}$ as random variables over different images $\mathbf{I}$ for a fixed $\mathbf{J}'$. The dependence between these two random variables is well fit with a straight line $c_{\mathbf{I},\mathbf{J}'} = \lambda\hat{c}_{\mathbf{I},\mathbf{J}'} + c_0$. Because the distribution of the deviation from the linear fit does not seem to vary with $\hat{c}_{\mathbf{I},\mathbf{J}'}$ (see Fig. 11), we make a simplifying assumption that the conditional probability

$$\Pr(c_{\mathbf{I},\mathbf{J}'} - \lambda\hat{c}_{\mathbf{I},\mathbf{J}'} - c_0 = x | \hat{c}_{\mathbf{I},\mathbf{J}'}) \approx f_{\mathbf{J}'}(x), \tag{44}$$

is independent of $\hat{c}_{\mathbf{I},\mathbf{J}'}$.

When $\mathbf{I}$ was used by Eve in the multiplicative forgery, due to the additional common signal $\Xi_{\mathbf{I}}$, the correlation $c_{\mathbf{I},\mathbf{J}'}$ increases to $\beta c_{\mathbf{I},\mathbf{J}'}$, where $\beta$ is the following multiplicative factor derived in [12, 23]

[1] Equation (43) holds independently of whether or not $\mathbf{I}$ was used by Eve.

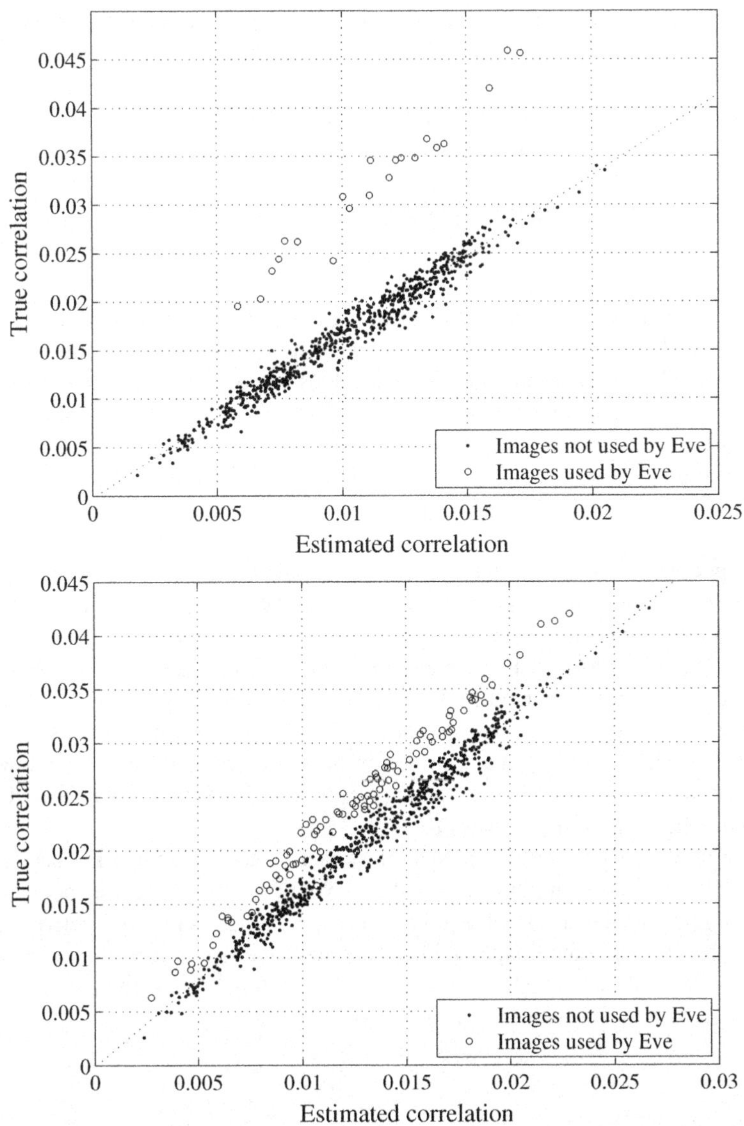


**Fig. 11** True correlation $c_{\mathbf{I},\mathbf{J}'}$ versus the estimate $\hat{c}_{\mathbf{I},\mathbf{J}'}$ for image no. 5. Eve's fingerprint was estimated from $N = 20$ images (*top*) and $N = 100$ (*bottom*)

$$\beta = 1 + \frac{\alpha}{N} \frac{\sum_b a_{\mathbf{J}',b} \overline{\mathbf{J}'_b} \left\| \boldsymbol{\Xi}_{\mathbf{I},b} \right\|^2}{\sum_b a_{\mathbf{I},b} a_{\mathbf{J}',b} \overline{\mathbf{I}_b \mathbf{J}'_b} \left\| \mathbf{K}_b \right\|^2}. \tag{45}$$

Notice that the percentual increase is proportional to the fingerprint strength $\alpha$ and the energy of the common noise component $\Xi_{\mathbf{I},b}$; it is inversely proportional to $N$.

Alice now runs the following composite binary hypothesis test for every candidate image $\mathbf{I}$ from her set of $N_c$ candidate images:

$$\begin{aligned} H_0 &: \ c_{\mathbf{I},\mathbf{J}'} - \lambda \hat{c}_{\mathbf{I},\mathbf{J}'} - c_0 \sim f_{\mathbf{J}'}(x), \\ H_1 &: \ c_{\mathbf{I},\mathbf{J}'} - \lambda \hat{c}_{\mathbf{I},\mathbf{J}'} - c_0 \nsim f_{\mathbf{J}'}(x). \end{aligned} \tag{46}$$

The reason why (46) cannot be turned into a simple hypothesis test is that the distribution of $c_{\mathbf{I},\mathbf{J}'}$ when $\mathbf{I}$ is used for forgery is not available to Alice and it cannot be determined experimentally because Alice does not know the exact actions of Eve. Thus, we resort to the Neyman-Pearson test and set our decision threshold $\tau$ to bound the probability of false alarm,

$$\Pr(c_{\mathbf{I},\mathbf{J}'} - \lambda \hat{c}_{\mathbf{I},\mathbf{J}'} - c_0 > \tau | H_0) = P_{FA}. \tag{47}$$

The pdf $f_{\mathbf{J}'}(x)$ is often very close to a Gaussian but for some images $\mathbf{J}'$, the tails exhibit a hint of a polynomial dependence. Thus, to be conservative, we used Student's $t$-distribution for the fit.

Note that, depending on $\mathbf{J}'$, the constant of proportionality $\lambda > 1$, which suggests the presence of an unknown multiplicative hidden parameter in (41) most likely due to some non-periodic NUAs that were not removed using zero-meaning as described in Sect. 3.1. The quality of Alice's fingerprint, $q$, can be considered unknown (or simply set to 1) as different $q$ will just correspond to a different $\lambda$ (scaling of the $x$ axis in the diagram of $c_{\mathbf{I},\mathbf{J}'}$ versus $\hat{c}_{\mathbf{I},\mathbf{J}'}$).

Alice now has two options. She can test each candidate image $\mathbf{I}$ separately by evaluating its p-value, and thus on a certain level of statistical significance, identify those images that were used by Eve for estimating her fingerprint. Alternatively, Alice can test for $N_c$ candidate images $\mathbf{I}$ all at once whether $c_{\mathbf{I},\mathbf{J}'} - \lambda \hat{c}_{\mathbf{I},\mathbf{J}'} - c_0 \sim f_{\mathbf{J}'}(x)$. This "pooled test" will be a better choice for her for large $N$ where the reliability of the triangle test for individual images becomes low.

## 6.2 Experiments

In this chapter, we report the results of experiments when testing individual images. The original publications [12, 23] contain much more detailed experimental evaluation including the pooled test and another case when multiple forged images are analyzed.

In the experiment, the signals entering the triangle test were preprocessed by zero-meaning. Wiener filtering, as described in Sect. 3.1 to suppress the NUAs, was only applied to $\mathbf{W}_{\mathbf{J}'}$ and not to $\mathbf{W}_{\mathbf{I}}$ to save computation time. The camera $C'$ is the 4-megapixel Canon PS A520 while $C$ is Canon PS G2, which has the same native resolution. Both cameras were set to take images at the highest quality JPEG compression and the largest resolution. The picture taking mode was set to "auto."

The fingerprint estimation algorithm and the forging algorithm depend on a large number of parameters, such as the parameters of the denoising filter. In presenting our test results, we intentionally opted for what we consider to be the most advantageous setting for Eve and the hardest one for Alice. This way, the results will be on the conservative side. Furthermore, to obtain a compact yet comprehensive report on the performance of the triangle test, the experiments were designed to show the effect of only the most influential parameters. To estimate her fingerprint $\hat{\mathbf{K}}_E$, Eve uses the most accurate estimator she can find in the literature (7) implemented using the denoising filter $F$ described in [43] with the wavelet-domain Wiener filter parameter $\sigma = 3$ (valid for 8 bit per channel color images). From our experiments, the reliability of the triangle test is insensitive to the denoising filter or the mismatch between the filters used by Eve and Alice.

Then, Eve forges a 24-bit color image $\mathbf{J}$ from camera $C'$ to make it look as if it came from camera $C$. She first slightly denoises $\mathbf{J}$ using the same denoising filter $F$ (with its Wiener filter parameter $\sigma = 1$) to suppress the fingerprint from camera $C'$ and possibly other artifacts introduced by $C'$. The filter is applied to each color channel separately. Then, Eve adds the fingerprint to $\mathbf{J}$, using (37), and saves the result as JPEG with quality factor 90, which is slightly smaller than the typical qualities of the original JPEG images $\mathbf{J}$. The fingerprint strength factor $\alpha$ is determined so that the response of the generalized matched filter (Equation (11) in [11]) matches its prediction obtained using the predictor as described in Sect. 4.4. The predictor was implemented as a linear combination of intensity (31), texture (33), and flattening features (32), and their second-order terms. The coefficients of the linear fit were determined from 20 images of natural scenes using the least square fit. Note that because Eve needs to adjust $\alpha$ so that the JPEG-compressed $\mathbf{J}'$ elicits the same GMF value as the prediction, the proper value of $\alpha$ must be found, e.g., using a binary search. Finally, the true fingerprint $\mathbf{K}$ was estimated from 300 JPEG images of natural scenes.

On the defense side, Alice estimates her fingerprint $\hat{\mathbf{K}}_A$ from $N_A = 15$ blue-sky raw images (fingerprint quality was $q = 0.56$). Surprisingly, the quality of $\hat{\mathbf{K}}_A$ has little impact on the triangle test. Tests with $N_A = 70$ produced essentially identical results. In particular, it is not necessary for Alice to work with a better quality fingerprint than Eve! In all experiments, the block size was $128 \times 128$ pixels. The triangle test performed equally well for blocks as small as $64 \times 64$ and as large as $256 \times 256$.

We now evaluate the performance of the triangle test when applied to each candidate image individually. By far the most influential element is the number of images used by Eve, $N$, and the content of the forged image $\mathbf{J}$. Six randomly selected test images $\mathbf{J}$ shown in Fig. 12 were tested with the values of $N \in \{20, 50, 100, 200\}$. To give the reader a sense of the extent of Eve's forging activity, in Table 1 we report the PSNR between $\mathbf{J}$ and $\mathbf{J}'$ before it is JPEG compressed. The PSNR between $\mathbf{J}'$ and $F(\mathbf{J}')$ measures the total distortion that includes the slight denoising, $F(\mathbf{J})$, and

**Fig. 12** Six original images $\mathbf{J}$ from a Canon PS A520 numbered by rows (#1, #2, #3); (#4, #5, #6)

**Table 1** PSNR between the original image $\mathbf{J}$ and the forgery $\mathbf{J}'$ before JPEG compression for six test images

| # | $PSNR(F(\mathbf{J}), \mathbf{J}')$ (dB) | | | | $PSNR(\mathbf{J}, \mathbf{J}')$ (dB) | | | |
|---|---|---|---|---|---|---|---|---|
| $N$ | 20 | 50 | 100 | 200 | 20 | 50 | 100 | 200 |
| 1 | 48.8 | 51.8 | 53.2 | 53.9 | 47.6 | 49.5 | 50.3 | 50.7 |
| 2 | 49.0 | 51.8 | 53.1 | 53.8 | 47.8 | 49.8 | 50.6 | 50.9 |
| 3 | 50.1 | 51.8 | 52.9 | 53.4 | 48.7 | 49.8 | 50.5 | 50.8 |
| 4 | 54.5 | 56.4 | 57.5 | 58.7 | 49.5 | 50.0 | 50.2 | 50.4 |
| 5 | 49.5 | 52.2 | 53.2 | 54.0 | 47.7 | 49.3 | 49.7 | 50.1 |
| 6 | 50.8 | 53.2 | 54.3 | 55.1 | 49.3 | 51.0 | 51.6 | 52.1 |

quantization to 24-bit colors after adding the fingerprint. The PSNR between $\mathbf{J}'$ and the slightly denoised $F(\mathbf{J})$ measures the energy of the PRNU term only.

To accurately estimate the probability distribution $f_{\mathbf{J}}(x)$, we used 358 images from camera $C$ that were for sure not used by Eve. All these images were taken within a period of about 4 years. In practice, depending on the situation, statistically significant conclusions may be obtained using a much smaller sample.

Figure 11 presents a typical plot of $c_{\mathbf{I},\mathbf{J}'}$ versus $\hat{c}_{\mathbf{I},\mathbf{J}'}$ for $N = 20$ and $N = 100$. As expected, the separation between images used by Eve and those not used deteriorates with increasing $N$. When applying the triangle test individually to each candidate image, after setting the decision threshold to satisfy a desired probability of false alarm, $P_{FA}$, the probability of correct detection $P_D$ in the hypothesis test (46) is shown in Table 2. Each value of $P_D$ was obtained by running the entire experiment as explained in Sect. 6.1 and evaluating the p-values for all images used by Eve.

The lower detection rate for image no. 4 is due to the low energy of the fingerprint (see the corresponding row in Table 1) dictated by the predictor. Because the image has smooth content, which is further smoothened by the denoising filter, the fingerprint PSNR in the noise residual $\mathbf{W}$ is higher than for other images. Consequently, a low fingerprint energy is sufficient in matching the predicted correlation. Image no.

**Table 2** Detection rate in percents for six text images

| # | $P_D$ (%) for $P_{FA} = 10^{-3}$ | | | | $P_D$ (%) for $P_{FA} = 10^{-4}$ | | | |
|---|---|---|---|---|---|---|---|---|
| $N$ | 20 | 50 | 100 | 200 | 20 | 50 | 100 | 200 |
| 1 | 100 | 92 | 63 | 15 | 100 | 80 | 44 | 6 |
| 2 | 100 | 84 | 40 | 5 | 100 | 74 | 26 | 0 |
| 3 | 95 | 78 | 35 | 4 | 95 | 66 | 14 | 0 |
| 4 | 95 | 64 | 21 | 3 | 95 | 42 | 8 | 1 |
| 5 | 100 | 90 | 56 | 11 | 100 | 82 | 41 | 2 |
| 6 | 100 | 94 | 59 | 14 | 100 | 90 | 40 | 2 |

3 also produced lower detection rates, mostly due to the fact that 26% of the image content is overexposed (the entire sky) with fully saturated pixels. The attenuation factor $a_b$ in (39) is thus effectively equal to zero for such blocks $b$, while it is estimated in (43) under $H_1$ as being relatively large due to the absence of the noise term $\Xi_{\mathbf{J}',b}$. A possible remedy is to apply the triangle test only to the nonsaturated part of the image. However, then we experience a lower accuracy again due to a smaller number of pixels in the image. At this point, we note that if the attacker makes the forgery using (37) without attenuating the PRNU in saturated areas, the fingerprint will be too strong there, which could be used by Alice to argue that the fingerprint has been artificially added and the image did not come from her camera.

We conclude this section with the statement that while it is possible to maliciously add a sensor fingerprint to an image to frame an innocent victim, adding it so that no traces of the forging process are detectable appears rather difficult. The adversary, Eve, will likely have to rely on images taken by Alice that she decided to share with others, for example on her Facebook site. However, the estimation error of the camera fingerprint estimated from such images will contain remnants of the entire noise residual from all images used by Eve. This fact is the basis of the triangle test using which Alice can identify the images that Eve used for her forgery and, in doing so, prove her innocence.

As a final remark, we note that while the reliability of the single image triangle test breaks up at approximately $N = 100$, the test in its pooled form (which is described in [12, 23]) can be applied even when Eve uses a high quality fingerprint estimated from more than 300 images. The reliability in general decreases with increasing ratio $N_c/N$.

The reliability of the triangle test can be somewhat decreased by modifying the fingerprint estimation process to limit outliers in the noise residuals of images used for the estimation [5, 39]. This will generally suppress the remnants of the noise residual used by the triangle test. More extensive tests, however, appear necessary to investigate the overall impact of this counter–counter-counter measure on the reliability of the identification algorithm when the fingerprint is not faked.

## 7 Temporal Forensics

The goal of temporal forensics is to establish causal relationship among two or more objects [42, 44]. In this section, we show how pixel defects can be used to order images by their acquisition time given a set of images from the same camera whose time ordering is known.[2] Even though temporal data is typically found in the EXIF header, it may be lost when processing or editing images offline. Additionally, since the header is easily modifiable, it cannot be trusted. Developing reliable methods for establishing temporal order among individual pieces of evidence is of interest for multiple reasons. Such techniques can help reveal deception attempts of an adversary or a criminal. The causal relationship also provides information about the whereabouts of the photographer.

Strong point defects (hot/stuck pixels) occur randomly in time and space on the sensor independently of each other [13, 36–38], which makes them useful for determining an approximate age of digital photographs. New defects appear suddenly and with a constant rate (It is a Poisson process.), which means that the time among the onsets of new defects follows the exponential distribution. Once a defect occurs, it becomes a permanent part of the sensor and does not heal itself.

The main cause of new pixel defects is environmental stress, primarily due to impacting cosmic rays. In general, smaller pixels are more vulnerable to point defects than larger pixels. Sensors age both at high altitudes and at the sea level. They do age faster at high altitudes or during airplane trips where cosmic radiation is stronger. Consequently, in real life the defect accumulation may not be linear in time.

The main technical problem for using point defects for temporal forensics is that such defects may not be easily detectable in individual images, depending on the image content, camera settings, exposure time, etc. In the next section, we describe a method for estimating pixel defect parameters and use it for determining an approximate age of a digital photograph using the principle of maximum likelihood.

### 7.1 Estimating Point Defects

Even though sensor defects can be easily estimated in a laboratory environment by taking test images under controlled conditions, a forensic analyst must work with a given set of images taken with camera settings that may be quite unfavorable for estimating certain defects. For example, the dark current is difficult to estimate reliably from images of bright scenes taken with a short exposure time and low ISO.

Let $\mathbf{Y}_k$, $k = 1, \ldots, d$, be $d$ images of regular scenes taken at known times $t_1, \ldots, t_d$. As in the previous section, we work with noise residuals $\mathbf{W}_k = \mathbf{Y}_k - F(\mathbf{Y}_k)$ obtained using a denoising filter $F$. Since hot pixels (and stuck pixels with a large offset) are spiky in nature, denoising filters of the type [43] that extract additive white

[2] Temporal ordering of digital images was for the first time considered in [42].

Gaussian noise are likely a poor choice for estimating point defects. Instead, nonlinear filters, such as the median filter, are more likely to extract the spike correctly.

We use the model of pixel output (3):

$$\mathbf{W}_k(i) = \mathbf{I}_k(i)\mathbf{K}(i) + \tau_k \mathbf{D}(i) + \mathbf{c}(i) + \Xi_k(i), \tag{48}$$

where $k$ and $i$ are the image and pixel indices, respectively. For simplicity, we model $\Xi_k(i)$, $k = 1, \ldots, d$, as an i.i.d. Gaussian sequence with zero mean and variance $\sigma^2(i)$. The known quantities in the model are $\mathbf{I}_k \approx F(\mathbf{Y}_k)$ and $\tau_k$; $\mathbf{W}_k$ are observables.

From now on, all derivations will be carried out for a fixed pixel $i$. This will allow us to drop the pixel index and make the expressions more compact. The unknown vector parameter $\boldsymbol{\theta} = (\mathbf{K}, \mathbf{D}, \mathbf{c}, \sigma)$ (for a fixed pixel, $\boldsymbol{\theta} \in \mathbb{R}^4$) can be estimated using the Maximum Likelihood (ML) principle:

$$\hat{\boldsymbol{\theta}} = \underset{\boldsymbol{\theta}}{\arg\max}\, L(\mathbf{W}_1, \ldots, \mathbf{W}_d | \boldsymbol{\theta}), \tag{49}$$

where $L$ is the likelihood function

$$L(\mathbf{W}_1, \ldots, \mathbf{W}_d | \boldsymbol{\theta}) = (2\pi\sigma^2)^{-d/2} \exp\left( -\frac{1}{2\sigma^2} \sum_{k=1}^{d} (\mathbf{W}_k - \mathbf{I}_k\mathbf{K} - \tau_k\mathbf{D} - \mathbf{c})^2 \right). \tag{50}$$

Because the modeling noise $\Xi_k$ is Gaussian, the ML estimator becomes the least squares estimator [31]. Additionally, the model linearity allows us to find the maximum in (54) in the following well-known form:

$$(\hat{\mathbf{K}}, \hat{\mathbf{D}}, \hat{\mathbf{c}}) = (\mathbf{H}\mathbf{H}')^{-1}\mathbf{H}'\mathbf{W}, \tag{51}$$

$$\hat{\sigma}^2 = \frac{1}{d} \sum_{k=1}^{d} \left( \mathbf{W}_k - \mathbf{I}_k\hat{\mathbf{K}} - \tau_k\hat{\mathbf{D}} - \hat{\mathbf{c}} \right)^2, \tag{52}$$

where

$$\mathbf{H} = \begin{pmatrix} \mathbf{I}_1(i) & \tau_1 & 1 \\ \mathbf{I}_2(i) & \tau_2 & 1 \\ \ldots & \ldots & \ldots \\ \mathbf{I}_d(i) & \tau_d & 1 \end{pmatrix} \tag{53}$$

is a matrix of known quantities and $\mathbf{W} = (\mathbf{W}_1, \ldots, \mathbf{W}_d)'$ the vector of observations (noise residuals).

### 7.2 Determining Defect Onset

We now extend the estimator derived above to the case when we have observations (pixel intensities) across some time span during which the pixel becomes defective. Our goal is to estimate $\boldsymbol{\theta}$ before and after the onset, $\boldsymbol{\theta}^{(0)}, \boldsymbol{\theta}^{(1)}$, and the onset time $j$. The estimator derived in the previous section easily extends to this case

$$(\hat{\boldsymbol{\theta}}^{(0)}, \hat{\boldsymbol{\theta}}^{(1)}, \hat{j}) = \arg \max_{(\boldsymbol{\theta}^{(0)}, \boldsymbol{\theta}^{(1)}, j)} L_j(\mathbf{W}_1, \ldots, \mathbf{W}_d | \boldsymbol{\theta}^{(0)}, \boldsymbol{\theta}^{(1)}), \tag{54}$$

where

$$L_j(\mathbf{W}_1, \ldots, \mathbf{W}_d | \boldsymbol{\theta}^{(0)}, \boldsymbol{\theta}^{(1)}) = L(\mathbf{W}_1, \ldots, \mathbf{W}_j | \boldsymbol{\theta}^{(0)}) \times L(\mathbf{W}_{j+1}, \ldots, \mathbf{W}_d | \boldsymbol{\theta}^{(1)}) \tag{55}$$

is the likelihood function written in terms of (50). Because of the form of (55), the maximization in (54) factorizes:

$$(\hat{\boldsymbol{\theta}}^{(0)}, \hat{\boldsymbol{\theta}}^{(1)}, \hat{j}) = \arg \max_j \left\{ \arg \max_{\boldsymbol{\theta}^{(0)}} L(\mathbf{W}_1, \ldots, \mathbf{W}_j | \boldsymbol{\theta}^{(0)}) \times \right.$$
$$\left. \arg \max_{\boldsymbol{\theta}^{(1)}} L(\mathbf{W}_{j+1}, \ldots, \mathbf{W}_d | \boldsymbol{\theta}^{(1)}) \right\}, \tag{56}$$

which converts the problem of onset estimation to the problem of estimating pixel defect addressed in the previous section.

### 7.3 Determining Approximate Acquisition Time

We are now ready to develop the algorithm for placing a given image $\mathbf{I}$ under investigation among other $d$ images, $\mathbf{I}_1, \ldots, \mathbf{I}_d$, whose time of acquisition is *known*, monotone increasing, and whose pixel defects are known, including the onset time and the parameters before and after the onset. This problem is again addressed using the ML approach. This time, only the time index $j$ of $\mathbf{I}$ is the unknown as the parameters of all defective pixels are already known. Denoting the set of all defective pixels $\mathcal{D}$, the estimator becomes:

$$\hat{j} = \arg \max_j \prod_{i \in \mathcal{D}} L(\mathbf{W}_{\mathbf{I}}(i) | \boldsymbol{\theta}^{[j > j(i)]}(i)) \tag{57}$$

written in terms of (50). The superscript of $\boldsymbol{\theta}$ is the Iverson bracket.

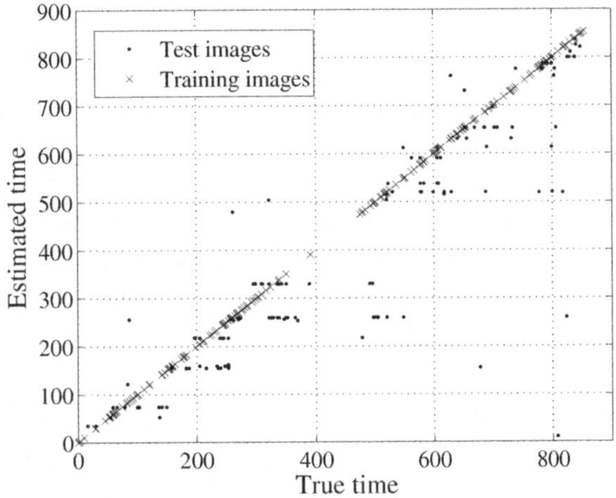


**Fig. 13** True acquisition date versus date estimated using (57) based on detecting pixel defects. The average absolute error between the estimated and true date was 61.56 days

## *7.4 Performance Example*

We applied the approach outlined above to images from a Canon PS sd400 digital camera. A total of $d = 329$ images with known acquisition times spanning almost 900 days were used to estimate the defect parameters for 46 point defects. Then, the acquisition times were estimated for 159 other images. The accuracy with which one can estimate the time obviously depends on how many new defects appear during the entire time span.

The noise residual $\mathbf{W}$ was extracted using a $3 \times 3$ median filter. Figure 13 shows the true date versus the estimated date for all 159 images. The circles on the diagonal correspond to the training set of 329 images with known dates. They show the temporal distribution of training images. Note that no training images appear between time 400 and 500, thus limiting our ability to date images within this time interval.

## *7.5 Confidence Intervals*

In forensic setting, the analyst will likely be interested in statements of the type "the probability that image $\mathbf{I}$ was taken in time interval $[t, t']$ is at least $p$." The approach outlined above allows us to quantify the results in this way because the conditional probabilities $\Pr(\mathbf{W_I}|j) = \prod_{i \in \mathcal{D}} L(\mathbf{W_I}(i)|\theta^{[j>j(i)]}(i))$ are known for each $j$. From the Bayes formula,

$$\Pr(j|\mathbf{W_I}) = \Pr(\mathbf{W_I}|j)\frac{\Pr(j)}{\Pr(\mathbf{W_I})}. \tag{58}$$

Thus, the probability that $\mathbf{I}$ was taken in time interval $[t, t']$ is

$$\Pr(j \in [t, t'] | \mathbf{W_I}) = \frac{\sum_{k \in [t,t']} \Pr(\mathbf{W_I}|k) \Pr(k)}{\sum_k \Pr(\mathbf{W_I}|k) \Pr(k)}. \quad (59)$$

Depending on the situation at hand, the prior probabilities $\Pr(k)$ may be known from other forensic evidence or may be completely unknown. In the absence of any information about the priors, one could resort to the principle of maximum uncertainty and assume the least informative prior distribution—that the owner of the camera was taking images at a uniform rate, leading to $\Pr(k) = 1/(t_k - t_{k-1})$ for all $k$, where $t_k$ is the time when the $k$th training image was taken.

## 8 Summary

Every imaging sensor contains defects and imperfections that can be utilized for a variety of forensic tasks. The photo-response non-uniformity (PRNU) plays the role of a fingerprint that survives processing well and can thus be used for identifying images taken by a camera whose fingerprint is known. Its absence in individual regions testifies about malicious processing applied to an image (forgery detection). The fingerprint can also serve as a template to recover previously applied geometrical processing. The methodology applies to CCD as well as CMOS sensors and to digital still and video cameras as well as scanners [16, 25, 34].

Although it is possible for an adversary to superimpose a camera fingerprint onto an image from a different camera and thus frame an innocent victim (the fingerpint-copy attack), it is not easy to do this without leaving detectable traces. The so-called triangle test can reveal when a fingerprint has been maliciously added and one can even identify the images that the adversary utilized in the attack.

Besides PRNU, sensors also contain the so-called point defects, examples of which are hot and stuck pixels and pixels with abnormal sensitivity. Such defects occur randomly on the sensor and randomly in time. This makes them useful for determining an approximate time when an image was taken. This chapter outlines a maximum-likelihood estimator of time of acquisition that basically detects the presence of point defects with a known onset in an image.

The performance of all forensic methods introduced here is briefly demonstrated on real images. Throughout the text, references to previously published articles guide the interested reader to more detailed technical information.

### *8.1 Related Publications*

When the camera identification technique is deployed on a large scale, one quickly runs into complexity issues. For example, the seemingly routine task of matching an image (or fingerprint) to a large database of fingerprints stored in a database may be quite time consuming already when the database holds merely hundreds of fingerprints. This is because correlating each database fingerprint with the query signal may take hours even in the simplest case when no search for geometrical transformation is carried out. To address this problem, researchers [24] introduced the concept of a fingerprint digest and sparse data structures to cut down the searching time significantly.

A similar task of matching a large database of images to a small number of fingerprints was proposed in [2]. Here, the complexity was decreased using a hierarchical binary search.

For completeness, we note that there exist approaches combining sensor noise defects with machine-learning classification [6, 25, 33]. An older version of this forensic method was tested for cell phone cameras in [6] and in [47] where the authors show that a combination of sensor-based forensic methods with methods that identify camera brand can decrease false alarms. The improvement reported in [47], however, is unlikely to hold for the newer version of the sensor noise forensic method presented in this chapter as the results of [47] appear to be influenced by uncorrected non-unique artifacts discussed in Sect. 3.1. The problem of pairing of a large number of images was studied in [3] using an ad hoc approach. A large-scale experimental evaluation of camera identification on over one million images from over 6,800 cameras covering 150 models appears in [19]. Identification of printed images is the subject of [20]. Anisotropy of image noise for classification of images into scans, digital camera images, and computer art appeared in [33]. The effect of denoising filters on the performance of sensor-based camera identification was studied in [1].

## *8.2 Real-life Applications*

Camera identification based on sensor fingerprints described in this chapter passed the Daubert challenge http://en.wikipedia.org/wiki/Daubert_standard in the State of Alabama in July 2011. In March 2009, Miroslav Goljan testified in Scotland as an expert witness in a high-profile case that involved child abuse crimes by a pedophile ring. See the article "Operation Algebra" at http://p10.hostingprod.com/@spyblog.org.uk/blog/2009/05/. Cases involving movie piracy and illegal acquisition and distribution of movies taped inside a movie theater form another area of possible applications of this technology. The sensor fingerprint can also be used to reverse engineer in-camera geometrical processing, such as digital zoom or correction of lens distortion, and to provide a blind estimate of the focal length at which an image was taken [22].

**Acknowledgments** The work on this chapter was partially supported by the NSF grant number CNF-0830528. The U.S. Government is authorized to reproduce and distribute reprints for Governmental purposes notwithstanding any copyright notation there on. The views and conclusions contained herein are those of the authors and should not be interpreted as necessarily representing the official policies, either expressed or implied, of Air Force Office of Scientific Research or the U.S. Government. The author would like to thank Miroslav Goljan for useful discussions.

## References


1. Amerini I, Caldelli R, Cappellini V, Picchioni F, Piva A (2009) Analysis of denoising filters for photo-response non-uniformity noise extraction in source camera identification. In: Proceedings of the 16th international conference on Digital Signal Processing, pp 511–517
2. Bayram S, Sencar HT, Memon N (2010) Efficient techniques for sensor fingerprint matching in large image and video databases. In: Memon N, Dittmann J (eds) Proceeedings of SPIE electronic media forensics and security XII, vol 7541, pp 09-01–09-12
3. Bloy GJ (March 2008) Blind camera fingerprinting and image clustering. IEEE Transactions on Pattern Analysis and Machine Intelligence 30(3):532–534
4. Böhme R, Kirchner M (2009) Synthesis of color filter array pattern in digital images. In: Memon N, Dittmann J (eds) Proceedings of SPIE media forensics and security XI, vol 7254. San Jose, CA, pp 0K–0L
5. Caldelli R, Amerini I, Novi A (2011) An analysis on attacker actions in fingerprint-copy attack in source camera identification. In: Proceedings of IEEE WIFS, Iguazu Falls, Brazil
6. Çeliktutan O, Avcibas I, Sankur B (2007) Blind identification of cellular phone cameras. In: Delp EJ, Wong PW (eds) Proceedings of SPIE electronic imaging, steganography, security, and watermarking of multimedia contents IX, vol 6505, pp H1–H12
7. Celiktutan OB, Sankur B, Avcibas I (2008) Blind identification of source cell-phone model. IEEE Transactions on Information Forensics and Security 3(3):553–566
8. Chen M, Fridrich J, Goljan M (2007) Digital imaging sensor identification (further study). In: Delp EJ, Wong PW (eds) Proceedings of SPIE electronic imaging, steganography, security, and watermarking of multimedia contents IX, vol 6505, pp 0P–0Q
9. Chen M, Fridrich J, Goljan M (2007) Imaging sensor noise as digital x-ray for revealing forgeries. In: Furon T et al (eds.) Proceedings of ninth information hiding workshop, vol 4567 of LNCS, Springer-Verlag, Saint Malo, France, pp 342–358
10. Chen M, Fridrich J, Goljan M (2007) Source digital camcorder identification using ccd photo response nonuniformity. In: Delp EJ Wong PW (eds) Proceedings of SPIE electronic imaging, steganography, security, and watermarking of multimedia contents IX, vol 6505, pp 1G–1H
11. Chen M, Fridrich J, Goljan M (March 2008) Determining image origin and integrity using sensor noise. IEEE Transactions on Information Forensics and Security 1(1):74–90
12. Chen M, Fridrich J, Goljan M (2010) Defending against fingerprint-copy attack in sensor-based camera identification. IEEE Trans Inform Security Forensics, submitted
13. Dudas J, Wu LM, Jung C, Chapman GH, Koren Z, Koren I (2007) In: Martin RA, DiCarlo JM, Sampat N (eds) Identification of in-field defect development in digital image sensors, vol 6502. SPIE, pp 65020Y
14. Filler T, Fridrich J, Goljan M (2008) Using sensor pattern noise for camera model identification. In: Proceedings of IEEE international conference on image processing (ICIP)
15. El Gamal A, Fowler B, Min H, Liu X (1998) Modeling and estimation of FPN components in CMOS image sensors. In: Proceedings of SPIE solid state sensor arrays: development and applications II, vol 3301-20, pp 168–177
16. Gloe T, Franz E, Winkler A (2007) Forensics for flatbed scanners. In Delp EJ, Wong PW (eds) Proceedings of SPIE electronic imaging, steganography, security, and watermarking of multimedia contents IX, vol 6505, pps 1I–1J
17. Gloe T, Kirchner M, Winkler A, Boehme R (2007) Can we trust digital image forensics? In: Proceedings of the fifteenthth international conference on multimedia, Multimedia '07, ACM, pp 78–86
18. Goljan M, Chen M, Fridrich J (2007) Identifying common source digital camera from image pairs. In: Proceedings of IEEE international conference on image processing (ICIP)
19. Goljan M, Filler T, Fridrich J (2009) Camera identification-large scale test. In: Memon N, Dittmann J (eds) Proceeedings of SPIE electronic media forensics and security XI, vol 7254, pp 0I-01–0I-12

20. Goljan M, Fridrich J (2008) Camera identification from printed images. In: Delp EJ, Wong PW, Memon N, Dittmann J (eds) Proceeedings of SPIE electronic imaging, security, forensics, steganography, and watermarking of multimedia contents X, vol 6819, pp OI1-OI12
21. Goljan M, Fridrich J (2008) Camera identification from scaled and cropped images. In: Delp E, Wong PW, Memon N, Dittmann J (eds) Proceeedings of SPIE electronic imaging, security, forensics, steganography, and watermarking of multimedia contents X, vol 6819, pp OE1-OE13
22. Goljan M, Fridrich J (2012) Sensor-fingerprint based identification of images corrected for lens distortion. In: Memon ND, Alattar A, Delp EJ (eds) Proceedings SPIE electronic imaging, media watermarking, security and forensics January 22–26, 2012
23. Goljan M, Fridrich J, Chen M (2010) Sensor noise camera identification: Countering counter-forensics. In: Memon N, Dittmann J, Alattar A (eds) Proceedings of SPIE electronic imaging, steganography, security, and watermarking of multimedia contents XII, vol 7541, pp 0S-01–0S-12, January 17–21 2010
24. Goljan M, Fridrich J, Filler T (2010) Managing a large database of camera fingerprints. In Memon N, Dittmann J, Alattar A (eds) Proceeedings of SPIE electronic media forensics and security XII, vol 7541, pp 08-01–08-12, January 2010
25. Gou H, Swaminathan A, Wu M (2007) Robust scanner identification based on noise features. In Delp EJ, Wong PW (eds) Proceedings of SPIE electronic imaging, steganography, security, and watermarking of multimedia contents IX, vol 6505, pp 0S–0T
26. Healey G, Kondepudy R. Radiometric CCD camera calibration and noise estimation
27. Healey GE, Kondepudy R (March 1994) Radiometric CCD camera calibration and noise estimation. IEEE Trans Pattern Anal Mach Intell 16(3):267–276
28. Holst GC (1998) CCD Arrays, Cameras, and Displays, 2nd edn. JCD Publishing & SPIE Press, USA
29. Holt CR (1987) Two-channel detectors for arbitrary linear channel distortion. IEEE Trans Acoust Speech Signal Process ASSP-35(3):267–273
30. Janesick JR (2001) Scientific charge-coupled devices, vol PM83. SPIE Press Monograph
31. Kay SM (1998) Fundamentals of statistical signal processing, vol I: Estimation Theory, vol II. Prentice Hall, Upper Saddle River, NJ
32. Kay SM (1998) Fundamentals of statistical signal processing, Detection Theory, vol II. Prentice Hall, Englewood Cliffs
33. Khanna N, Mikkilineni AK, Chiu GTC, Allebach JP, Delp EJ (2007) Forensic classification of imaging sensor types. In: Delp EJ, Wong PW (eds) Proceedings of SPIE electronic imaging, steganography, security, and watermarking of multimedia contents IX, vol 6505, pp U1–U9
34. Khanna N, Mikkilineni AK, Chiu GTC, Allebach JP, Delp EJ Scanner identification using sensor pattern noise. In: Delp EJ, Wong PW (eds) Proceedings of SPIE electronic imaging, security, steganography, and watermarking of multimedia contents IX, vol 6505
35. Kurosawa K, Kuroki K, Saitoh N (1999) CCD fingerprint method - identification of a video camera from videotaped images. In: Proceedings of IEEE international conference on image processing (ICIP), pp 537–540
36. Leung J, Chapman GH, Koren I, Koren Z (2008) Automatic detection of in-field defect growth in image sensors. In: DFT '08: Proceedings of the 2008 IEEE international symposium on defect and fault tolerance of VLSI systems, IEEE Computer Society, Washington, DC, pp 305–313
37. Leung J, Chapman GH, Koren I, Koren Z (2009) Characterization of gain enhanced in-field defects in digital imagers. In: IEEE international symposium on defect and fault-tolerance in VLSI systems, pp 155–163
38. Leung J, Dudas J, Chapman GH, Koren I, Koren Z (2007) Quantitative analysis of in-field defects in image sensor arrays. In: 22nd IEEE international symposium on defect and fault-tolerance in VLSI systems, 2007. DFT '07, pp 526–534
39. Li C-T (2009) Source camera identification using enhanced sensor pattern noise. In: Proceedings of IEEE ICIP, pp 7–11
40. Lukáš J, Fridrich J, Goljan M (2005) Determining digital image origin using sensor imperfections. In: Delp EJ, Wong PW (eds) Proceedings of SPIE, electronic imaging, security, steganography, and watermarking of multimedia contents VII, San Jose, CA, pp 249–260

41. Lukáš J, Fridrich J, Goljan M (June 2006) Digital camera identification from sensor pattern noise. IEEE Transactions on Information Forensics and Security 1(2):205–214
42. Mao J, Bulan O, Sharma G, Datta S (2009) Device temporal forensics: an information theoretic approach. In: Proceedings of IEEE international conference on image processing (ICIP), vol 1
43. Mihcak MK, Kozintsev I, Ramchandran K, Moulin P (1999) Low-complexity image denoising based on statistical modeling of wavelet coefficients. IEEE Signal Process Lett 6(12):300–303
44. Rosa AD, Uccheddu F, Costanzo A, Piva A, Barni M (2010) Exploring image dependencies: a new challange in image forensics. In: Memon N, Dittmann J, Alattar A (eds) Proceeedings of SPIE electronic media forensics and security XII, vol 7541, pp 0X-1–0X-12
45. Rosenfeld K, Sencar HT (2009) A study of the robustness of PRNU-based camera identification. In : Memon N, Dittmann J, Alattar A (eds) Proceedings of SPIE electronic imaging, steganography, security, and watermarking of multimedia contents XI, vol 7254, pp 0M–0N
46. Steinebach M, Liu H, Fan P, Katzenbeisser S (2010) Cell phone camera ballistics: attacks and countermeasures. In: Proceedings of SPIE, multimedia on mobile devices 2010, vol 7542, San Jose CA, pp 0B–0C
47. Sutcu Y, Bayram S, Sencar HT, Memon N (2007) Improvements on sensor noise based source camera identification. In: IEEE international conference on multimedia and expo, pp 24–27

# Source Attribution Based on Physical Defects in Light Path

**Ahmet Emir Dirik**

**Abstract** Recent studies in multimedia forensics show that digital images contain intrinsic patterns, traces, and marks generated by imaging pipeline components (sensor) and processes (demosaicing and color adjustment). Some of these patterns and marks, such as photo response non-uniformity noise (PRNU), are unique to individual component characteristics of imaging system. Similar to PRNU noise, physical defects in imaging pipeline such as dust particles in DSLR camera chamber, scratches on flatbed scanners also generate unique patterns in image output. Due to unique and random nature of these patterns, they can be utilized in digital image forensics problems. In this chapter, we will give an overview of state-of-the-art camera identification techniques which utilize such defects and patterns.

## 1 Introduction

Frequent and widespread use of digital images in our daily lives brings some serious issues and problems such as authenticity and integrity of digital images. Today, digital images are commonly used in military, finance, mainstream media, medical records, etc. They are also presented in court proceedings as legal evidence. When a digital image is introduced as evidence in a court proceeding, its veracity and authenticity should be guaranteed with reliable tools and methods. In this context, establishing a link between a digital image and its source (camera, scanner, smart phone, etc) is vital to assess the authenticity of an image before admitting as legal evidence.

Up to now, several blind forensic techniques (not requiring any watermark) have been proposed to identify image source class, model, or device. Most of these techniques are based on intrinsic component characteristics in image pipeline, sensor

A. E. Dirik (✉)
Faculty of Engineering and Architecture,
Department of Electronics Engineering, Uludag University,
16059 Nilufer, Bursa, Turkey
e-mail: edirik@uludag.edu.tr

H. T. Sencar and N. Memon (eds.), *Digital Image Forensics*,
DOI: 10.1007/978-1-4614-0757-7_7,

defects, and sensor imperfections [1, 2, 8–10, 16, 17]. In recent years, alternative individual source identification approaches have also been proposed based on physical defects in imaging system [4–6]. Here, the term "individual" refers to exact source "device", not model, brand, or class of imaging device. "Physical defects" refer to physical changes and imperfections caused by misuse and/or aging in image pipeline components such as dust contamination over lens, optical sensor, [4, 5] or platen of flatbed scanners [6].

Recent studies in multimedia forensics show that digital images contain intrinsic patterns, traces, and marks generated by imaging pipeline components (sensor) and processes (demosaicing and color adjustment) [13]. Some of these patterns and marks, such as photo response non-uniformity noise (PRNU) [2], are unique to individual component characteristics of imaging system. Due to their consistent nature, these marks and patterns can be used as a natural camera fingerprint to identify individual image source device. Similar to PRNU noise, dust particles in DSLR camera chamber, scratches on flatbed scanners also generate unique patterns which are suitable to be used in individual camera identification.

For an accurate and reliable source camera identification, decision error rates (false alarm and miss rate) should be in acceptable ranges. False alarm in source identification refers to the case where an image is matched with another camera which is not the original source device. Miss rate gives the probability of not identifying original source when an image is questioned with its exact source device. In individual source identification, there are two main challenges: First, imperfections and defects in imaging pipeline components cause minute differences in image output. Detection of these artifacts is challenging especially for busy image content. Second, these minute artifacts and their characteristics may not be consistent in nature and vary in time and excessive camera use. Thus, for an accurate and reliable source identification, these issues should be taken into account.

In this chapter, we will give an overview of state-of-the-art camera identification techniques which utilize physical defects as intrinsic device fingerprint. The organization of this chapter is as follows: Sensor dust problem in DSLR cameras is introduced and studied in Sect. 2. Recent image forensic techniques which utilize sensor dust traces are introduced in the same section. Section 3 studies the individual scanner identification based on platen scratches. Finally, Sect. 4 concludes this chapter.

## 2 Forensic Use of Sensor Dust

### *2.1 Dust Problem*

In recent years DSLR cameras showed a significant growth in digital camera market. Despite their large dimensions and weights, users tend to buy DSLR cameras due to their attractive features such as noticeably good image quality, error-free

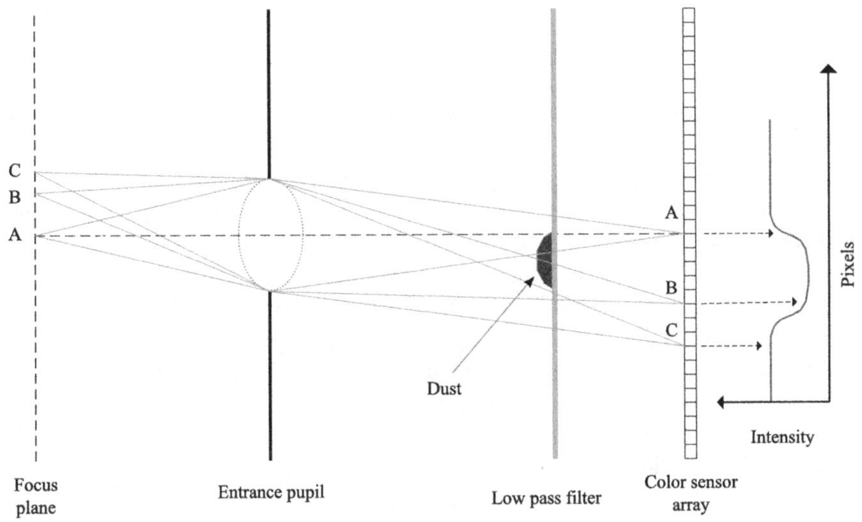


**Fig. 1** Intensity loss due to the dust in DSLR chamber and dust spot formation

viewing of the scenery, less shutter lag, good depth-of-view control, and interchangeable lenses. Today, not surprisingly, DSLR cameras take the highest places in most popular camera ranking lists reported in popular image sharing websites such as www.flickr.com.

Different from compact cameras, lenses of DSLR can be detached and interchanged with other lenses. Nevertheless, this popular feature brings a serious issue. During the mounting/demounting process of DSLR lens, the inner chamber of DLSR is exposed to the outside environment. When DSLR lens is removed, dust, dirt, and other small particles in the air are attracted to the sensor element due to electrostatic forces inside of the DSLR chamber. These attracted particles settle on the protective component (dichroic mirror or low-pass filter) in front of the optical sensor surface. Surprisingly, these small specks of dirt, dust, lint, or hair remain on the sensor without changing their positions unless the sensor surface is cleaned out physically. These dust particles with their fixed positions compromise the image quality and they appear as dark blemishes and blotches in taken images. As a general term we will refer these artifacts as *dust spots* in the rest of the chapter.

Dust and dirt in front of the sensor obstruct the incident light and absorb most of the light causing local intensity degradations at the same positions in every taken image (Fig. 1). The amount of intensity degradation on the picture depends on both dust particle and optical camera parameters such as focal length, aperture, etc. Blemishes in taken images due to sensor dust become noticeable especially for low apertures and high f-number settings (F/#). F-number is the ratio between the camera focal length and the aperture. For low apertures, dust spots become softer and less visible.

Today, most DSLR cameras are equipped with internal anti-dust mechanisms. Generally, these mechanisms are attached to the optical sensor and they shake dust particles by vibrations and/or ultrasonic waves. For instance, some DSLR cameras, such as Sony's Alpha A10, use an anti-dust coating on the sensor element with a vibrating mechanism to prevent dust contamination. However, anti-dust systems do not work satisfactorily and do not remove dust and dirt from the sensor surface completely. Popular DSLR manufacturers, e.g., Nikon, Canon also offer software tools to remove dust spots as a complementary solution to dust problem.

Although vibration-based dust removal mechanisms cannot clean the sensor surface completely, they can move dust particles and affect their positions. However, in practice dust shifts due to vibrations are minute. For instance, after running the built-in cleaning mechanism of Canon EOS-40D 25th times, 97% of dust particles remained at their positions and the maximum dust spot position shift was observed as 5.83 pixels for $800 \times 532$ pixels resolution [5]. We can infer from these results that the effect of vibration mechanisms on dust positions can be omitted without any significant error on dust positions.

Although the main cause of sensor dust problem is interchanging lenses, in some cases, brand new cameras can be shipped with sensor dust. Thus, even DSLR users who do not detach their lenses may notice dust specks in their images taken with DSLR. Besides, non-professional DSLR users may not even know whether their sensor has dirt on them if dust particles are tiny and not many.

Sensor dust can be cleaned physically by using air blower, wiping it with alcohol, or other chemical solutions. Nevertheless, careless wiping can damage the sensor. Thus, it is not recommended for nonprofessionals. Another alternative to clean out dust spots is to locate them visually in image and paint them with an image editing software using cloning tool. This is the least effective dust cleaning method and should be performed manually for every image. Since physical dust removal is a cumbersome process, it is mostly preferred by professionals. It is also not practical to detect and clean dust spots manually in hundreds or thousands of images. Even dust spots can be removed with cloning tool, there is still a chance to determine deleted dust spots by searching cloning and copy-move operation traces with image forgery detection techniques [7, 12]. It should be noted that cloning tool can fail in irregular backgrounds and highly textured content.

Although dust spots in images are annoying for users, they generate an intrinsic pattern unique to DSLR camera. Intrinsic dust spot pattern is random in nature and do not change with time unless the sensor is cleaned physically [5]. Thus, sensor dust pattern can be considered as an intrinsic camera fingerprint and can be used in forensic investigations to keep track of original camera source from a given image. However, the lack of sensor dust pattern does not always mean the image is not taken with the questioned camera.

The effects of dust over optical lens to the image output were investigated in [18]. Willson et al. in [18] introduced an optical model which simulates the size and appearance of dust spots caused by dust and dirt in front of optical lens. The introduced optical artifact model was tested on machine vision system used in 2003 Mars Exploration Rovers. Experimental results given in the paper showed that using camera parameters (focal length, f-number, distance between lens and entrance pupil) it is possible to model dust artifacts realistically for particles with various opacity values.

The proposed model in [18] can be used to simulate the effect of illuminated dusts which appear as bright spots in image output. Such bright artifacts can be seen in images acquired with flatbed scanners as tiny bright spots due to glass scratches over scanner platen. These bright spots can be used in scanner device identification, since the relative positions of these glass scratches are fixed and do not vary by time [6].

## *2.2 Dust Spot Modeling and Removal*

Modeling and detection of dust spots in images is essential for forensic analysis of dust spots and its use in source identification. It is also important for users and camera manufacturers to remove annoying dust artifacts in taken images. In recent years, several techniques have been proposed to model, detect, and remove sensor dust artifacts in images taken with DSLR cameras [3, 14, 15, 19, 20]. Although the authors in these papers did not discuss the forensic use of sensor dust artifacts, the proposed optical dust models and dust spot detection techniques can be utilized in forensic analysis and investigations.

Zhou et al. in [20] proposed a dust artifact model utilizing contextual information and color consistency in dust spot regions for automatic dust detection and removal. Since sensor dust stays on the low-pass filter component in front of the sensor, there is a distance between dust particle and sensor array. Due to this small distance, size, position, and appearance of dust spots (the shadow of the dust on the sensor) vary depending on camera settings. The formation of dust spot not only depends on its size and opacity but also on image geometry of DSLR. In general, dust spots reveal themselves as dark blemishes due to the attenuation of scene irradiance at dust locations. One of the main challenges in dust spot detection is determining whether a dark spot in a given image is sensor dust or it is a part of the image content. In [20], the proposed dust detection method does not make hard decision on the presence of dust spots; instead it outputs a list of likely candidate dust spots with specific metrics [20]. These metrics are utilized to evaluate how much a given dark spot resembles an actual sensor dust. Sorting out the dust metrics values it is possible to eliminate false dust detections.

In another application, dust detection algorithm introduced in [20] can be run automatically on a large image database. Coinciding dust spot positions detected in several images can be tagged to find images taken with the same DSLR.

Dust spot formation in [20] is modeled as follows: Dust in front of imaging sensor is assumed as a light attenuator which degrades the image quality according to:

$$I(x) = U(x) \times \alpha(x) \tag{1}$$

where, $I(x)$ is the image intensity at position $x$, $U(x)$ is the original intensity value before light attenuation due to sensor dust, and $\alpha(x)$ is the attenuation factor of sensor dust at $x$. Light diffractions and scattering are assumed as little and omitted in dust spot modeling. In the formula above, $\alpha(x)$ is related to the transparency ($\beta$) of sensor dust at position $x$ and this is dependent with camera settings such as f-number and camera aperture. Let the distance between dust and optical sensor be $d$ and camera focal length be denoted with $f$. The relation between dust transparency function $\beta$ and intensity degradation function $\alpha(x)$ is modeled with convolution operator as:

$$\alpha = \beta * \Pi_{\tau} \tag{2}$$

where, $\Pi_{\tau}$ is the rectangle function which models the smooth transitions on dust edges. The size and shape of dust spot vary with camera parameters. This relation is modeled with the scale of the rectangle function $\tau$, which is defined as:

$$\tau = \frac{d \times A}{f} \tag{3}$$

where $A$ is aperture and $\frac{f}{A}$ is the camera's f-number, which is generally recorded in EXIF field. The function space of dust artifact model $\alpha$ depending on $\tau$ is expressed as:

$$S_{\tau} = \{\alpha : \beta * \Pi_{\tau} = \alpha, 0 \leq \beta \leq 1\} \tag{4}$$

Hence, the image intensity at dust region $D$ can be written as:

$$I_D = U_D \times (\beta * \Pi_{\tau}) + n \tag{5}$$

where $n$ is additive image noise. Here, $U_D$ refers to the dust-free appearance of taken image. In [20], authors also utilize contextual information outside the dust region to estimate $U_D$ and remove dust artifact. Dust detection algorithm proposed in [20] uses intensity attenuation factors to locate likely dust spots. However, false positives of dust spot detection is relatively high. Thus, after dust spot detection, a list of candidate dust positions are shown to users and let users select dust spots for image cleaning. It is also stated in [20] that unknown camera parameters such as $d$ can be estimated more accurately by using large number of images in dust analysis. Hence a more reliable candidate dust spot detection can be achieved.

In [3], authors proposed a method to model sensor dust appearance in images taken with DSLR. The authors also introduced and evaluated a technique to detect and correct dust spots. In [3], effects of dust artifacts and related optical modeling was given as follows: Let $(u, v)$ represent the coordinate system of input surface and $(x, y)$ denote the coordinate system of image plane. The input plane is the surface of low-pass filter component where dust and dirt are attracted on it. Let the width of filter component be $t_w$ and the distance between exit pupil and sensor array be $P_e$. Let the principal ray of the beam intersect the input surface at $(u, v)$ and reach to the position $(x, y)$ on image plane. The relation between $(u, v)$ and $(x, y)$ can be written as:

$$x = (1 - \frac{t_w}{P_e}) \times u \tag{6}$$

$$y = (1 - \frac{t_w}{P_e}) \times v \tag{7}$$

It is assumed in this model that the ratio of $(t_w/P_e)$ is considerably small. Let the scene irradiance at $(x, y)$ coordinate on sensor array be $I(x, y)$ for clean sensor surface case. It is assumed that dust and dirt on optical sensor completely absorbs and reflects the incoming beam. In other words, the transparency of dust particles is omitted. Let the image intensity at position $(x, y)$ be degraded by the presence of a dust particle. Thus, the irradiance at $(x, y)$ can be written as:

$$I'(x, y) = I(x, y) \times \frac{A_{\text{total}} - A_{\text{blocked}}}{A_{\text{total}}} \tag{8}$$

where $A_{\text{total}}$ is the area of the light beam cone intersecting with the dust particle and $A_{\text{blocked}}$ is the area of the dust region which blocks the light beam cone converging to the image point. The area $A_{\text{total}}$ parameter and pixel values at dust region depend on lens f-number $f/\#$ and low-pass filter component $t_w$. In general, the intersection of the conic beam with dust surface is an ellipse. The intersection area of the beam cone is directly proportional to the camera aperture. Thus, for large apertures, dust spots become more blurry and less prominent.

The ratio of $t_w/P_e$ determines the position of the dust spot. Since dust artifacts are shadows of dust particles on filter component, their positions are dependent on camera parameters $P_e$ and $t_w$. This relation can be formulated as:

$$h = \frac{h_0}{1 + \frac{t_w}{P_e}} \cong h0 = 1 - \frac{t_w}{P_e} \tag{9}$$

where $h_0$ is the actual position of dust and $h$ is the position of dust shadow in image plane. For telecentric illumination $P_e$ is infinitive and $h$ converges to $h_0$. In this formulation the value of exit pupil distance $P_e$ varies with different lens and f-numbers, whereas the filter width is fixed for a particular type of DSLR.

In [3] two techniques were introduced to determine dust locations of a given DSLR. The first technique involves taking a special photograph which makes dust spots appear sharp and dark in a flat background. The second technique utilizes a batch of images to determine dust positions employing summation and averaging of taken images by the same DSLR. Corcoran et al. also reported that the combination of these two methods could be employed together to find dust spot locations more accurately. These two techniques assume that DSLR camera is available at hand and camera settings such as f-number can be obtained reliably from EXIF.

Extending the work in [3], authors in [19] studied sensor dust problem and proposed an optical model for the appearance (size, position, shape, and transparency)

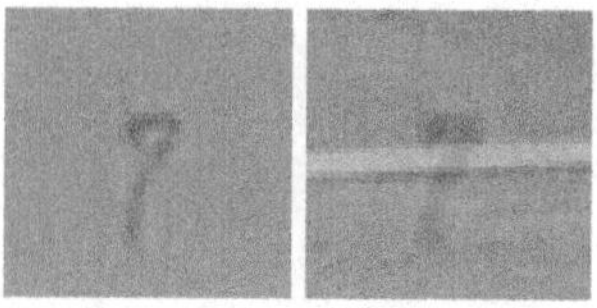


**Fig. 2** hair/lint artifacts in images taken with different f-numbers (Nikon D50)

of dust particles utilizing camera parameters. Based on this model, Zamfir et al. proposed a dust spot removal algorithm which involves estimation of DSLR optical parameters. To achieve camera parameter estimation, several calibration images should be taken with same DSLR under specific conditions. Considering optical parameters of DSLR and estimated positions, shapes, and transparency values of dust spots, dust spot removal is realized. It is stated in [19] that dust shadow positions can vary in different images due to different camera settings. These shifts are dependent on the exit pupil distance, the width of the filter component, and the distance of the dust spot to the center of the image. Dust spot shifts due to different camera settings are more noticeable at regions close to the image corners than the regions close to the image center. It is seen from the experimental results in [19] that the relation between relative dust shift and focal length value is almost linear.

In the presence of DSLR camera, dust spot positions can be determined reliably by taking calibration images under special settings. However, if DSLR is not available and given images are taken with large apertures it is hard to notice dust spots in images. In [4, 5], the authors proposed a technique to highlight dust spots for visual examinations when they are not noticeable easily. To make dust spots more apparent, the image is applied to local histogram equalization within a fixed window size. After local histogram equalization is applied, tiny dark spots which are noticeably different from their backgrounds in the resulting image can be considered as dust spots.

## 2.3 Dust Spot Characteristics

Tiny dust particles in front of DSLR sensor generally appear as round blemishes in images (Fig. 4). Due to their dark color and unique shapes large particles in front of sensor are easily noticeable and more likely to be cleaned out either physically or manually by software. Considering this hypothesis, Dirik et al. developed a dust spot model in [5] with the assumption that most of dust spots are tiny and have round shapes. In this model blemishes in images caused by large particles such as hair or lint (see Fig. 2) are omitted.

The optical dust spot formation model used in [5] is depicted in Fig. 3. It should be noted that the dust spot in the figure has a circular shape. In the Fig. 3, $A$, $f$, $t$, $D$, and $S$ are aperture, focal length, low-pass filter width, dust diameter, and dust shadow diameter, respectively. The size of the dust shadow $S$ is depicted with $s_0$

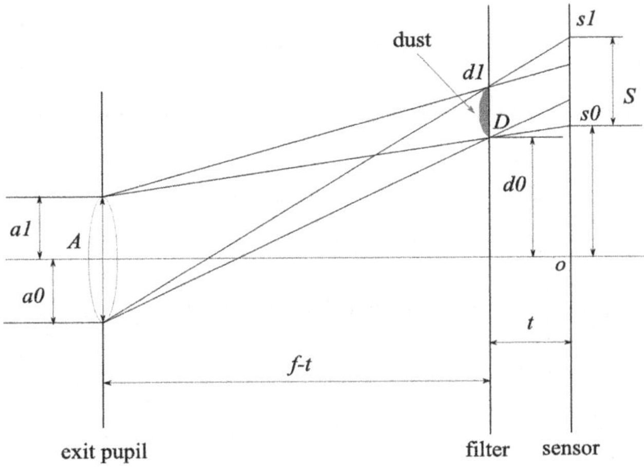


**Fig. 3** Optical dust spot model

**Fig. 4** Same dust spot for different f-numbers, focal length is fixed to 55 mm (Nikon D50)

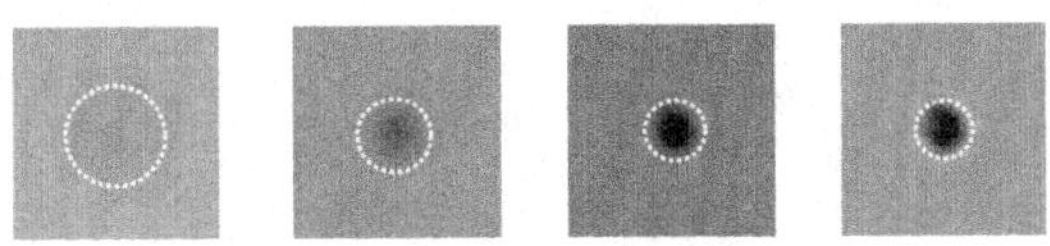


and $s_1$ points. Let the diameter of actual dust $D$ be defined with $d_0$ and $d_1$ points. Using the similarity of triangles $s_0$ and $s_1$ can be computed in terms of aperture, focal length, and position of actual dust measured from the optical image center $o$. Considering the circular dust shape, $D$ and $S$ can be written as $d_1 - d_0$ and $s_1 - s_0$, respectively. Hence, dust shadow position in terms of $s$ points becomes:

$$s_0 = \frac{f \times d_0 - t \times a_1}{f - t} \tag{10}$$

$$s_1 = \frac{f \times d_1 + t \times a_0}{f - t} \tag{11}$$

Hence, dust spot size in the image becomes:

$$S = D\frac{f}{f - t} + A\frac{t}{f - t} \tag{12}$$

It is seen from (12) that there is a direct relation between dust spot size and DSLR aperture $A$. The relation between dust spot shape and camera aperture can also be seen in Fig. 4, where same dust spot is shown with four different apertures (f-numbers). In the figure, the focal length is fixed to 55 mm. As it is seen from the Fig. 4, large

**Table 1** Same dust with different f-numbers (image resolution: 1504 × 1000 pixels)

| F-number | Aperture (mm) | Max. intensity degradation | Dust spot diameter (pixels) |
|---|---|---|---|
| 5.6 | 9.82 | 11 | 20 |
| 10 | 5.5 | 46 | 14 |
| 20 | 2.75 | 61 | 11 |
| 36 | 1.53 | 83 | 10 |

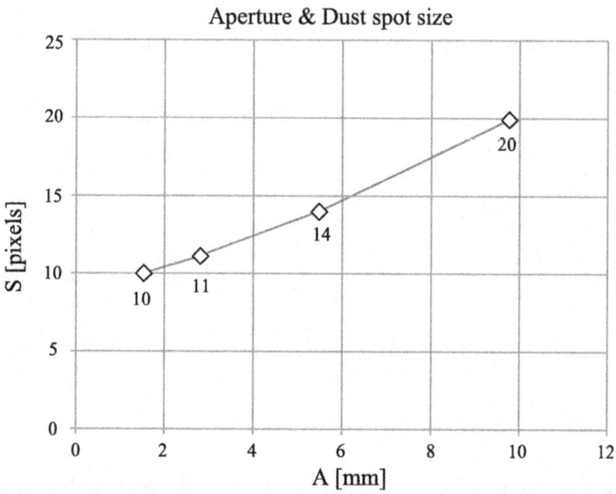


**Fig. 5** Size of dust spots for different f-numbers

f-numbers and small apertures cause sharp and strong dust shadows because at small apertures light source can be assumed as a pin point source yielding a relatively narrow light cone and most of the light energy in the light cone is blocked by dust particles. On the other hand, for large aperture and low f-numbers, the light cone becomes wider and most of the light beams pass around the particles causing blurry, faint, and a larger dust blemish in the image. Dust spot properties and corresponding DSLR camera parameters are also given in Table 1. The graphical representation of table is provided in Fig. 5.

In [5], the position of dust spot (dust shadow) is modeled as follows: Let the center of circular dust spot be $\rho = \frac{s_0+s_1}{2}$. Substituting the (10) and (11) into the central point definition, $\rho$ is obtained as:

$$\rho = \frac{f(d_0 + d_1) + t(a_0 - a_1)}{2(f - t)} \tag{13}$$

$$\rho = \frac{d_0 + d_1}{2} \frac{f}{f - t} \tag{14}$$

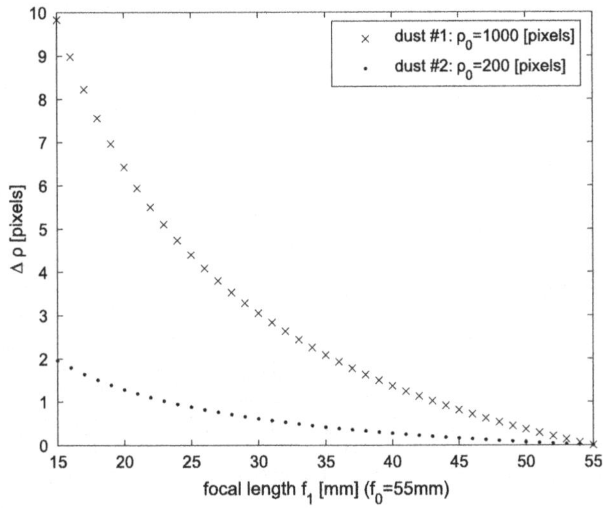


**Fig. 6** Radial shifts of two dust spots. $t$ and $f_0$ are fixed. $f_1$ is changed

where, $a_0$ and $a_1$ gives the aperture size of DSLR. It is seen from (14) that dust spot position is dependent with focal length but not aperture. Indeed, in Fig. 4, dust spots positions remain same with different aperture settings. However, for different f-numbers, dust spot positions may shift little. Let dust spot shift $\Delta\rho$ be the distance between the centers of two same dust spots in two different images taken with different DSLR settings. The shift is defined as: $\Delta\rho = \rho_1 - \rho_0$, where $\rho_0$ and $\rho_1$ are the central points of dust spots $f_0$ and $f_1$, respectively. Substituting (14) into dust shift definition, we get:

$$\Delta\rho = \frac{(f_0 - f_1)t}{(f_1 - t)(f_0 - t)} \frac{d_0 + d_1}{2} \tag{15}$$

$$\Delta\rho = \frac{f_0 - f_1}{f_0(f_1 - t)} \times t \times \rho_0 \tag{16}$$

It is seen in (16), the amount of dust spot shift is directly proportional with the filter width $t$ and the radial position of the dust particle. To visualize the relation between dust spot shift and camera parameters ($f$ and $t$), (16) was evaluated for different $f$ and $t$ values. The resulting shift values are depicted in Fig. 6 and Fig. 7.

The amount of dust spot shift due to focal length change is also dependent on dust particle position. The farther the dust particle ($d_0$) is from optical center (0), the larger the dust spot shifts along the radial axis. It should be noted that shift vectors lie along the radial axes (see Fig. 8). Actual dust shift observation results given in Table 2 also agree with the proposed dust shift model (Eq. 16). The measurements in Table 2 were obtained with a Nikon D50. To measure dust spot shifts, two pictures (with 1504 × 1000 resolution) were taken at focal lengths of 18 mm (F/18) and

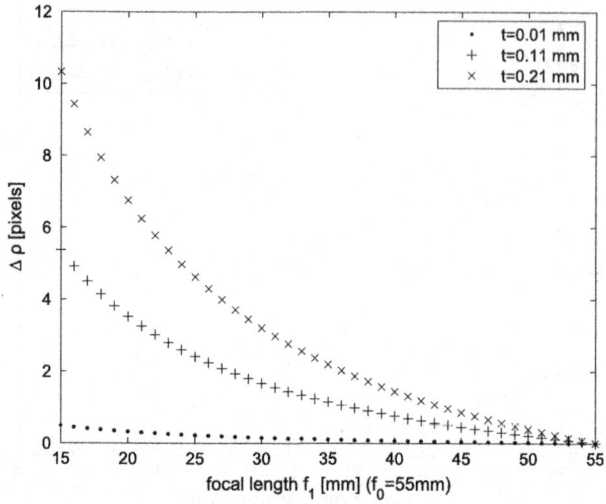


**Fig. 7** Dust spot shifts for different filter width ($t$) values

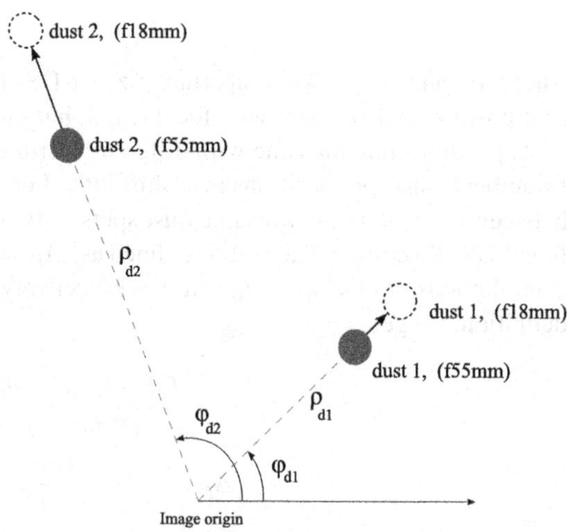


**Fig. 8** Dust spot (shadow) shifts due to focal length $f$ change

55 mm (F/36). In these two images, 4 dust spots were tagged and their positions were computed in polar coordinates.

## 2.4 Forensic Use of Dust Spots

Dust spots in images can be potentially useful to forensic investigators in several aspects such that image source type and individual image source device can be

**Table 2** Dust spot shift observations for different focal lengths, (Nikon D50).

| Dust no | 1st position $f = 55$ mm | | 2nd position $f = 18$ mm | | Shift vector | |
|---|---|---|---|---|---|---|
| | $\rho$ (pixels) | $\varphi$ (deg.) | $\rho$ (pixels) | $\varphi$ (deg.) | $\Delta\rho$ | $\Delta\varphi$ |
| $dust_1$ | 389.9 | 80.5 | 392.8 | 80.5 | 3.0 | 0.1 |
| $dust_2$ | 451.3 | 72.5 | 457.0 | 72.4 | 5.7 | −0.2 |
| $dust_3$ | 617.2 | 50.4 | 626.6 | 50.5 | 9.4 | 0.1 |
| $dust_4$ | 817.7 | 147.7 | 830.0 | 147.7 | 12.2 | 0.0 |

identified based on dust marks. If optical sensor is not cleaned physically, it may accumulate additional dust and dirt particles. New dust particles do not affect the positions of old dust and dirt over the sensor. Therefore, this dust accumulation can be used to date images taken with DSLR since new dust particles should not appear in old photographs while the old ones should be seen in new images taken with the same DSLR. Since most people do not take sensor dust as a serious problem and let optical sensor accumulate dust, old dust positions will remain unchanged for a long time and can be used by forensic investigators to assess authenticity of images taken with DSLR.

In general, sensor dust traces has a potential to answer the following questions in forensic examinations:

- Was given image X taken with a DSLR camera or not?
- Did image X originate from DSLR source Y as claimed?
- Are the date and the time of image X in a group of images consistent with dust contamination history?
- Has image X undergone any digital tampering or not?

In this context, Dirik et al. proposed an individual source camera identification method in [4]. The proposed forensic method in [4] relied on several assumptions such as tiny dust specks can be modeled with round and dark blemishes. Based on these assumptions, the authors proposed a dust detection technique using match filtering and round dust contour analysis. With the help of proposed dust spot detection algorithm, reference dust spot patterns of three DSLR cameras were generated. Then, these dust spot patterns were tested on 20 images taken with same DSLR and 60 images taken with other cameras. In the paper, the authors reported 92% average camera identification accuracy with no false positives.

Sensor dust problem and its use for forensic identification of image source camera were also investigated by Oliver in [11]. In this paper, the author discussed the question whether dust marks can be used in source camera identification. With the light of this question, dust spot appearance and its relation with several camera settings were studied. Based on these investigations, the author proposed a dust pattern matching technique and analyzed the identification accuracy in terms of false positive rate.

In [5], previous source camera identification work in [4] has been extended. Similar to [4], the proposed source identification technique in [5] has two main steps: dust pattern generation and image-DSLR source matching. In both of these steps, dust spot detection was employed to determine likely dust spots. In dust spot detection, intensity loss due to dust particles was modeled as a 3D bell shape. Considering this model, dust spots were detected searching a gaussian dust pattern in a given image by normalized cross correlation. As it is expected, some dark regions (not dust spots) of the image content could be tagged as dust spots. Thus, to reduce the false positive rate and increase the camera identification accuracy, dust pattern matching algorithm in [5] was devised as follows:

After the locations of likely dust spots are found, the locations are compared with the actual dust spot locations in camera dust pattern. If DSLR camera is available, the camera dust pattern can be generated easily by searching gaussian dust spot model over blank calibration images (the calibration image can be a plain background or a clear sky without any cloud). Once DSLR camera dust pattern is generated, dust spot locations in the pattern are compared with detected dust spot positions in a given image (see Fig. 9). To reduce false positives, each matched dust spot is then analyzed with three metrics. These metrics together validate the result of dust detection algorithm. These metrics are as follows:

- *Dust occurrence metric,* $v_1$, checks if there is a strong and noticeable dust mark at that particular region. Higher values of $v_1$ refer to salient dust spots.
- *Smoothness metric,* $v_2$, examines the smoothness of the region at which a dust spot is detected. Since busy regions can yield false detections, dust spots in flat image parts are given more weights than detected dusts in complex regions.
- *Shift validity metric,* $v_3$, measures the validity of dust spot shift. If there is a shift in detected dust spot position, it should be along radial axis. Besides, its orientation and magnitude should be consistent with dust shift model. If a computed shift vector is away from the radial axis the detected dust spot is ignored and not used for source matching.

These three metrics are computed for each dust spot matched with dust template and then they are combined to compute over all dust spot match confidence metric as:

$$C = \sum_{i=1}^{N} f(v_1(i)/\alpha) + v_2(i) + v_3(i) \times u(N-2) \tag{17}$$

$$u(x) = \begin{cases} 0, & x < 0; \\ 1, & x \geq 0. \end{cases} \tag{18}$$

where $N$ is the number of matched dust spots with dust template, $f$ is a normalization function (Gauss error function), $\alpha$ is a scaling parameter, and $u$ is a step function defined in (18), respectively. The higher the confidence value $C$ is,

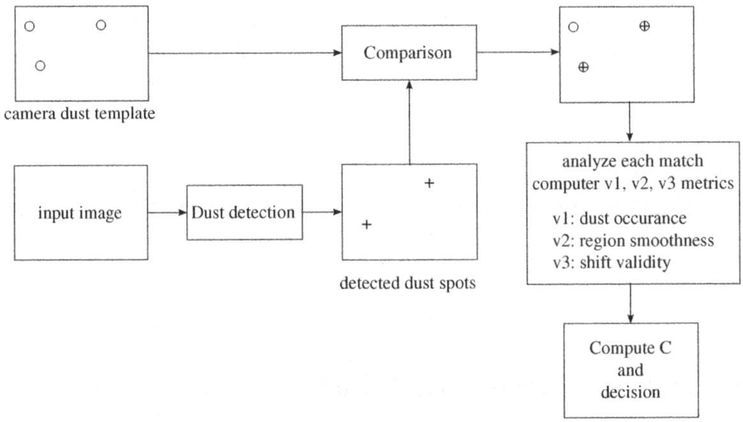


**Fig. 9** Source identification with sensor dust

the more likely the given image is taken with the questioned DSLR. On the other hand, low confidence value does not imply that the image is not taken with the tested DSLR.

In the experiments given in [5], dust spot/dirt positions over imaging sensor were determined using the Gaussian dust spot model for two DSLR cameras (Nikon D50 and Canon EOS digital rebel). As a result, two DSLR fingerprints were generated. Then, 100 images were taken with two DSLR cameras to create genuine image sets. To estimate false positives, 1000 images taken with different cameras were used in the experiments. For each image in genuine and other image sets, dust spot positions were detected and compared with two camera's fingerprints. DLSR camera identification accuracies for this experiment were 0.963 for Nikon D50 and 0.991 for Canon EOS, respectively and the corresponding false positive rate was 0.002. It was also reported in [5] that the proposed source identification method was robust to image resizing and Jpeg compression.

If questioned image is geometrically transformed, it is also possible to estimate geometrical transform parameters using dust spot graph information. The most challenging issue here is that dust spots become almost invisible for low f-number settings. It is also hard to detect dust spots in highly textured backgrounds. These are still open questions in the digital forensic literature.

In a different forensic scenario, images taken with DSLR can be in printed form. In this case, sensor dust-based source identification proposed in [5] can be applied to the printed images. For source identification, first printed photographs are scanned with a proper resolution. Once they are scanned, dust spot detection and fingerprint matching can be employed. Comparing dust spot positions detected in printed image with questioned camera's dust spot pattern, printed image authentication can be realized. Here, we make an assumption that when an image taken with DSLR is printed out, dust spots in the original image will likely be available in the printed photograph. In this approach, it is likely that printed and scanned image pixel positions do not match with its original form due to geometrical distortions during print-scan operation. However, it is expected that dust spot displacements in printed image should not be large. Therefore, it is feasible to search for possible dust spot matches in DLSR fingerprint.

The presence of dust spots is also a good indicator that the source camera type is DSLR. However, to use the presence of dust spots false positive rate of dust spot detection should be very low. To reduce false positive rate, dust shape, shadow, and their relations with DSLR camera parameters can be used altogether in image source identification.

## 3 Forensic Use of Scratches in Scanners

Similar to DSLR cameras, optical scanners also suffer from dust, dirt, smudges, fingerprints, and scratches on scanner platen. This problem has been known for a long time and up to now many technologies and means have been developed for detection and correction of undesired artifacts appear in scanned images, documents, or films. Scanners differ from digital cameras in the sense that in flatbed scanners, the paper to be scanned is first fixed and then a one-dimensional optical sensor array moves along the document. In some scanners, document to be scanned moves over a fixed one-dimensional sensor pixel array.

Dust and dirt over platen are in main focal plane and degrades scanned image quality directly. However, the appearance of dust particles in DSLR chamber depends on several camera parameters such as f-number and aperture. Dust and dirt positions over scanner platen are not as consistent as sensor dust in DSLR. Document or image placing over platen can change existing dust positions and add new dust and debris to the scanner surface. Excessive and careless use of scanners can cause small scratches over platen and once they exist, they cannot be removed with cleaning. These tiny scratches over scanner platen can appear as bright spots in scanned images. Since their relative positions do not change, they generate a unique random pattern.

Inspired from [5], a novel work for scanner identification based on scratches, dirt, and dust over scanner platen was proposed in [6]. It was shown in [6] that scratches and dirt on scanner platen could be used as an intrinsic fingerprint of scanner device. Even if scanner platen is cleaned, the scratches over the platen do not disappear. Using scanner fingerprints, it is possible to find unique source scanner device. In [6] it was reported that the proposed scanner identification method was robust to Jpeg compression. Besides, the introduced forensic method has a potential to be extended for identification of all-in-one printer-scanner devices, fax machines, and even photocopiers.

## 4 Conclusion

Due to ubiquitous use of digital cameras and imaging sensors, digital images are generated easily and commonly used in our daily lives. Nevertheless, these new technologies bring some new issues such as integrity and authenticity of acquired images. In this chapter we presented recent image forensic techniques focusing on image source identification problem. We can infer from current studies in digital image forensics that source device identification from a single image is possible under certain conditions. However, decision error rates of these recent techniques are still quite high to use them in court proceedings. Thus, more research efforts are still needed in this field to achieve higher decision accuracy and lower error rates.

## References


1. Bayram S, Sencar HT, Memon N (2008) Classification of digital cameras based on demosaicing artifacts. J Digit Investig 5:46–59
2. Chen M, Fridrich J, Goljan M, Lukáš J (2008) Determining image origin and integrity using sensor noise. IEEE Trans Inf Secur Forensics 3(1):74–90
3. Corcoran P, Zamfir M, Buzuloiu V, Steinberg E, Zamfir A, SRL, F (2005) Automated Self-Calibrating Detection and Removal of Blemishes in Digital Images. In: Conference Proceedings-Pervasive Signal Processing (ISBN 0-9728718-2-9), GSPx, pp 24–27
4. Dirik AE, Sencar HT, Memon N (2007) Source Camera Identification Based on Sensor Dust Characteristics. In: IEEE Workshop on Signal Processing Applications for Public Security and Forensics, 2007. SAFE '07, pp 1–6
5. Dirik AE, Sencar HT, Memon N (2008) Digital single lens reflex camera identification from traces of sensor dust. IEEE Trans Inf Forensics Secur 3(3):539–552
6. Dirik AE, Sencar HT, Memon N (2009) Flatbed scanner identification based on dust and scratches over scanner platen. In: IEEE Conference on Acoustic, Speech and Signal Processing (ICASSP), pp. 1385–1388
7. Fridrich J, Soukal D, Lukàš J (2003) Detection of copy-move forgery in digital images. In: Digital Forensics Research, Workshop
8. Geradts ZJ, Bijhold J, Kieft M, Kurosawa K, Kuroki K, Saitoh N (2001) Methods for identification of images acquired with digital cameras. In: SPIE, Enabling Technologies for Law Enforcement and Security 4232:505–512
9. Kurosawa K, Kuroki K, Saitoh N (1999) Ccd fingerprint method—identification of a video camera from videotaped images. In: ICIP' 99, Kobe, Japan, pp 537–540
10. Lukáš J, Fridrich J, Goljan M (2006) Digital camera identification from sensor noise. IEEE Trans Inf Secur Forensics 1:205–214
11. Oliver MS (2008) Using sensor dirt for toolmark analysis of digital photographs. In: Ray I, Shenoi S (eds) Advances in Digital Forensics IV, vol 285. Springer, Boston, pp 193–206
12. Popescu AC, Farid H (2004) Exposing digital forgeries by detecting duplicated image regions. Department of Computer Science, Dartmouth College, Technical Report TR2004-515
13. Sencar HT, Memon N (2008) Overview of state-of-the-art in digital image forensics. In: Indian Statistical Institute Platinum Jubilee Monograph series titled Statistical Science and Interdisciplinary Research. World Scientific Press
14. Steinberg E, Prilutsky Y, Corcoran P, et al (2005) Method of detecting and correcting dust in digital images based on aura and shadow region analysis. US Patent 0,068,448,:A1
15. Steinberg E, Prilutsky Y, Corcoran P, Bigioi P, Zamfir A, Buzuloiu V, Ursu D, Zamfir M, et al (2009) Automated statistical self-calibrating detection and removal of blemishes in digital

images based on determining probabilities based on image analysis of single images. US Patent 7,536,061
16. Swaminathan A, Wu M, Liu KJR (2007) Non intrusive forensic analysis of visual sensors using output images. IEEE Trans Inf Forensics Secur 2(1):91–106
17. Swaminathan A, Wu M, Liu KJR (2008) Digital image forensics via intrinsic fingerprints. IEEE Trans Inf Forensics Secur 3:101–117
18. Willson R, Maimone M, Johnson A, Scherr L (2005) An optical model for image artifacts produced by dust particles on lenses. In: 8th International Symposium on Artificial Intelligence, Robotics and Automation in Space (i-SAIRAS)
19. Zamfir A, Drimbarean A, Zamfir M, Buzuloiu V, Steinberg E, Ursu D (2007) An optical model of the appearance of blemishes in digital photographs. In: Proceedings of SPIE, Digital Photography III, 65020I, vol 6502, pp 0I1–0I12
20. Zhou C, Lin S (2007) Removal of image artifacts due to sensor dust. In: CVPR07

# Part III
# Techniques Verifying the Integrity and Authenticity of Image Evidence

# Natural Image Statistics in Digital Image Forensics

**Siwei Lyu**

*Look deep into nature, and then you will understand everything better.*

Albert Einstein

**Abstract** The fundamental problem in digital image forensics is to differentiate tampered images from those of untampered ones. A general solution framework can be obtained using the statistical properties of natural photographic images. In the recent years, applications of natural image statistics in digital image forensics have witnessed rapid developments and led to promising results. In this chapter, we provide an overview of recent developments of natural image statistics, and focus on three applications of natural image statistics in digital image forensics as (1) differentiating photographic images from computer-generated photorealistic images, (2) generic steganalysis; (3) rebroadcast image detection.

## 1 Introduction

Thanks to the increasing availability and sophistication of digital imaging technology (digital cameras, computers, and photo-editing software) and the popularity of the Internet and digital images have become our main information source. However, concomitant with the ubiquity of digital images is the rampant problem of digital forgeries, which has seriously debased the credibility of photographic images as definite records of events. Accordingly, digital image forensics that aims to reveal tampering operations in digital images has emerged as a new research field.

One fundamental problem in digital image forensics is to differentiate images that have undergone tampering operations from those of untampered images. In the

S. Lyu (✉)
LI-67A, 1400 Washington Avenue, Albany, NY 12222, USA
e-mail: lsw@cs.albany.edu

H. T. Sencar and N. Memon (eds.), *Digital Image Forensics*,
DOI: 10.1007/978-1-4614-0757-7_8,

typical forensic analysis scenario, however, we do not have access to the original image. On the other hand, we can solve the problem using the general properties of untampered images. There have been several different methodologies developed in the recent literature, the details of which can be found in the other chapters of this book. In this chapter, we focus on the use of general *statistical* properties of natural photographic images in digital image forensics.

The rest of this chapter is organized as follows. In Sect. 2, we briefly introduce statistical properties of natural images in multi-scale image representations, and a type of image feature composed of several statistics of natural images in multi-scale image representations that can capture properties of natural images. In Sect. 3, we apply this image features to three problems in digital image forensics. We conclude this chapter with a discussion in Sect. 4.

## 2 Natural Image Statistics

The biological perceptual systems have evolved to adapt and operate in the natural environment. Among all types of sensory data that probe the surrounding environment, visual sensory signals, i.e., images, provide the most comprehensive information critical to survival. Accordingly, human vision is most sensitive to images that resemble scenes likely to be encountered by an imaging device (an eye or a camera) in the physical world. Such images are usually referred to as natural images.

However, natural images correspond only to a tiny fraction in the space of all possible images [1, 2, 12]. For instance, there are more than $10^{1,500}$ different 8-bit gray-scale images of size as small as $25 \times 25$ pixels, while the current estimated total number of atoms in the whole universe is only about $10^{80}$. Yet, if we uniformly sample from this colossal image space by randomly choosing intensity values from $\{0, 1, \cdots, 255\}$ for each pixel, most of the time a noise pattern lack of any consistent structure will appear, yet with a very rare chance such a procedure will produce a natural image. The fact that such a uniform sampling seldom produces a natural image confirms that natural images are sparsely distributed in the space of all possible images.

Although relatively small in number, natural images exhibit strong regularities that distinguish themselves from the sea of all possible images. Particularly, natural images are not simply a collection of independent pixels. The visual structures making them look natural are the result of strong correlations among pixels. Such statistical properties hold the key for a fundamental understanding of the biological mechanisms enabling visual perception, and play equally critical roles in building machine vision systems to simulate biological visual functions.

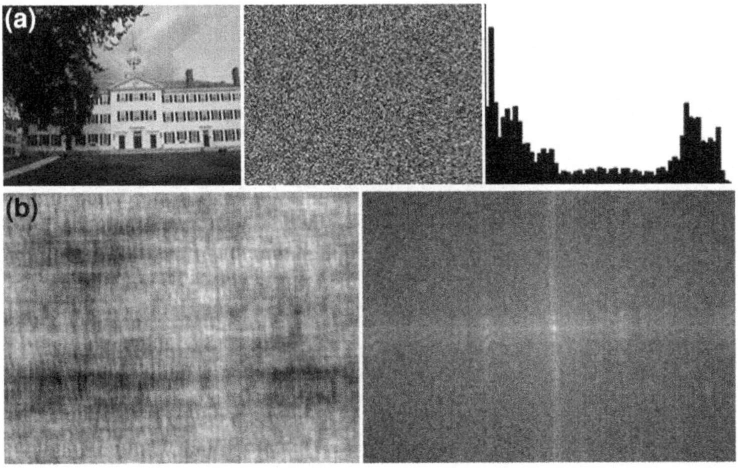


**Fig. 1** The second and the fourth images have the same pixel intensities or the frequency energies as the natural images shown in the first column. Note the difference in the visual appearance of these images. Shown in the second and the fifth columns are the corresponding pixel intensity histograms and power spectrum density

## *2.1 Image Representation*

To study the statistical properties of natural images, the essential first step is to choose a proper image representation. Image representation provides the domain where image signals are analyzed, processed, and synthesized. As such, the choice of representation can strongly affect our basic understanding of images, and is of fundamental importance to any scientific or engineering field centering around the analysis and processing of images. There are, of course, many different image representations to choose from. However, the choice should be made based on their effectiveness in revealing statistical regularities in natural images. Once a proper image representation is chosen, the next step is to make observations on an ensemble of natural images for regular statistical properties.

The simplest representation, for example, is an intensity-based approach, where the representation is simply the original intensity values. An $n \times m$ grayscale image is considered as a collection of $mn$-independent samples of intensity values. Similarly, an $n \times m$ RGB color image is represented as a collection of $mn$-independent 3D vectors. However, it is generally difficult to capture the statistical regularities of natural images in the intensity domain. Especially, two images with the same set of intensity values can have very different visual appearance, one example is shown in Fig. 1.

Another popular image representation is based on the global Fourier decomposition, where an image is decomposed as a linear combination of complex exponentials with varying frequency and orientation as

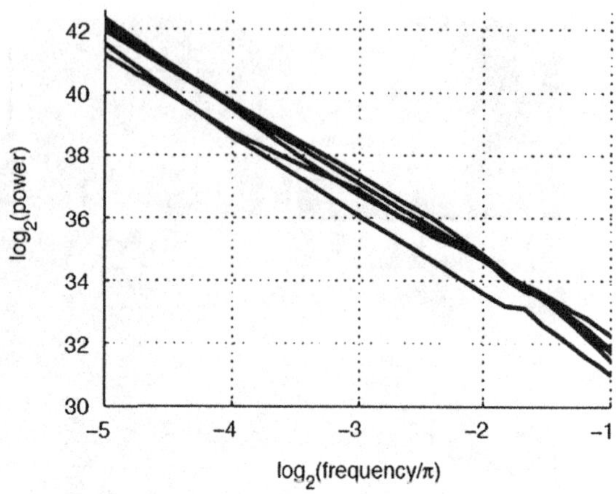


**Fig. 2** Power law behavior of the power spectrum density of natural images as shown in the log–log domain. Each curve corresponds to one natural image

$$F(\omega_x, \omega_y) = \sum_{x,y} I(x, y) \exp(-\imath[\omega_x x + \omega_y y]),$$

where $I(x, y)$ is a grayscale image, and $F(\omega_x, \omega_y)$ is its Fourier transform (each channel of a color image is independently represented in the same way) for spatial frequency of $(\omega_x, \omega_y)$. It has been observed [13] that the magnitudes of Fourier transform of natural images, $|F(\omega_x, \omega_y)|$, can be well modeled with a "power law" model as

$$|F(\omega_x, \omega_y)| \propto \frac{1}{(\omega_x^2 + \omega_y^2)^\gamma},$$

where $\gamma$ is a constant determined on each individual image. Such a power law model appears as straight lines in the log–log domain, as shown in Fig. 2. Such a power law behavior reflects the scale invariance of natural images. However, such statistical properties are not sufficient to specify natural images, and one example is shown in Fig. 1.

The intensity- and Fourier-based representations are, in some ways, at opposite ends of a spectrum of representations. The basis functions for the pixel-based representation are perfectly localized in space, but are infinite in terms of their frequency coverage. On the other hand, the basis functions for a Fourier-based representation are perfectly localized in frequency, but are infinite in the spatial domain. Image representations based on multi-scale image decomposition (e.g., wavelets [3]) decompose an image with basis functions partially localized in both space and frequency, and thus offer a compromise between these representations. As natural images are characterized by spatial-varying localized structures such as edges, these representations are generally better than intensity- or Fourier-based representations at describing natural images. Within the general framework of multi-scale image decomposition, there exist many different implementations, each having its own advantage and effective in different problems.

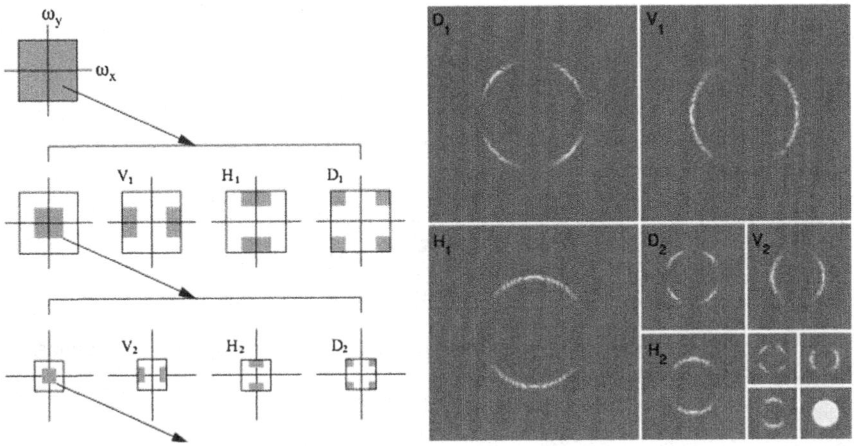


**Fig. 3** **a** An idealized frequency domain decomposition with a three-scale QMF pyramid decomposition. Shown, from *top* to *bottom*, are scales 0, 1, and 2, and from *left* to *right* are the low-pass, vertical, horizontal, and diagonal subbands. **b** The magnitude of a three-scale QMF pyramid decomposition of a "disc" image. For the purpose of display, each subband is normalized into range [0, 255]

One particular effective multi-scale image representation is based on the separable quadrature mirror filters (QMF) [5, 6], which minimizes aliasing from the reconstructed image, making it suitable for the purpose of image analysis. Shown in Fig. 3 is the idealized frequency domain decomposition with a three-scale QMF pyramid (it is idealized as using finite support filters it is not possible to achieve the sharp cutoff in frequency domain as shown). The QMF pyramid decomposition splits the image frequency space into three different scales, and within each scale, into three orientation subbands (vertical, horizontal, and diagonal). Visually each subband captures the local orientation energy in an image. The resulting vertical, horizontal, and diagonal subbands at scale i are denoted by $V_i(x, y)$, $H_i(x, y)$, and $D_i(x, y)$, respectively. The first scale subbands are the result of convolving the image with a pair of 1D $2N + 1$-tap finite impulse response (FIR) low-pass and high-pass QMF filters, denoted as $l(\cdot)$ and $h(\cdot)$, respectively. The vertical subband is generated by convolving the image, $I(x, y)$, with the low-pass filter in the vertical direction and the high-pass filter in the horizontal direction as:

$$V_1(x, y) = \sum_{m=-N}^{N} h(m) \sum_{n=-N}^{N} l(n) I(x - m, y - n).$$

The horizontal subband is generated by convolving the image with the low-pass filter in the horizontal direction and the high-pass filter in the vertical direction as:

$$H_1(x, y) = \sum_{m=-N}^{N} l(m) \sum_{n=-N}^{N} h(n) I(x - m, y - n).$$

The diagonal subband is obtained by convolving the image with the high-pass filter in both directions as:

$$D_1(x, y) = \sum_{m=-N}^{N} h(m) \sum_{n=-N}^{N} h(n) I(x - m, y - n).$$

Finally, convolving the image with the low-pass filter in both directions generates the residue low-pass subband as:

$$L_1(x, y) = \sum_{m=-N}^{N} l(m) \sum_{n=-N}^{N} l(n) I(x - m, y - n).$$

The next scale is obtained by first downsampling the residual low-pass subband L1 and recursively filtering with $l(\cdot)$ and $h(\cdot)$ as

$$V_2(x, y) = \sum_{m=-N}^{N} h(m) \sum_{n=-N}^{N} l(n) I(\lfloor x/2 \rfloor - m, \lfloor y/2 \rfloor - n)$$

$$H_2(x, y) = \sum_{m=-N}^{N} l(m) \sum_{n=-N}^{N} h(n) I(\lfloor x/2 \rfloor - m, \lfloor y/2 \rfloor - n)$$

$$D_2(x, y) = \sum_{m=-N}^{N} h(m) \sum_{n=-N}^{N} h(n) I(\lfloor x/2 \rfloor - m, \lfloor y/2 \rfloor - n)$$

$$L_2(x, y) = \sum_{m=-N}^{N} l(m) \sum_{n=-N}^{N} l(n) I(\lfloor x/2 \rfloor - m, \lfloor y/2 \rfloor - n).$$

Subsequent scales are generated similarly by recursively decomposing the residual low-pass subband. The decomposition of a RGB color image is performed by decomposing each color channel independently.

There are two distinct regularities of natural images that have been widely observed in the multi-scale representations. First, as shown in Fig. 4a, the marginal distributions of the QMF pyramid subband coefficients exhibit a highly non-Gaussian shape, and usually have high positive kurtosis, characterized by a sharp peak at zero and long symmetric tails. An intuitive explanation is that natural images contain large smooth regions and abrupt transitions (e.g., edges). The smooth regions, though dominant, produce small coefficients near zero, while the transitions generate large

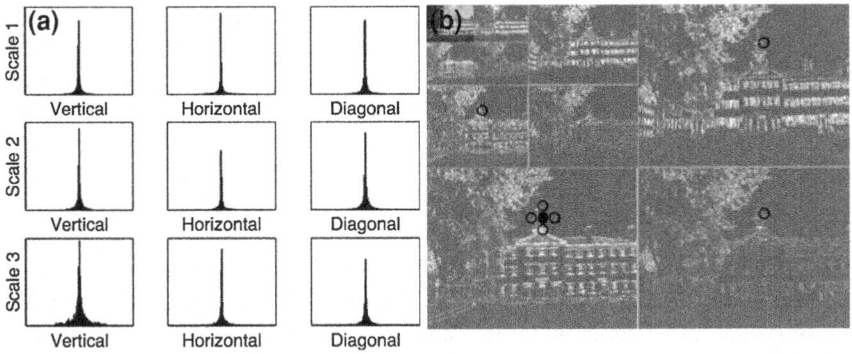


**Fig. 4** **a** The histograms of the coefficients of the three-scale QMF decomposition of a natural image. **b** The magnitudes of the three-scale QMF pyramid decomposition of the same natural image

coefficients. This property holds for the QMF pyramid decomposition as well as many other multi-scale image decompositions (e.g., wavelets).

Furthermore, for natural images, the coefficients in each subband and across different scales and orientations are correlated as the result of salient image features (e.g., edges and junctions). Such image features tend to orient spatially in certain directions and extend across multiple scales, which result in substantial local energy, measured by the magnitude of the decomposition coefficient, across many scales, orientations, and spatial locations, Fig. 4b. As such, a coefficient with a large magnitude in a horizontal subband (an indication that the corresponding pixel is on or near an edge having horizontal orientation) suggests that the left and right spatial neighbors in the same subband may also have a large magnitude. Similarly, if there is a coefficient with a large magnitude at scale $i$, it is likely that its "parent" at scale $i + 1$ will also have a large magnitude.

## 2.2 Features Based on Regularities of Natural Images

In general, the statistical descriptions of natural images play three important roles: (a) prior or regularization terms for probabilistic inference about natural images; (b) constraints for sampling procedures for special class of natural images, such as textures; (c) discriminative/descriptive features for differentiating or describing natural images.

For the purpose of digital image forensics, where our main objective is to differentiate natural and unnatural images, we would like to extract useful features based on regularities of natural images. Since the QMF-based multi-scale image representation has advantage of revealing statistical properties of natural images, we focus here on the statistical features in this domain that can capture the two statistical properties.

First, we use the first four cumulants (i.e., the mean, variance, skewness, and kurtosis) of the coefficients in each subband of all orientations, scales, and color channels to characterize the marginal distributions of the coefficients [15]. The cumulants determine the distribution indirectly—distributions sharing similar cumulants will have similar shapes. Furthermore, we capture the higher order statistical correlation of coefficient magnitudes (Fig.4b) by another set of image statistics, which are collected from the linear prediction errors of coefficient magnitudes. Particularly, the coefficient magnitudes near image features follow a simple first-order linear dependency [14]; therefore, it is possible to build a linear predictor of magnitudes from the magnitudes of neighboring coefficients for natural images. The prediction errors from this linear predictor provide a measure of correlations among neighboring coefficients. For the purpose of illustration, consider a coefficient in the vertical subband at scale $i$, $V_i(x, y)$. A linear predictor of its magnitude is built from a fixed subset from its all possible spatial, orientation, and scale neighbors. Formally, the linear predictor of the coefficient magnitudes is formed as a linear combination of the neighboring magnitudes:

$$
\begin{aligned}
|V_i(x, y)| = {} & w_1|V_i(x|1, y)| + w_2|V_i(x + 1, y)| + w_3|V_i(x, y|1)| \\
& + w_4|V_i(x, y + 1)| + w_5|V_{i+1}(x/2, y/2)| + w_6|D_i(x, y)| \\
& + w_7|D_{i+1}(x/2, y/2)|,
\end{aligned}
$$

where $|\cdot|$ denotes the magnitude operator and $w_k, k = 1, \ldots, 7$ are scalar weights associated with each type of neighbors. Interpolated values (e.g., rounding) are used when $x/2$ or $y/2$ is noninteger. When evaluated throughout the whole subband (and assuming homogeneity), the above linear predictor can be expressed more compactly in form of matrix and vectors as:

$$\mathbf{v} = Q\mathbf{w},$$

where the column vector $\mathbf{v}$ is formed by stacking the magnitudes of all coefficients in $V_i$, and each column of the matrix $Q$ contains the magnitudes of each type of neighboring coefficients. The unknowns, $\mathbf{w} = (w_1, \cdots, w_7)^T$, are determined by a least squares estimation, which minimizes the following quadratic error function:

$$E(\mathbf{w}) = \|\mathbf{v} - Q\mathbf{w}\|^2.$$

This error function is an over-constrained least squares problem, which can be minimized and the solution is given by

$$\mathbf{w} = (Q^T Q)^{-1} Q^T \mathbf{v}$$

Given the large number of constraints (one per coefficient in the subband) in only seven unknowns, it is generally safe to assume that the $7 \times 7$ matrix $Q^T Q$ is invertible. Similar linear predictors are formed on all other orientation subbands—the linear predictor for horizontal and diagonal subbands.

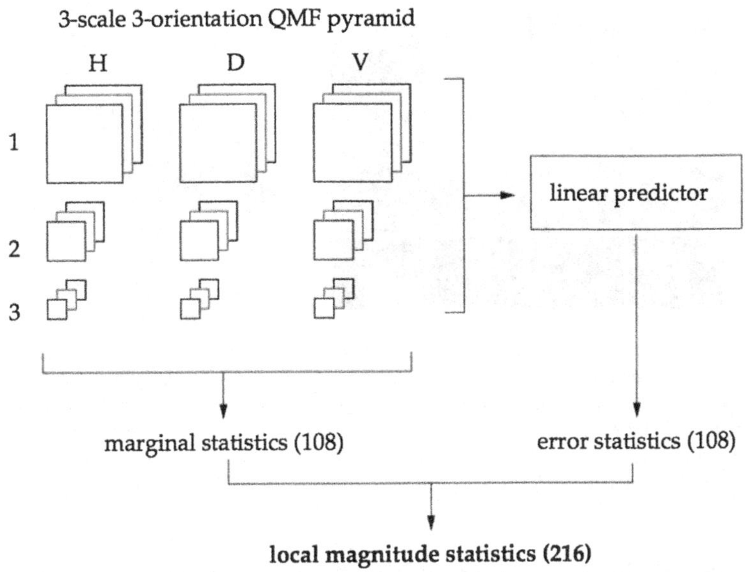


**Fig. 5** Overall process of collecting local magnitude statistics from a RGB color image with a three-scale QMF pyramid

With the linear predictor, the log errors between the actual and the predicted coefficient magnitudes are computed as $\log(\mathbf{v}) - \log(|Q\mathbf{w}|)$. The log error quantifies the correlations of the coefficients in a subband with their neighbors, and natural images tend to have a specific distribution for these errors. Following the same rationale as the coefficient statistics, the same statistics (i.e., the mean, variance, skewness and kurtosis) are collected to characterize the error distributions of each scale and orientation subbands.

For a QMF pyramid of n scales, the coefficient marginal statistics is collected for scales $i = 1, \cdots, n|1$, with a total number of $12(n|1)$ (the mean, variance, skewness, and kurtosis for the vertical, horizontal, and diagonal subbands in each scale). Similarly, the error statistics is collected at scales $i = 1, \cdots, n|1$ which also yields a total number of $12(n|1)$. Combining both types of statistics results in a grand total of $24(n|1)$ statistics from a grayscale image. If we consider the three color channels of a color image, the total number of statistics is tripled to $72(n|1)$ in a QMF pyramid decomposition of $n$ scales. The overall process of collecting this image feature types is illustrated in Fig. 5.

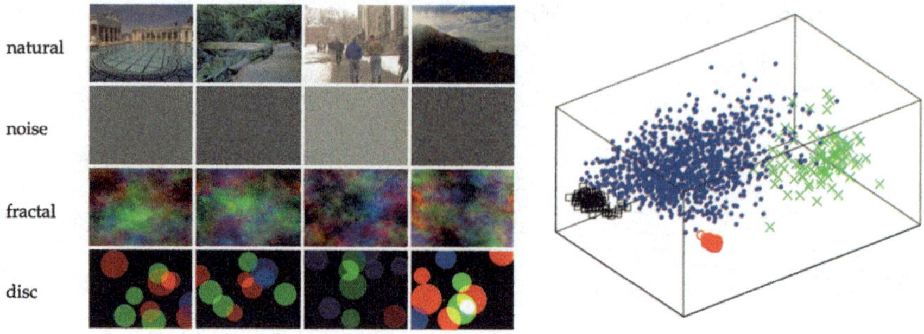


**Fig. 6** *Left* Natural and synthetic images. From *top* to *bottom* are examples of the 1,000 natural images, 300 noise patterns with the same intensity histograms, 300 fractal patterns with the same frequency magnitude envelopes, and 300 disk images with similar phase statistics as the corresponding natural images. *Right* Projection of the 216 color local magnitude image statistics for 1,000 natural image (•), and the synthetic images (noise (×), fractal (◇), and disk (○), onto the *top* three principal components

## 2.3 Natural Versus Synthetic Using Image Features

We need to confirm that the described image features capture certain non-trivial statistical regularities of natural images. Ideally, under these image statistics, natural images will show more similarity beyond their difference in content, and dissimilarities between natural and unnatural images will also be more distinct. More specifically, in the space of all possible feature vectors consisting of the proposed image statistics, we would like to see that natural images cluster together and separate from unnatural images. It is, however, even harder to define an "un-natural" image, and in our experiments, we employed some synthetic images of different types, usually considered "un-natural" as they rarely appear in natural environment.

We collected 1,000 natural images from photographs taken with a range of different films, cameras, and lenses, and digitally scanned. They are RGB color images, spanning a range of indoor and outdoor scenes, JPEG compressed with an average quality of 90%, and typically $600 \times 400$ pixels in size (on average, 85.7 KB). Three different types of synthetic images, each type of these synthetic images preserving certain statistical properties of natural images, were also generated. Shown in the second to the fourth row in Fig. 6 are examples of: (1) noise patterns, which were created by scrambling the pixels of the corresponding natural images and thus had the same intensity histograms; (2) fractal patterns which kept the $\frac{1}{(\omega_x^2+\omega_y^2)^\gamma}$, for $\gamma \in [1, 2)$ magnitude envelope of the corresponding natural image in the frequency domain but were with random phases; and (3) disc images that were formed by overlapping anti-aliased discs of variable size radii. About 300 of each type of synthetic images were generated in our experiment.

From these images, natural and synthetic alike, the local magnitude statistics from the QMF pyramid decomposition was extracted to form image features. To visualize the distribution of these image feature vectors, in all three cases, the high-dimensional feature vectors were projected onto a three-dimensional linear subspace spanned by the top three principal components as a result of the principal component analysis of all image features. Shown in the right panel of Fig. 6 are the projected feature vectors of the 1,000 natural and the 900 synthetic images onto the top three principal components. For the 216 local magnitude statistics the three top principal components capture over 75% of the total variance in the original data set. By reducing the dimensionality, a significant fraction of information was discarded, but it allows us to visually inspect the distribution of the feature vectors for natural and synthetic images. The proposed image statistics for natural images form a relatively tight cluster. More importantly, the synthesized images are well separated from the ensemble of natural images. The similar statistical regularities in each type of synthetic images are also reflected by their own clusters. These indicate that the proposed image statistics is capable of capturing statistical regularities in natural images not present in the synthetic images.

## 3 Application to Digital Image Forensics

Statistical descriptions of natural images also have important applications in the burgeoning field of digital image forensics. The popularity of digital images also makes it vulnerable to tampering. An image not only tells a story, but may also tell a secret. An innocent-looking natural image may conceal a hidden message generated by a computer graphics program. Such manipulations, along with tampering with image contents, have become increasingly easy due to the development of technologies. Digital image forensics are techniques aiming to shed lights on such secrets beneath digital images, and is therefore of the great interest to the law-enforcement and national security agencies.

The image features based on natural image statistics are particularly useful for three applications in digital image forensics:

- **Photographic or photorealistic**: Sophisticated computer graphics softwares can generate highly convincing photorealistic images able to deceive the human eyes. Differentiating between these two types of images is an important task to ensure the authenticity and integrity of photographs.
- **Generic image steganalysis**: Image steganography hides messages in digital images in a non-intrusive way that is hard to detect visually. The task of generic steganalysis is to detect the presence of such hidden messages without the detailed knowledge of the embedding methods. Because of the potential use of steganography as a covert communication method, steganalysis is of interest to the law-enforcement, counter-intelligence, and anti-terrorism agencies.
- **Live or rebroadcast**: Biometrics-based (e.g., face, iris, or voice) authentication and identification systems are vulnerable to the broadcast attacks. An example is the defeat of a face recognition system using a high-resolution photograph of a

human face. An effective protection against such an attack is to differentiate a live image (captured in real time by a camera) and a broadcast one (a photograph).

Common to these problems is the task of differentiating natural photographic images from some other classes of images: in photographic versus photorealistic, the other class is the computer-generated photorealistic images; in steganalysis, they are images with steganographic messages; and in live versus rebroadcast, they are the prints of natural images. Instead of seeking an individual solution to each problem, we propose to solve them in a unified framework. First, statistics capturing certain aspects of natural images are collected. Using these statistics to form discriminative features, classification systems are built to determine the class of an unknown image. In the following, we describe the application of the image features to these three different problems in detail.

## 3.1 Photographic Versus Photorealistic

In an age prevailing with digital media, it is no longer true that seeing is believing. This is partly attributed to the development of sophisticated computer graphics rendering algorithms and softwares that can generate remarkably photorealistic images [16]. These technologies have started challenging our long-held notion of photorealism.Somehow unexpectedly, this technology also bears legal implications. In 1996, the United States Congress passed *The Child Pornography Prevention Act*, which in part prohibited any image that appears to be or conveys the impression of someone under 18 engaged in sexually explicit conduct. This law made illegal the computer-generated images that only appear to show minors involved in sexual activity. In 2002, however, the United States Supreme Court struck down portions of this law in their 6–3 ruling in *Ashcroft v. Free Speech Coalition*—the court said that the language in the 1996 *The Child Pornography Prevention Act* was unconstitutionally vague and far-reaching. This ruling essentially legalized the computer-generated child pornographic images and makes it considerably more difficult for law-enforcement agencies to prosecute such crimes—anyone trafficking these illegal images may always claim that they are computer generated to avoid punishment. Therefore, the law-enforcement agencies are in a great need of methods that can reliably differentiate between true photographic images and computer-generated photorealistic (hereafter, photorealistic) images.

One promising methodology to solve this problem is to take the advantage of the statistical regularities in natural photographic images. No matter how visually resembling a photographic image, a photorealistic image is created from a fundamentally different process. A photographic image is the result of the physical world projected on the image sensors in imaging devices, such as the film in an optical camera or the CCD (charge-coupled device) in a digital camera. On the other hand, photorealistic images are produced by rendering algorithms that simulate this imaging process. The rendering algorithms can only roughly model the highly complex and subtle inter-

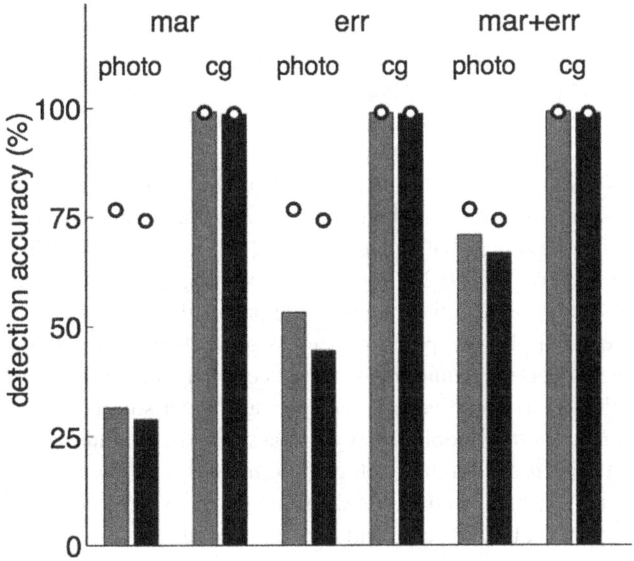


**Fig. 7** Classification accuracy for a nonlinear SVM trained on (from *left* to *right*) the 108 color coefficient marginal statistics only, the 108 magnitude linear prediction error statistics only, and the 216 local magnitude statistics including both coefficient marginal and magnitude linear prediction error statistics. The dots correspond to a no-linear SVM trained on the complete set of 216 color image statistics. The gray bars correspond to the accuracies on the training set and the black bars correspond to the accuracies on the testing set, for photographic (photo) and photorealistic (cg) images

actions between the physical world and the imaging device. Such discrepancy will well reveal themselves in image statistics. Differentiating photographic and photorealistic images then proceeds as a binary classification problem, where the type of an unknown image is automatically determined based on the proposed image statistics.

We formulate the problem of differentiating photographic and photorealistic images as a binary classification, where a classifier is trained to determine if an image is photographic or photorealistic. In building the classifier, the training data are essential, as we want to avoid learning accidental difference between photorealistic and photographic images in color, texture, or other aspects of image contents. To this end, we need to train the classifiers on a considerably large set of images with contents as diverse as possible, so as to integrate out superficial difference in image contents. For the photographic images, we used the 40,000 natural images. For the photorealistic images, 6,000 were downloaded from www.raph.com and www.irtc.org. Shown in Fig. 5.2 are eight samples from the 6,000 photorealistic images. The relative fewer number of photorealistic images reflects the fact that photorealistic images, especially those of high quality, require more effort to create. All photorealistic images are color (RGB), JPEG compressed (with an average quality of 90%), and typically in the order of 600–400 pixels in size. Visually, these photorealistic images span a range of contents (e.g., landscapes and city scenes) and imaging conditions (e.g., indoor and outdoor lighting, close up, far away views, etc.), and have different levels of photorealism. They were created from popular computer graphics software packages (e.g., 3D Studio Max, Maya, SoftImage 3D, PovRay, Lightwave 3D, and Imagine).

From the 40,000 photographic images and 6,000 photorealistic images, 32,000 photographic and 4,800 photorealistic images were randomly chosen to form the training set for a nonlinear SVM classifier [11]. To accommodate different image sizes, only the central $256 \times 256$ region of each image was analyzed. During the training phase, the false negative rate (i.e., the probability of a photorealistic image being classified as a photographic image) was controlled to be less than 1%. This specific setting reflects the requirement in practice, where the cost of a false negative is much higher than a false positive, and also sets the comparisons henceforth described on a fair ground. Using a nonlinear SVM classifier, the nonlinear SVM classifier affords a 74.3% accuracy on the photographic images, as shown in Fig. 7. We also compare different types of statistics in the same figure, which shows that to achieve the best performance, both type of statistics are needed.

### *3.2 Generic Image Steganalysis*

The goal of steganography is to hide messages in an innocuous cover medium in an unobtrusive way, so as to evade inspection [17–20]. Steganography can be used as a covert communication method by criminals, terrorists, and spies. Malicious message can be embedded into an innocuous-looking image, and posted on the Internet or sent in an e-mail without being suspected. Therefore, it is not surprising that with the emergence of steganography, that the development of a counter technology, steganalysis, has also emerged [21]. The goal of steganalysis is to determine if an image (or other carrier medium) contains an embedded message. As this field has developed, determining the length of the message and the actual contents of the message are also becoming an active area of research.

Though visually hard to differentiate, the statistical regularities in the natural image as the steganography cover are disturbed by the embedded message. For instance, changing the LSBs of a grayscale image will introduce high frequency artifacts in the cover images. We use the statistical image features for a general framework of generic image steganalysis, based on nonlinear SVM classification. Without the knowledge of the embedding algorithm, this method detects steganography based on the abnormality in the statistics of the stego images.

Our experiment is based on 40,000 natural images, from which 40,000 stego images (1,600 per embedding message type and per steganography tool) were generated by embedding random noise messages of various sizes into the full-resolution cover images. The messages were of sizes 6.0, 4.7, 1.2, 0.3 KB, corresponding to an average of 100, 78, 20, and 5% of the total steganography capacity of the cover images, respectively. These messages were embedded using Jsteg [7], Outguess [10],

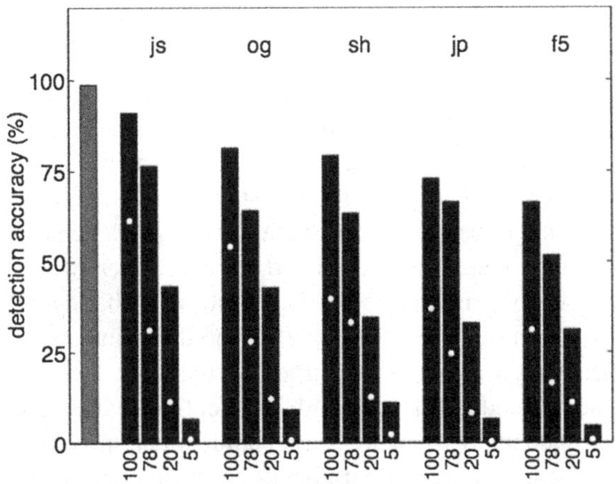


**Fig. 8** Classification accuracy with color image statistics for nonlinear SVMs with a 1% training false positive rate. The left-most *gray* bar corresponds to the false positive rate (a clean image classified as stego), by subtracting it from 100%. Each group of four bars corresponds to different steganography embedding programs (jsteg (js); outguess (og); steghide (sh); jphide (jp); and F5 (f5)). The numeric values on the horizontal axes correspond to the message size as an average percentage of the steganography capacity of the covers

Steghide [8], Jphide [9], and F5 [4]. Each stego image was generated with the same quality factor as the original cover image so as to minimize double JPEG compression artifacts.

We trained nonlinear support vector machines with radial basis function (RBF) kernels based on the collected image statistics. The training sets for SVM consisted of image statistics from the 32,000 randomly chosen natural images and 32,000 randomly chosen stego images (6,400 per embedding program tested). The image statistics of the remaining cover and stego images was used to test the classifiers obtained throughout; results from the testing stage are presented. In the training phase, the false positive rate (i.e., the probability of a cover image being incorrectly classified as a stego image) was controlled to be less than 1%. This specific setting is the practical requirement of applying steganalysis, where the cost of a false positive is much higher than a false negative. As shown in Fig. 8, the average detection accuracy of the nonlinear SVM classifier is 78.2, 64.5, 37.0, and 7.8% with a maximum/minimum detection accuracy of 91.1%/66.4%, 76.4%/51.8%, 43.2%/31.3%, and 11.2%/4.8% for each embedding types.

### *3.3 Live or Rebroadcast*

In recent years, biometric-based authentication (e.g., face, iris, voice or fingerprint) is increasingly gaining popularity in a large spectrum of applications, ranging from governmental programs (e.g., national ID card and visa) to commercial applications such as logical and physical access control. Compared to the traditional password-based authentication systems, the biometrics-based systems have the advantages of easier management (no need to memorizing the passwords and changing them periodically) and better security (biometrics confirm the identity of the user).

Surprisingly, however, even the most sophisticated biometric-based authentication systems may be vulnerable to a simple broadcast attack. For instance, to break an iris recognition-based authentication system, a malicious intruder can find a photograph of the face of an authenticated person, printed out on a paper (with sufficiently high printing quality), cut the iris images out and paste it on his eyelids. When being presented to the system, instead of the live captured iris image, the system is fed with the rebroadcast iris image. For the purpose of recognition, these images are just the same, with the rebroadcast images with different noise characteristics from the printing process. Since many biometric-based authentication systems are built to be robust in image noises, such a simple trick will, unfortunately, work for the intruder purpose. A key feature for biometric-based authentication systems withstanding the "rebroadcast" attacks is to enable the systems to differentiate between a rebroadcast and a live image. This problem is again formulated as a binary classification where classifiers of live and rebroadcast images were built based on the proposed image statistics.

For this task, the simpler image feature comprised of the 72 grayscale local magnitude statistics and linear classifier (LDA) are sufficient for a good performance [22]. In our experiment, we randomly chose 1,000 photographic images. From these images, 200 "rebroadcast" images were generated by printing the grayscale converted photograph images on a laser printer and then scanned with a flatbed scanner. This is to simulate the condition when the printed-out images are captured by the camera in the authentication system. For consistency, printing and scanning were both done at 72 dpi (dots per inch) to reduce loss of image information as much as possible. Next, the 72 grayscale local magnitude image statistics were collected on all the 1,200 images, from which 750 photograph images and 150 rebroadcast images were randomly chosen to form the training set. The remaining 250 photograph images and 50 rebroadcast images were included in the testing set. In the training stage, the LDA classifier correctly classified 99.5% of the live and 100% of the rebroadcast images. A threshold was selected so as to afford a less than 0.5% false positive rate. In the testing stage, 99.5% of the live and 99.8% of the rebroadcast images were correctly classified. To avoid reporting results pertaining to a specific training/testing split, these values were the average of results over 100 random training/testing splits. The relatively easy classification of live and rebroadcast images attributes to the artifacts introduced by the printing and scanning process, which the simple grayscale local magnitude statistics and linear classification suffice to capture.

## 4 Conclusion

Natural images are ubiquitous, yet they are relatively rare in the sea of all possible images. More importantly, natural images have statistical regularities that has been employed by the biological vision systems. Capturing such statistical regularities of natural images are essential for many applications in image processing, computer vision, and, especially, digital image forensics.

In this chapter, we have presented a set of image statistics based on the multi-scale image decompositions, which correspond to image representations with basis functions localized in both spatial and frequency domains. Such image representations are shown to be suitable to reveal statistical regularities in natural images. Specifically, our image statistics consists of the local magnitude statistics from a quadrature mirror filters (QMF) pyramid decomposition and the local phase statistics from a local angular harmonic decomposition. Empirically, we showed that these image statistics can differentiate natural images from some simple "un-natural" synthetic images. We then demonstrated their effectiveness, combined with nonlinear classification techniques, in applications in digital image forensics on: (1) differentiating photographic and computer-generated photorealistic images (photographic versus photorealistic); (2) detecting hidden messages in photographic images (generic steganalysis); and (3) identifying printed copies of photographic images to protect the biometric-based authentication systems from the rebroadcast attack (live vs. rebroadcast). All these techniques follow a unified framework of image classification: first feature vectors based on the proposed image statistics are collected on a set of labeled images, then a classifier is trained on these feature vectors and used to determine the class of an unknown image.

In conclusion, we believe that statistical characterization of natural images is at the heart of many challenging problems in image processing and analysis, computer vision, and digital image forensics. The work described in this chapter represents just a small fraction of the tremendous potential of these image statistics in digital image forensics. Though a well-posed starting point for future exploration in this direction, there are still a lot of room for future improvement.

## References

1. Daugman JG (1989) Entropy reduction and decorrelation in visual coding by oriented neural receptive fields. IEEE Trans Biomed Eng 36(1):107–114
2. Kersten D (1987) Predictability and redundancy of natural images. J Opt Soc America A, 4(12):2395–2400
3. Mallat SG (1989) A theory for multiresolution signal decomposition: the wavelet representation. IEEE Trans Pattern Mach Intell 11:674–693
4. Westfeld A, F5. www.wrn.inf.tu-dresden.de/westfeld/f5
5. Vaidyanathan PP (1987) Quadrature mirror filter banks, M-band extensions and perfect reconstruction techniques. IEEE ASSP Mag 4(3):4–20
6. Vetterli M (1987) A theory of multirate filter banks. IEEE Trans ASSP, 35(3):356–372

7. Upham D, Jsteg. http://ftp.funet.fi
8. Hetzl S, Steghide. http://steghide.sourceforge.net
9. Latham A, Jpeg Hide-and-Seek. http://linux01.gwdg.de/alatham/stego
10. Provos N, Honeyman P (2002) Detecting steganographic content on the internet. In ISOC NDSSf02, San Diego, CA
11. Shawe-Taylor J, Cristianini N (2004) Kernel methods for pattern analysis. Cambridge
12. Ruderman DL, Bialek W (1994) Statistics of natural image: scaling in the woods. Phys Rev Lett 73(6):814–817
13. Field DJ (1987) Relations between the statistics of natural images and the response properties of cortical cells. J Opt Soc Am A 4(12):2379–2394
14. Buccigrossi RW, Simoncelli EP (1999) Image compression via joint statistical characteri- zation in the wavelet domain. IEEE Trans Image Process 8(12):1688–1701
15. Gluckman J (2003) On the use of marginal statistics of subband images. In IEEE international conference on computer vision, Nice, France
16. Foley J, van Dam A, Feiner S, Hughes J (1997) Computer graphics: principles and practice. Addison-Wesley
17. Anderson RJ, Petitcolas FAP (1998) On the limits of steganography. IEEE J Sel Areas Commun 16(4):474–481
18. Johnson N, Jajodia S (1998) Exploring steganography: seeing the unseen. IEEE Comput. 31(2):26–34
19. Kahn D (1996) The history of steganography. In proceedings of information hiding, First international workshop, Cambridge, UK
20. Petitcolas EAP, Anderson RJ, Kuhn MG (1999) Information hiding—a survey. Proc IEEE 87(7):1062–1078
21. Fridrich J, Goljan M, Hogea D (2003) New methodology for breaking steganographic techniques for JPEGs. In SPIE Symposium on Electronic Imaging, Santa Clara, CA
22. Duda RO, Hart PE, Stork DG (2000) Pattern classification. 2nd edn. Wiley, New York

# Detecting Doctored Images

**Micah K. Johnson**

**Abstract** With the availability of powerful image-editing software, digital cameras, and a wealth of online imagery, almost anyone can doctor an image. Consequently, image hoaxes are now commonplace and people even expect that images of celebrities have been retouched. For many, seeing is no longer believing. While most doctored images are made for entertainment or artistic purposes, they have also shown up in courtrooms as evidence and in scientific publications. The presence of manipulated images in these settings is troubling and the field of digital image forensics has emerged to address this growing problem. This chapter provides an overview of current tools for detecting doctored images and discusses trends that have emerged in the field. Several tools are described in detail and references are provided for related techniques.

There are a growing number of tools for detecting doctored images. These tools tend to fall into two different categories: watermarking or forensics. The watermarking approach embeds information into the image at the time of recording—this information can be extracted at a later date to verify authenticity. Watermarking requires specialized cameras and is applicable in domains where the make and model of cameras can be controlled, such as law enforcement. By contrast, the forensics approach focuses on information that is inherent to images in general, and does not assume specialized cameras. These techniques often consider specific forms of tampering and how the tampering alters the image.

A fundamental assumption of image-forensic techniques is that images possess certain regularities, or invariants, that are disturbed by tampering. These invariants are often difficult to create synthetically and invisible to the inexperienced observer. Although tampering may not affect the image quality, changes to invariants are often measurable. These regularities can come from a variety of sources, including the world, the camera, or the image itself. In Fig. 1, a basic imaging pipeline is shown

M. K. Johnson (✉)
Computer Science and Artificial Intelligence Laboratory,
Massachusetts Institute of Technology, Cambridge, MA 02139, USA
e-mail: kimo@csail.mit.edu

H. T. Sencar and N. Memon (eds.), *Digital Image Forensics*,
DOI: 10.1007/978-1-4614-0757-7_9,

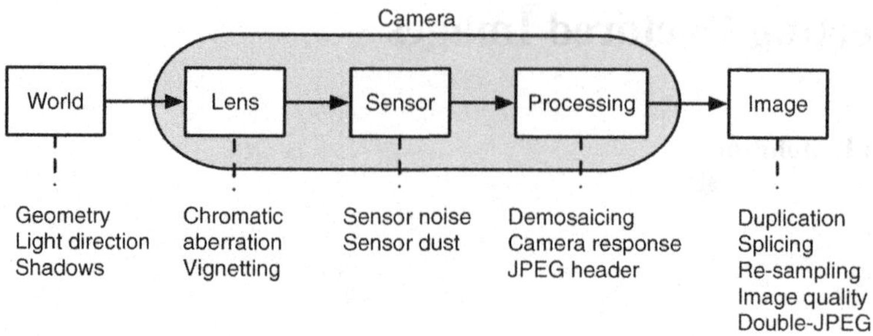


**Fig. 1** Sources of regularities in the imaging process and current forensic tools that exploit these regularities

along with forensic techniques that exploit invariants along each stage in the pipeline. The common approach taken by these techniques is to measure the regularities and detect differences in the measurements. Most of the current forensic tools target specific types of tampering since a single manipulation may disturb only some of the regularities. While there is no single tool that can detect all types of tampering, the tools used together are a powerful way to detect forgeries.

This chapter reviews techniques for detecting doctored images. The techniques are grouped into categories based on the imaging property that they exploit: world, lens, sensor, post-processing, or the image itself. For any technique, there are often variations that relax the assumptions or address related problems. This chapter focuses on the trends that have emerged in image forensics and presents the earliest, and often the simplest, version of an idea. Citations for related techniques are provided with a brief discussion of the differences.

## 1 World

Images are recordings of light from the world. Since rays of light obey certain physical laws, the first source of image regularities is the world itself. The regularities imposed by the world can be grouped into two basic categories: geometry and illumination. The geometry category includes geometric relationships, such as angles, lengths, areas, and describes how these relationships are derived from properties in the world. The illumination category includes invariants that are derived from the interaction of light with objects in the world. The effects of material are often modeled within illumination, but could be treated separately. The regularities of the world provide constraints on variation that can appear in images and provide a rich source for image-forensic techniques.

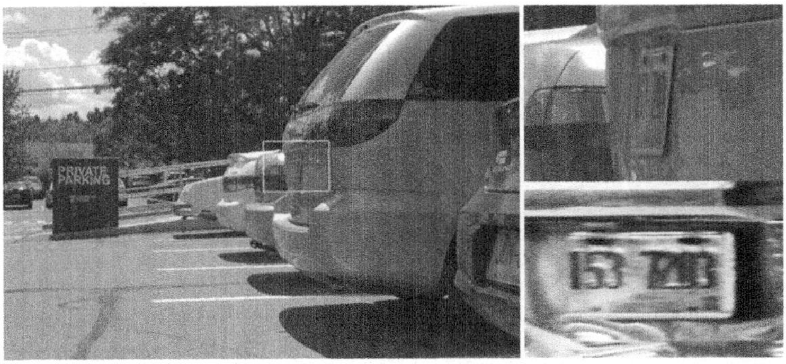


**Fig. 2** An image of a license plate (*left*) and a close up view (*top right*). The number on the plate is illegible due to the angle of the plate in the image. A planar rectification, followed by histogram equalization, reveals the number (*bottom right*)

## *1.1 Geometry*

The relationships between geometric properties in the world and image have been well studied in computer vision [13, 20, 48]. The imaging process, i.e., mapping light rays in the world to points in an image, can be modeled using projective geometry. Projective geometry describes relationships between the image and the world, providing constraints on features in the world due to features in the image. With knowledge of a few properties, such as lengths, angles, or length ratios, it becomes possible to measure image structures and undo projective image distortions. This section describes a simple application of geometric techniques to forensic image analysis. A more detailed overview of projective geometry for computer vision can be found in good books on the subject [13, 20, 48].

Consider the image of the license plate shown in Fig. 2. The number on the plate is illegible due to angle of the plate in the image. Since the dimensions of the license plate are known, the image can be rectified using techniques derived from projective geometry. The result (bottom right) shows the license plate as if it were viewed head-on and the number is now legible. This technique, known as planar rectification, is a simple but useful tool for forensic image analysis [25].

Under an ideal pinhole camera model, points $\mathbf{X}$ on a plane in the world coordinate system are mapped to points $\mathbf{x}$ on the image plane as follows:

$$\mathbf{x} = H\mathbf{X}, \tag{1}$$

where both points are homogeneous 3-vectors in their respective coordinate systems and $H$ is a $3 \times 3$ matrix. Note that under this model, straight lines in the world are imaged to straight lines in the image. Under a more realistic lens model, however, straight lines in the world are distorted and, therefore, not imaged as straight lines.

**Fig. 3** **a** An original image with a billboard and **b** a tampered version of the image

The distortions caused by real lenses and forensic techniques that exploit them are discussed in Sect. 2.

With equations involving homogeneous coordinates, there is typically an unknown scale factor. Equation 1 implies that the cross product is zero and this expression provides a set of equations that are valid independent of the unknown scale factor:

$$\mathbf{x} \times H\mathbf{X} = \mathbf{0}\,,$$

$$\begin{bmatrix} x_1 \\ x_2 \\ x_3 \end{bmatrix} \times \begin{bmatrix} h_1 & h_2 & h_3 \\ h_4 & h_5 & h_6 \\ h_7 & h_8 & h_9 \end{bmatrix} \begin{bmatrix} X_1 \\ X_2 \\ X_3 \end{bmatrix} = \begin{bmatrix} 0 \\ 0 \\ 0 \end{bmatrix}, \tag{2}$$

where $\times$ denotes the cross product.

The cross product in Eq. 2 can be expanded into three equations that are linear in $h_i$, the coefficients of the matrix $H$. Re-ordering the terms yields a system of linear equations for the components of the matrix $H$, expanded into a column vector $\mathbf{h}$:

$$M\mathbf{h} = \mathbf{0}. \tag{3}$$

This system of equations seems to provide three constraints in the nine unknowns of $\mathbf{h}$. However, the rows of the matrix $M$ are not linearly independent (the third row is a linear combination of the first two rows). As such, this system provides only two constraints in the nine unknowns. In order to solve for the projective transformation matrix $H$, four or more points with known coordinates $\mathbf{X}$ and $\mathbf{x}$ are packed into the rows of the matrix $M$.[1] The vector $\mathbf{h}$ that satisfies $\min_{\mathbf{h}} \|M\mathbf{h}\|$, such that $\|\mathbf{h}\|^2 = 1$, is the minimal eigenvalue eigenvector of $M^T M$.

Once the projective transform $H$ is estimated, the inverse transform can be applied to rectify the image region, e.g., Fig. 2. With additional knowledge, such as a known length, measurements can be made in the image plane.

In addition to making measurements in an image, various manipulations may be difficult to perform in the presence of perspective projection and inconsistencies in perspective distortion may be evidence of tampering. As an example, consider the

[1] With four points, the rank of matrix $M$ is 8, one less than the number of unknowns in $\mathbf{h}$. This is sufficient as we only can recover $\mathbf{h}$ to within a scale factor.

problem of manipulating the text or image on a billboard, e.g., Fig. 3. Any manipulation will need to take into account perspective projection. Such manipulations are difficult because small errors in perspective are not perceptually salient [12].

One work has considered the problem of analyzing text under perspective projection [6]. The goal is to determine if the text in the image obeys the laws of perspective projection or if there are inconsistencies. The technique begins by identifying stable features on the text in the image using SIFT [32]. The SIFT features are used to robustly estimate the homography between the text and a fronto-parallel version of the same text in the same font. This process is simple if the font is known, but for unknown fonts, a large pool of common fonts are considered and the best font is selected using the random sample consensus (RANSAC) algorithm [14]. After estimating the homography, the text in the image is warped according to the homography and an image-based error metric is used to estimate authenticity. This approach can detect tampering with 88% accuracy at a false alarm rate of 2% [6].

Planar rectification is one of the basic techniques for exploiting geometric invariants in images, but there are many related techniques, depending on what information is available in the image. For example, in structured environments it is often possible to establish relationships between measurements on parallel planes or to measure the height of a person from the ground plane [7]. The study of techniques for measuring geometric properties of objects from images is known as photogrammetry and there is a vast literature available [46].

## 1.2 Illumination

The fact that light rays obey physical laws during the imaging process gave rise to geometric constraints on the image data. But an image is more than a collection of light rays. Images typically depict scenes and the scene can impose further constraints on the image data. For example, in many cases the source of illumination is constant over a scene, i.e., all objects are illuminated by the same light sources. This property imposes an additional constraint that is useful for image forensics: the illumination conditions on different objects in a scene should be consistent.

When making a composite image from different regions, it is difficult to ensure illumination consistency. The images used as source material have their own illumination conditions and it is challenging to modify illumination effects without knowledge of the light directions, 3D shapes of objects, and material properties. Consequently, relighting objects in images requires technical or artistic skills beyond those of the novice forger. In addition, illumination inconsistencies are not perceptually salient, perhaps due to the adaptation of the human visual system to the wide range of illumination conditions present in the natural world [47]. To the extent that illumination properties can be measured from images, inconsistencies in these measurements are possible evidence of tampering.

Much of the work on estimating illumination properties, in particular illumination direction, from images uses simple illumination and reflectance models, such as point light sources and Lambertian reflectance [5, 23, 37, 39]. For real-world images, illumination conditions can be arbitrarily complex—lights can occur in any number of positions. Although lighting can be complex, its effect on diffuse surfaces can be approximated by a low-dimensional model [2, 45]. This observation allows for illumination analysis with less restrictive assumptions about illumination conditions.

Under the assumption of distant lighting with no cast shadows or interreflections, the intensity at a position $\mathbf{x}$ in an image can be described through the brightness equation [21]:

$$I(\mathbf{x}) = R(\mathbf{N}(\mathbf{x})) , \tag{4}$$

where $\mathbf{N}(\mathbf{x})$ is the surface normal at the position $\mathbf{x}$. The function $R$ maps a surface normal to a scalar intensity and can be viewed as a function on the unit sphere.

Spherical functions can be expressed in terms of spherical harmonics. Spherical harmonics form an orthonormal basis for piecewise continuous functions on the sphere and are analogous to the Fourier basis on the line or plane. For Lambertian reflectance, the function $R$ in Eq. 4 can be approximated with a small number of basis functions [44]:

$$R(\mathbf{N}) \approx \sum_{n=0}^{2} \sum_{m=-n}^{n} \hat{r}_n l_{n,m} Y_{n,m}(\mathbf{N}) . \tag{5}$$

The parameters $\hat{r}_n$ are constants related to the Lambertian reflectance function, the parameters $l_{n,m}$ are the lighting environment coefficients, and the functions $Y_{n,m}$ are the spherical harmonic basis functions. Equations 4 and 5 together imply that image intensities can be approximated as the linear combination of nine basis functions with coefficients $l_{n,m}$. With $p \geq 9$ image intensities and corresponding surface normals the lighting environment coefficients can be estimated as the solution to the following system of linear equations [26]:

$$\begin{bmatrix} \pi Y_{0,0}(\mathbf{N}(\mathbf{x}_1)) & \frac{2\pi}{3} Y_{1,-1}(\mathbf{N}(\mathbf{x}_1)) & \dots & \frac{\pi}{4} Y_{2,2}(\mathbf{N}(\mathbf{x}_1)) \\ \pi Y_{0,0}(\mathbf{N}(\mathbf{x}_2)) & \frac{2\pi}{3} Y_{1,-1}(\mathbf{N}(\mathbf{x}_2)) & \dots & \frac{\pi}{4} Y_{2,2}(\mathbf{N}(\mathbf{x}_2)) \\ \vdots & \vdots & \ddots & \vdots \\ \pi Y_{0,0}(\mathbf{N}(\mathbf{x}_p)) & \frac{2\pi}{3} Y_{1,-1}(\mathbf{N}(\mathbf{x}_p)) & \dots & \frac{\pi}{4} Y_{2,2}(\mathbf{N}(\mathbf{x}_p)) \end{bmatrix} \begin{bmatrix} l_{0,0} \\ l_{1,-1} \\ \vdots \\ l_{2,2} \end{bmatrix} = \begin{bmatrix} I(\mathbf{x}_1) \\ I(\mathbf{x}_2) \\ \vdots \\ I(\mathbf{x}_p) \end{bmatrix},$$

$$M\mathbf{v} = \mathbf{b}, \tag{6}$$

where $M$ is the matrix containing the sampled spherical harmonics, $\mathbf{v}$ is the vector of unknown lighting environment coefficients, and $\mathbf{b}$ is the vector of intensities at $p$ points. But this solution requires 3D surface normals, which are difficult to measure from a single image of objects with unknown geometry.

While arbitrary surface normals are difficult to estimate from objects in a single image, surface normals along the occluding contour are simple to measure. Under an assumption of orthographic projection, surface normals along an occluding contour are perpendicular to the line of sight, i.e., their $z$ components are zero. These normals

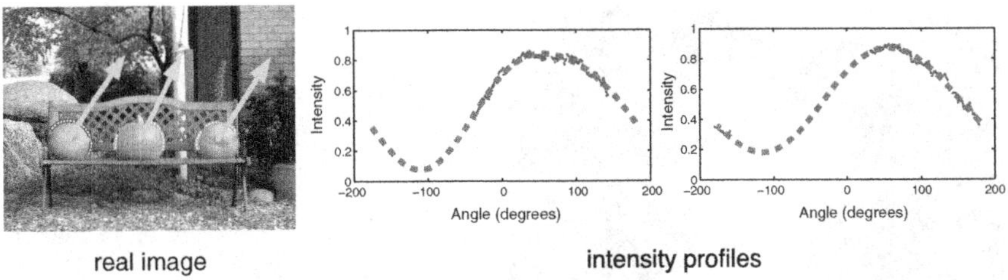


**Fig. 4** Illumination analysis of a real image with three pumpkins. The lighting environment model fit to the intensity profiles (*center* pumpkin not shown) reveals light directions of 55, 62, and 56 degrees

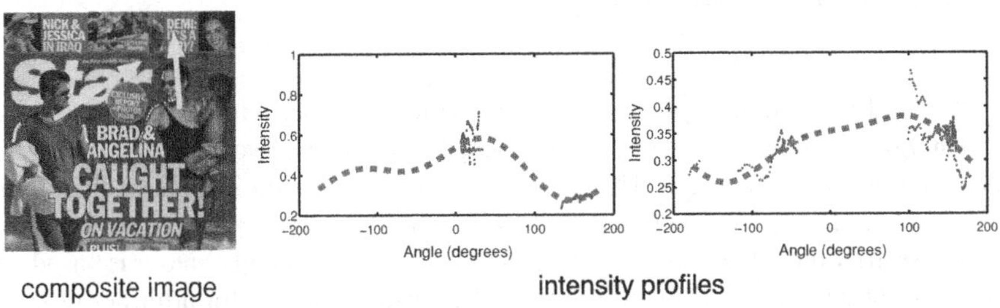


**Fig. 5** Illumination analysis of a composite image of two celebrities on vacation. The lighting environment model fit to the intensity profiles reveals light directions of 94 and 33 degrees

are essentially two dimensional and can be measured by fitting a curve to the occluding contour of the object. Using 2D surface normals simplifies Eq. 6 into a system of equations involving five unknown lighting environment coefficients. While providing less information than would be available with 3D normals, this assumption makes the problem tractable [37]. The resulting system of equations has the same form as Eq. 6, $M\mathbf{v} = \mathbf{b}$, and the solution is obtained through least-squares:

$$\mathbf{v} = \left(M^T M\right)^{-1} M^T \mathbf{b} \,. \tag{7}$$

In practice, it is necessary to use regularization when only parts of the occluding contour are available for analysis [26].

The coefficients in the vector $\mathbf{v}$ describe the lighting environment—measurements from different objects in the image should be similar if the lighting in the scene is assumed to be constant. For example, consider the image of three pumpkins on a bench shown in Fig. 4. Intensity samples from the occluding contours for two of the pumpkins are shown to the right of the image, with the lighting model plotted as the dashed (red) line. The peak of these curves can be interpreted as the dominant light direction. Analysis of the lighting on the three pumpkins reveals light directions of 55, 62, and 56 degrees.

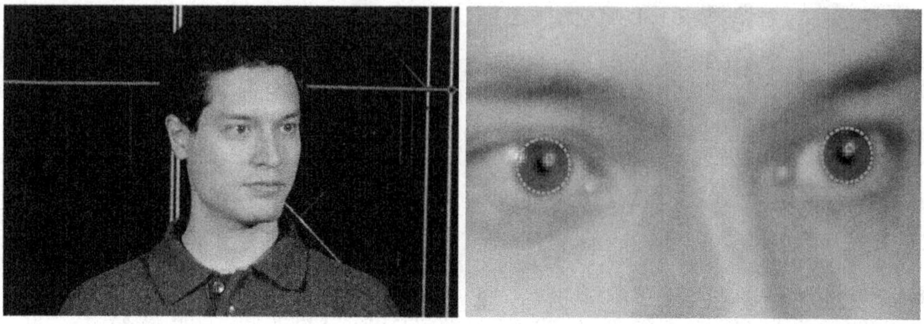

**Fig. 6** The 3D direction to the light source can be estimated from the elliptical shape of the iris and the location of the specular highlight on the eye

When objects come from different sources images, the lighting may be inconsistent. For example, consider the magazine cover depicting two celebrities walking along a beach shown in Fig. 5. The lighting analysis reveals dominant light directions of 94 and 33 degrees for the two people, suggesting that this image is a composite.

The restriction of the illumination estimation to the occluding contour is based on the assumption that it is difficult to obtain 3D surface normals for arbitrary objects in images. In images containing people, however, the human face can be used as a source of 3D surface normals and richer illumination models can be estimated [29]. One technique uses a 3D morphable face model derived from a set of 3D laser scans [3]. This face model is useful in computer graphics for creating realistic 3D models of faces from input images. The model requires a paired frontal and profile view of the person, or a single frontal view, as well as the manual selection of fiducial points. After the inputs are specified, the morphable model is registered to the face in the image by optimizing over the intrinsic and extrinsic camera parameters. Once the model is registered with the image, the full illumination model, Eq. 6, can be fit using 3D surface normals. This illumination model allows for more accurate lighting environment estimates since the analysis is not limited to the occluding contour. However, this approach requires a human face or another object with a 3D shape that can be estimated from an image.

Another approach that considered forgeries of people looked at specular highlights on eyes and estimated 3D light directions from these highlights [27]. For example, consider the image of the man in Fig. 6. There are specular highlights in each of his eyes. A specular highlight is a reflection of a light source back to the camera. The physics of reflection constrains the position of the light source with respect to the plane of the eyes and location of the camera. In addition, the boundary of the iris is approximately circular and it appears in an image as an ellipse if the person is not directly looking at the camera. By measuring the shape of the iris, the person's gaze direction can be measured which allows the 3D light direction to be estimated from the position of the highlights on both eyes. While this technique requires high-resolution images of eyes, it provides another example of invariants in images that are measurable and may be overlooked when tampering with an image.

Shadow consistency can also be measured for image forensics [51]. The location, size, and shape of shadows are directly related to the illumination and 3D scene structure and may be difficult to synthesize in a physically correct way. At a more theoretical level, identities exist for verifying consistency of lighting and shading when the object shapes are known [35]. Future work may extend these identities to apply in a practical setting, when object shape is unknown and must be estimated from the image.

## 2 Lens

Computer vision algorithms often assume a pinhole camera model for simplicity. For example, the geometric relationships discussed in Sect. 1.1 all assume projection through an infinitesimally small principal point. While pinhole cameras exist, they do not gather enough light in a short time span to be practical in most scenarios. Lenses gather light over a larger area, so that exposure times can be short enough to freeze motion in the scene. But lenses are not perfect imaging devices and they introduce a variety of artifacts into an image. From a forensic point-of-view, these artifacts are another property of an image that can be measured and used to detect tampering or to authenticate an image.

In a perfect imaging system, light rays from a point in the world that pass through a lens are all imaged to the same point on the sensor. Real optical systems, however, fail to perfectly focus light of all wavelengths. Different wavelengths of light can be focused to different focal planes or to different positions in the same focal plane. These distortions are known as longitudinal and lateral chromatic aberration, respectively. To a first-order approximation, longitudinal aberrations amount to slight differences in blur and magnification between the color channels. Lateral chromatic aberration can be modeled as an expansion/contraction pattern of the color channels about the optical center. Since the expansion/contraction pattern has a simple parametric form, the parameters of the pattern can be estimated from a single image. Large inconsistencies in estimates of the parameters from different locations in an image could indicate tampering.

In optics, the refraction of light at the boundary between two different media is described by Snell's Law:

$$n_a \sin(\theta_a) = n_b \sin(\theta_b) \tag{8}$$

where $n_a$ and $n_b$ are the refractive indices of the different media. The refractive index of glass depends on the wavelength of the light that traverses it. In a camera, this dependency causes light rays of different wavelengths to be imaged at different locations on the sensor. The shift in position in 1D results in an expansion/contraction pattern in 2D, as shown in Fig. 7.

The expansion/contraction pattern can be approximated using a low-parameter model. For example, let $(x_b, y_b)$ and $(x_r, y_r)$ be coordinates in the blue and red channels, respectively. The model has the form:

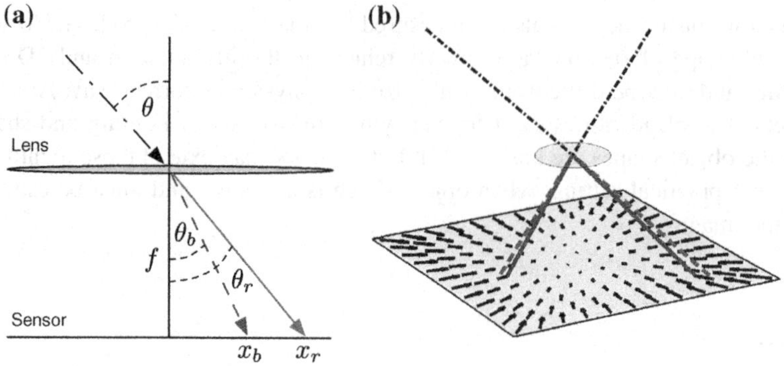


**Fig. 7** Lateral chromatic aberration in 1D (**a**) and 2D (**b**). Polychromatic light enters the lens and emerges at an angle that depends on the wavelength. Consequently, different wavelengths of light will be imaged at different points on the sensor. In 2D, the distortion can be modeled as an expansion/contraction pattern

$$x_r = \alpha(x_b - x_0) + x_0 \tag{9}$$

$$y_r = \alpha(y_b - y_0) + y_0 \tag{10}$$

where $(x_0, y_0)$ is the center of the aberration. In real lenses, the center of optical aberrations is often different from the image center due to the complexities of multi-lens systems [49].

In [24], a coarse-to-fine optimization using mutual information is proposed to estimate the three parameters of the model, Eq. 10. This optimization is run between two pairs of color channels, red to green and blue to green. The optimization finds a global estimate of the expansion/contraction pattern. To detect tampered regions in an image, the optimization can be performed on separate blocks in the image, provided the block has sufficient contrast to yield a reliable estimate. The average angular error between the expansion/contraction vectors in the global versus local estimates is used as evidence of tampering.

As an example, consider the image in Fig. 8a. A forgery is created by moving the ship from the left side of the image to the right, Fig. 8b. But with this manipulation, the local expansion/contraction pattern around the ship becomes inconsistent with the global pattern. As a result, the tampered region can be detected using the method described in [24].

A more recent work has considered a more general model of chromatic aberration that includes the effects of both lateral chromatic aberration and purple fringing [50]. Both of these artifacts increase in strength with distance from the image center and become more pronounced in regions of high contrast. Therefore, local estimates of this aberration provide constraints on the image center. In addition, estimates from high contrast image regions are generally more reliable than estimates from low contrast regions. Accordingly, this technique measures local purple fringing events and assigns a reliability measure to each event. The combined information from

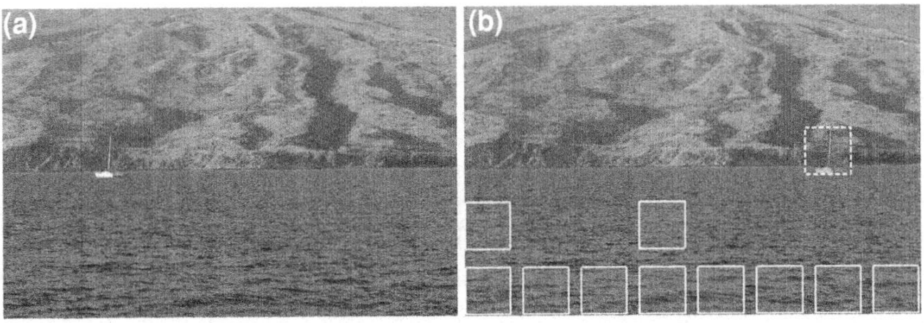


**Fig. 8** Detecting image forgeries using chromatic aberration. On the left is an original image and on the right is a tampered version, where the ship has been moved to the opposite side of the image. On the right, the red (*dashed outline*) blocks denote regions that are inconsistent with the global aberration estimate. The green (*solid outline*) blocks denote regions that are consistent with the global estimate

all events is used to estimate the image center. By using a more general model of aberration and pooling many local estimates, this approach is more robust and accurate than the original technique, described in [24]. The new approach is also able to detect cropped images when the estimated image center does not align with the numerical image center.

Another lens distortion that has been exploited for forensic purposes is vignetting. Vignetting causes a gradual darkening toward the image periphery and it is attributed to several effects related to light paths within the lens [16]. The method described in [34] models vignetting with a parametric function and uses maximum likelihood estimation to find the parameters of the model. The estimated vignetting functions are effective for classifying various lens models.

While high-end lens manufacturers strive to design lenses with fewer aberrations, the majority of images are being captured with cellphone cameras. These cameras often have inexpensive optics, such as plastic lenses, and the images from these cameras suffer from various aberrations. Techniques for using these aberrations as invariants to detect tampering is an exciting direction for future research.

## 3 Sensor

Two types of sensors are commonly found in digital cameras: charge-coupled device (CCD) and complementary metal-oxide semiconductor (CMOS). Both types of sensors consist of an array of photo detectors, called pixels, which convert photons into electrons. The charge accumulates during the exposure time and is then amplified and processed to form the image. But due to imperfections in the manufacturing process, there are slight variations in the ability of each pixel to gather light. These variations are known as the photo response nonuniformity (PRNU). Once the PRNU has been estimated, it can be used for a variety of forensic tasks, including camera identification [33], camera identification under scaling and cropping [18], device identification and linking [17], and image forgery detection [4].

The PRNU is not the only source of sensor imperfections that can be exploited for forensic purposes. While discussed above as a lens imperfection, purple fringing is prevalent on CCD devices and is thought to be related to the sensor as well as the optics [50]. For example, CCD sensors can suffer from electron overflow where bright areas of a scene may cause blur in nearby edges. Also, micro lenses and filters near the CCD may contribute to the effect. Whatever the true source, sensor or optics, the purple fringing effect is an invariant that may be difficult to simulate when tampering with an image and can therefore be used for detecting doctored images.

Another sensor invariant comes from the observation that the sensors in digital single-lens reflex cameras often accumulate dust due to the interchangeable lens. If the sensor is not cleaned, the dust particles will leave a pattern in every image. The authors of [9] model the effect of sensor dust using a 2D Gaussian function. They describe a method for estimating a dust template from ten images and show how this template can be used for camera identification.

The sensitivity of the imaging sensor to photons governs the amount of time that is necessary to obtain a properly exposed image. Any motion during the exposure time, either due to moving objects in the world or due to the hand holding the camera, will create blur in the image. Blur that is due to camera motion will affect all pixels in the image and is therefore another invariant that can be measured to detect tampering [28].

## 4 Processing

After the exposure time has elapsed, the accumulated charge at every pixel is processed to form an image. This processing typically involves an interpolation step, called demosaicing, and a nonlinear tone mapping known as the camera response. In addition, images from most digital cameras are stored in JPEG format with parameters chosen by the camera manufacturer. Each of these processing stages has been used for forensic purposes. These three stages of the processing pipeline are shown in Fig. 9.

To capture a color image, most cameras have a color filter array (CFA) in front of the sensor. This filter array splits the incoming light rays according to wavelength and images them onto different pixels. Thus, a typical sensor only captures one of the three color channels at each pixel. To create RGB values for each pixel, the missing color channels are interpolated from neighboring pixels using a demosaicing algorithm. The authors of [43] show that the interpolation process introduces statistical correlations that are useful for forensics. To understand these correlations, consider the simple linear interpolation of a 1D signal $f(n)$ to yield a new signal $g(n)$ that has twice as many samples:

$$g(n) = \begin{cases} f(n/2) & \text{if } n \text{ is even,} \\ \frac{1}{2}f(\lfloor n/2 \rfloor) + \frac{1}{2}f(\lfloor n/2 \rfloor + 1) & \text{if } n \text{ is odd,} \end{cases} \quad (11)$$

where $\lfloor n/2 \rfloor$ denotes the floor operation.

In this example, all of the even-numbered samples in $g(n)$ are copies of samples from the signal $f(n)$ and the odd-numbered samples in $g(n)$ are linear combinations of neighboring samples. The odd-numbered samples are therefore correlated with neighboring samples and the interpolation scheme can be detected by measuring such correlations. While actual CFA interpolation schemes are more complex than linear interpolation, they introduce specific correlations into the image that are also periodic due to the periodic arrangement of color filters in the CFA.

If the interpolation scheme were known, then it would be possible to determine which subset of pixels is interpolated from its neighbors. Similarly, if the subset of interpolated pixels were known, it would be possible to model the correlations due to interpolation. In practice, neither the form of the correlations nor the interpolated pixels are known. In [43], the authors use the expectation–maximization algorithm to simultaneously fit a linear model of the correlation and estimate the pixels that are correlated with their neighbors. The presence of periodic correlation patterns over an image is evidence that it is interpolated directly from the sensor data (i.e., it has not been tampered with). Conversely, abrupt changes in the correlation patterns within an image region could suggest tampering.

Other stages of the post processing that have been used for forensics include the camera response function and JPEG encoding. The response of the individual pixels are close to linear in the amount of incoming light, since the charge accumulates linearly with the number of photons [8]. However, nonlinearities are often introduced to account for the response of the human visual system or to compress the dynamic range of the image [19]. The authors of [31] describe a method for estimating the response function from different image segments and show how these estimates can be used for splicing detection. A similar technique is described in [22].

The final stage of the processing pipeline in many digital cameras is JPEG compression. This compression scheme has several parameters that are not specified in the standard, allowing for flexibility to balance compression and image quality. As described in [10] and [11], the differences in JPEG parameters can be used to identify the make and model of the camera that captured an image.

## 5 Image

We have considered the stages of the imaging pipeline shown in Fig. 1 and have described forensic techniques that exploit invariants introduced by each of these stages. The final class of forensic techniques relies on properties of the image itself without making assumptions about the camera or world. These techniques describe how common image manipulations affect the pixel values in an image. They typi-

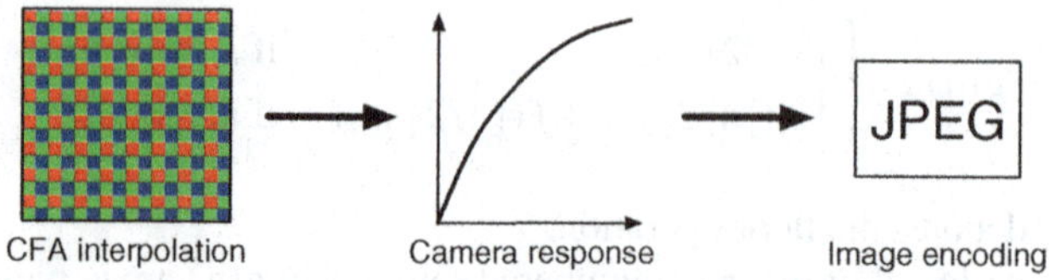


**Fig. 9** The post-processing pipeline. Most cameras use a color-filter array (CFA) to record information about the color of light at different positions on the imaging sensor. For cameras that record images in JPEG format, the raw sensor data is usually passed through a nonlinear camera response function and then quantized according to the JPEG specification. Each of these processing steps introduces regularities into the image that are useful for forensics

cally use simple models of the image invariants and detect tampering by measuring inconsistencies in the parameters of the models.

One of the most basic image manipulations is copy move or cloning. This manipulation is necessary if a forger needs to cover part of an image and it can be successful if a homogeneous texture is available (e.g., grass, sand, or water). Although different regions of a homogeneous texture may look similar qualitatively, it is highly unlikely that they will be exactly the same numerically. Two different forensic tools exploit this basic observation to detect cloning [15, 41].

The copy-move forgery creates two regions with identical pixels at a fixed offset. Intuitively, the forgery can be detected by searching for identical pairs of regions, but a naive search algorithm over all pairs of regions of arbitrary size would be prohibitively expensive. A more efficient algorithm is possible by dividing the image into fixed-size blocks and lexicographically sorting the blocks—blocks with exactly the same pixel values will be adjacent in the sorted list. To make this algorithm robust to noise and lossy compression, the authors of [41] use principal component analysis (PCA) to yield a reduced-dimension representation.

Suppose $\mathbf{x}_i$ is a $b^2 \times 1$ vector of pixels from the $i$th $b \times b$ block in the image. We can form a data matrix $X$ by placing vectors from $n$ different blocks in the image along the columns. The resulting data matrix will have size $b^2 \times n$.

The first step in PCA is to subtract the mean from the data matrix. Let $\hat{x}$ represent the mean:

$$\hat{x} = \frac{1}{n} \sum_{i=1}^{n} \mathbf{x}_i .$$

We can create a zero-mean data matrix $\hat{X}$ where the $i$th column of the matrix is $\mathbf{x}_i - \hat{x}$.

The goal of PCA is to find a orthogonal transformation of the zero-mean data matrix $\hat{X}$ that diagonalizes the covariance matrix $C = \hat{X}\hat{X}^T$. The eigenvalue decomposition of the covariance matrix $C$ provides such a relationship.

Let $V$ represent the eigenvectors of the covariance matrix, sorted by column in order of decreasing eigenvalue. After transforming the data by the orthogonal transformation $V$, the transformed coefficients will be ordered in terms of their ability

**Fig. 10** The copy-move forgery can conceal part of an image. **a** The original image. **b** A region of sand is copied over the people on the beach to create a tampered image. **c** The copy-move detection algorithm described in [41] can successfully identify the duplicated regions

to preserve the variance in the original signal. In other words, the first $p$ coefficients represent the best $p$-dimensional representation of the original data. Therefore, the data can be reduced to a $p$-dimensional representation by removing all coefficients beyond the $p$th coefficient. Here we assume that $p$ is smaller than $b^2$ and this reduced dimensional representation will improve runtime efficiency as well as robustness to noise [41].

The PCA coefficients are quantized and then sorted into lexicographic order. The offsets (i.e., spatial displacements along rows and columns) of adjacent blocks in the sorted list are tabulated and the algorithm detects duplicated regions by finding a large number of blocks with the same offset in the image.

As an example, consider the beach scene shown in Fig. 10a. The people on the beach were concealed by copying a region of sand from another location in the image, Fig. 10b. While the forgery is difficult to detect by visual inspection, it contains a large number of duplicate image blocks with the same offset. The algorithm described in [41] can detect this forgery, Fig. 10c; a similar algorithm is described in [15]. While both of these algorithms are robust to small amounts of additive noise and compression artifacts, they are not able to detect duplicated regions under geometric or illumination distortions. A newer method uses keypoints from the scale invariant feature transform (SIFT) to identify duplicated regions under various distortions [38].

When tampering with an image, it is often necessary to scale or rotate regions in an image. This process requires resampling the image region, which is an interpolation process that introduces periodic correlations, as discussed in Sect. 4. In this case, the periodic correlations due to resizing or rotation indicate tampering and they can be detected using an expectation–maximization algorithm [42]. Another approach uses image quality metrics detect resampling and other common image-processing operations [1]. Other approaches for detecting resampling are described in [30] and [36].

The workflow for creating a forgery involves opening an image in a photo editing application, making the desired changes, and resaving the image. If both the original and the tampered images use the JPEG format, the tampered image would have been compressed twice. The JPEG compression scheme quantizes the DCT coefficients of $8 \times 8$ blocks in the image. As described in [40], double quantization produces periodic artifacts in the histograms of DCT coefficients. The presence of these artifacts indicates that the image has been doubly compressed, which is suspicious, but not necessarily malicious—the user may have simply resaved the image with a different quality setting. However, the presence of these artifacts can warrant additional analysis by other forensic techniques.

## 6 Summary

This chapter has described the themes behind many of the current techniques for detecting doctored images. The core assumption of these techniques is that images possess certain regularities that are disturbed by tampering. These regularities may come from any stage in the imaging pipeline: the world, lens, sensor, post processing, or the image itself. Tampering can be revealed by measuring these regularities or deviations in them.

While not exhaustive, this chapter highlighted the themes in image-forensic research and described some of the techniques that have emerged in recent years. Interested readers can consult the original publications for more details about the various methods discussed in this chapter.

## References

1. Avcibas I, Bayram S, Memon N, Sankur B, Ramkumar M (2004) A classifier design for detecting image manipulations. In: 2004 International Conference on Image Processing, ICIP '04, vol 4. pp 2645–2648, 2004
2. Ronen B, Jacobs DW (2003) Lambertian reflectance and linear subspaces. IEEE Trans Pattern Anal Mach Intell 25(2):218–233
3. Blanz V, Vetter T (1999) A morphable model for the synthesis of 3D faces. In: Proceedings of the SIGGRAPH, pp 187–194, 1999
4. Chen M, Fridrich J, Goljan M, Lukáš J (2008) Determining image origin and integrity using sensor noise. IEEE Trans Inf Forensics Secur 3(1):74–90
5. Chojnacki W, Brooks MJ (1994) Revisiting Pentland's estimator of light source direction. J Opt Soc Am 11(1):118–124
6. Conotter V, Boato G, Farid H (2010) Detecting photo manipulations on signs and billboards. In: International Conference on Image Processing, pp 1–4, 2010
7. Criminisi A, Reid I, Zisserman A (2000) Single view metrology. Int J Comput Vis 40(2):123–148
8. Debevec PE, Malik J (1997) Recovering high dynamic range radiance maps from photographs. In: Proceedings of the 24th annual conference on Computer graphics and interactive techniques (SIGGRAPH), 1997
9. Dirik AE, Sencar HT, Memon N (2008) Digital single lens reflex camera identification from traces of sensor dust. IEEE Trans Inf Forensics Secur, 3(3):539–552
10. Farid H (2006) Digital image ballistics from JPEG quantization. Technical Report TR2006-583, Department of Computer Science, Dartmouth College, 2006
11. Farid H (2008) Digital ballistics from JPEG quantization: a followup study. Technical Report TR2008-638, Department of Computer Science, Dartmouth College, 2008

12. Farid H, Bravo MJ (2010) Image forensic analyses that elude the human visual system. In: SPIE Symposium on Electronic, Imaging, pp 1–10, 2010
13. Faugeras O (1993) Three-Dimensional Computer Vision. MIT Press, Cambridge
14. Fischler MA, Bolles RC (1981) Random sample consensus: a paradigm for model fitting with applications to image analysis and automated cartography. Communications of the ACM 24(6):381–395
15. Jessica F, David S, Lukáš J (2003) Detection of copy-move forgery in digital images, In: Proceedings of DFRWS, 2003
16. Goldman DB (2010) Vignette and exposure calibration and compensation. IEEE Trans Pattern Anal Mach Intell, 32(12):2276–2288
17. Goljan M, Chen M, Fridrich J (2007) Identifying common source digital camera from image pairs. In: Proceedings of thc IEEE ICIP, 2007
18. Goljan M, Fridrich J (2008) Camera identification from cropped and scaled images. In: Proceedings of the SPIE Electronic Imaging: Security, Forensics, Steganography, and Watermarking of Multimedia Contents X, vol 6819. 2008
19. Grossberg MD, Nayar SK (2004) Modeling the space of camera response functions. IEEE Trans Pattern Anal Mach Intell 26(10):1272–1282
20. Hartley R, Zisserman A (2004) Multiple View Geometry in Computer Vision. Cambridge University Press, Cambridge
21. Horn BKP (1970) Shape from shading; a method for obtaining the shape of a smooth opaque object from one view. PhD thesis, Massachusetts Institute of Technology
22. Hsu Y-F, Chang S-F (2010) Camera response functions for image forensics: An automatic algorithm for splicing detection. IEEE Trans Inf Forensics Secur 5(4):816–825
23. Johnson MK, Farid H (2005) Exposing digital forgeries by detecting inconsistencies in lighting. In: Proceedings of the 7th workshop on multimedia and security, MM&Sec '05, pp. 1–10, 2005
24. Johnson MK, Farid H (2006) Exposing digital forgeries through chromatic aberration. In: Proceedings of the 8th workshop on multimedia and security, MM&Sec '06, pp 48–55. ACM, 2006
25. Johnson MK, Farid H (2006) Metric measurements on a plane from a single image. Technical Report TR2006-579, Department of Computer Science, Dartmouth College, 2006
26. Johnson MK, Farid H (2007) Exposing digital forgeries in complex lighting environments. IEEE Trans Inf Forensics Secur 2(3):450–461
27. Johnson MK, Farid H (2008) Exposing digital forgeries through specular highlights on the eye. In: Information Hiding of Lecture Notes in Computer Science, vol 4567. pp. 311–325, 2008
28. Pravin K, Sudha N, Ser W (2011) Exposing digital image forgeries by detecting discrepancies in motion blur. IEEE Trans Multimedia 13(3):443–452
29. Kee E, Farid H (2010) Exposing digital forgeries from 3-D lighting environments. In: IEEE International Workshop on Information Forensics and Security, Seattle, WA, 2010
30. Kirchner M (2008) Fast and reliable resampling detection by spectral analysis of fixed linear predictor residue. In: Proceedings of the 10th workshop on multimedia and security, MM&Sec '08, pp 48–55. ACM, 2008
31. Zhouchen L, Rongrong W, Xiaoou T, Shum H-Y (2005) Detecting doctored images using camera response normality and consistency, In: IEEE Conference on Computer Vision and Pattern Recognition, 2005
32. Lowe DG (2004) Distinctive image features from scale-invariant keypoints. Int J Comput Vis 2(60):91–110
33. Lukáš J, Fridrich J, Goljan M (2006) Digital camera identification from sensor pattern noise. IEEE Trans Inf Forensics Secur 1(2):205–214
34. Lyu S (2010) Estimating vignetting function from a single image for image authentication. In: MM&Sec '10, pp 3–12, 2010
35. Mahajan D, Ramamoorthi R, Curless B (2008) A theory of frequency domain invariants: Spherical harmonic identities for BRDF/lighting transfer and image consistency. IEEE Trans Pattern Anal Mach Intell 30(2):197–213

36. Mahdian B, Saic S (2008) Blind authentication using periodic properties of interpolation. IEEE Trans Inf Forensics Secur 3(3):529–538
37. Nillius P, Eklundh J-O (2001) Automatic estimation of the projected light source direction. In: Computer Vision and Pattern Recognition, 2001
38. Pan X, Lyu S (2010) Region duplication detection using image feature matching. IEEE Trans Inf Forensics Secur 5(4):758–767
39. Pentland A (1982) Finding the illuminant direction. J Opt Soc Am 72(4):448–455
40. Popescu A, Farid H (2004) Statistical tools for digital forensics. In: 6th International Workshop on Information Hiding, 2004
41. Popescu AC, Farid H (2004) Exposing digital forgeries by detecting duplicated image regions. Technical Report TR2004-515, Department of Computer Science, Dartmouth College, 2004
42. Popescu AC, Farid H (2005) Exposing digital forgeries by detecting traces of re-sampling. IEEE Trans Signal Process 53(2):758–767
43. Popescu AC, Farid H (2005) Exposing digital forgeries in color filter array interpolated images. IEEE Trans Signal Process 53(10):3948–3959
44. Ramamoorthi R, Hanrahan P (2001) An efficient representation for irradiance environment maps. In: Proceedings of the 28th annual conference on Computer graphics and interactive techniques, SIGGRAPH '01, pp 497–500. ACM Press, 2001
45. Ramamoorthi R, Hanrahan P (2001) On the relationship between radiance and irradiance: determining the illumination from images of a convex Lambertian object. J Opt Soc Am A 18:2448–2559
46. Remondino F, El-Hakim S (2006) Image-based 3D modelling: A review. The Photogramm Rec 21(115):269–291
47. Sinha P (2000) Perceiving illumination inconsistencies. In: Investigative Ophthalmology and Visual Science, vol 41(4):1192, 2000
48. Szeliski R (2011) Computer Vision: Algorithms and Applications. Springer, New York, 2011
49. Willson RG, Shafer SA (1994) What is the center of the image? J Opt Soc Am A 11(11):2946–2955
50. Yerushalmy I, Hel-Or H (2011) Digital image forgery detection based on lens and sensor aberration. Int J Comput Vis 92(1):71–91
51. Zhang W, Cao X, Zhang J, Zhu J, Wang P (2009) Detecting photographic composites using shadows. In: IEEE Conference on Multimedia and Expo, pp 1042–1045, 2009

# Discrimination of Computer Synthesized or Recaptured Images from Real Images

**Tian-Tsong Ng and Shih-Fu Chang**

**Abstract** An image that appears to be a photograph may not necessarily a normal photograph as we know it. For example, a photograph-like image can be rendered by computer graphics instead of being taken by a camera or it can be a photograph of an image instead of a direct photograph of a natural scene. What is really different between these photographic appearances is their underlying synthesis processes. Not being able to distinguish these images poses real social risks, as it becomes harder to refute claims of child pornography as non-photograph in the court of law and easier for attackers to mount an image or video replay attack on biometric security systems. This motivates digital image forensics research on distinguishing these photograph-like images from true photographs. In this chapter, we present the challenges, technical approaches, system design and other practical issues in tackling this multimedia forensics problem. We will also share a list of open resources and the potential future research directions in this area of research which we hope readers will find useful.

## 1 Motivations

Since the ancient time of Greek and Roman, artists have been playing with special painting techniques for inducing visual illusion where objects in a painting appear to be real and immersed in the real surrounding. *Trompe l'oeil*, the name for such visual artistry, literally means *deceiving the eye*. For example, the painting entitled *Escaping Criticism* created by Pere Borrell del Caso in 1874 depicts a person climbing out of

T.-T. Ng (✉)
Institute for Infocomm Research, Singapore 138632, Singapore
e-mail: ttng@i2r.a-star.edu.sg

S.-F. Chang
Columbia University, New York, NY 10027, USA
e-mail: sfchang@ee.columbia.edu

H. T. Sencar and N. Memon (eds.), *Digital Image Forensics*,
DOI: 10.1007/978-1-4614-0757-7_10,

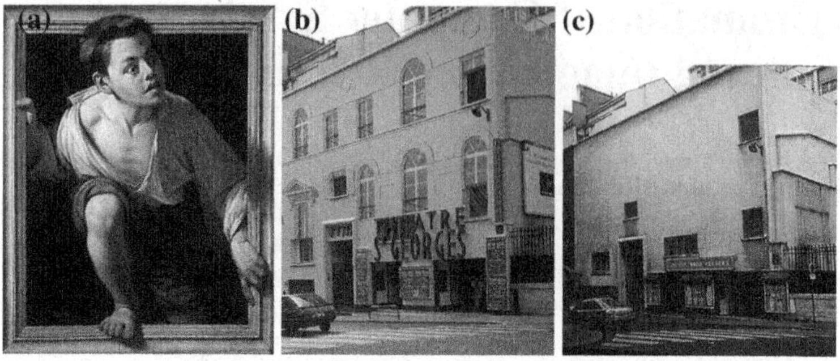


**Fig. 1** The art of visual deception, *trompel'oeil*. **a** A painting, *Escaping Criticism* created by Pere Borrell del Caso in 1874. **b** The facade of Saint-Georges Theater in Paris, France, created by the mural painter Dominique Antony. **c** The facade of Saint-Georges Theater before the mural painting

the painting with a make-believe quality (Fig. 1a); the mural painting on the facade of the Saint-Georges Theater created by a painter Dominique Antony induces an impression of balcony (Fig. 1b), while the facade is in fact just a flat wall (Fig. 1c).

*Trompe l'oeil* makes believe with not only the photorealistic quality in the painting, it exploits the visual gullibility of human observers through immersion into the real surrounding. Such adversarial nature of *trompe l'oeil* aptly mirrors that of digital image forensics where the intention to deceive is present. While the deceptive intent remains with human, the photorealism of an image which takes many years of practice for conventional artists to master can now be easily produced by laymen with modern technology. Physics-based computer graphics is capable of rendering photorealistic images that emulate images of real three-dimensional scenes, a great advancement from the two-dimensional graphics like cartoon which was popular at the early days of computer graphics. It has been shown that a scene with diffuse reflectance can be realistically rendered to the extent that the rendered scene radiance is close to that of a real scene and perceptually indistinguishable for human [43].

A photograph is photorealistic by definition. Recapturing a photograph with a camera when displayed in good quality on a paper or screen preserves the photorealism of the photograph if perspective distortion is minimized.[1] We refer to such type of image as *recaptured or rephotographed image*. The modern artist Richard Prince pioneered a unique art form of rephotographing advertisements from magazines where the artistic expression was conveyed through intensifying certain distinct characteristics of the reproduced image with various photographing techniques such as blurring, cropping and enlarging, while preserving photorealism of the original images.

As synthesizing photorealism gets easier, seeing a red apple can no longer immediately implies the actual presence of the apple, i.e., the link between the image

[1] The perspective distortion, in the form of planar homography, due to image recapturing can result in a non-zero skew in the camera internal parameters [24] and could appear visually unnatural to human observers.

of an object and the presence of the object is weakened. The weakened link poses real security risks. In the US, possession of child pornography is punishable as it implies abuse of minors. However, establishing the presence of minors from the child pornography is challenging on legal ground, as owners of child pornography can proclaim the images to be computer generated [13].[2] An important implication of the weakened link is the need for technology to recognize the underlying formation process of an image.

On the other hand, photorealistic rephotographed image is instrumental for *image/video replay attack* on biometric authentication systems [7]. In 2008, the Vietnamese Internet Security Center BKIS demonstrated the ease of breaching the face authentication login in commercial laptop computers using printouts of face images of a legitimate user [54]. Similar vulnerability is also shown for the face authentication login in a version of Andriod operating system known as Ice-cream Sandwich introduced in October 2011 [33]. The ease of accessing someone's face images on the Internet has made image replay attack problem looms larger.

Although computer graphics and rephotographed images can be as perceptually photorealistic as real photographs, their underlying image formation processes have distinctive characteristics. Such differences, though subtle, can be used for distinguishing these images as described in Sects. 4 and 5. Being able to recognize the underlying image formation process has far-reaching impacts. Such capability will make it harder for child pornography owners to get away with the computer graphic claim and harder for image/video replay attack to succeed. It will also make it harder for doctored images to escape detection through rephotographing, which can turn a doctored image into a quintessential photograph.

In essence, detecting rephotographed image is related to a fundamental problem in computer vison: *monocular depth perception*. It is important that a robot with monocular vision [45] does not confuse objects in a poster as real, as the two have greatly distinct semantics in a physical scene. Such monocular depth perception is also useful for algorithms that convert the conventional 2D movies into 3D content for display on 3D televisions.

General-class or specific object recognition has been widely researched for high-level computer vision. Object recognition is generally approached through appearance-based recognition, hence it does not differentiate a red apple for example from a red apple in a picture found in a scene. This is evidenced from the form of the public benchmark datasets for object recognition such as Caltech 101 [12, 16, 17, 39, 70], there is not an object class called poster for example. However, a versatile object recognition system ideally should be capable of recognizing a scene picture

[2] The ruling of the United States Supreme Court in 2002 on a clause in the 1996 Child Pornography Prevention Act (CPPA) defines child pornography as *any* visual depiction of explicit sexual conduct that involves a child. However, in 2002, the United States Supreme Court considered the broad definition of child pornography in CPPA that includes "virtual imagery" as unconstitutional and violating the freedom of speech as enacted in the First Amendment. As a result, the Court ruled that computer generated images including those with child pornography content are to be protected constitutionally.

such as a poster in the scene. The capability of detecting rephotographed image will augment the functionality of the current object recognition algorithms.

In Sect. 2, we show the evidence that distinguishing photorealistic computer graphics and rephotographed images from true photographs is challenging. In Sect. 3, we describe the desired characteristics in an algorithm for recognizing the underlying image formation of images when considering various fundamental and application-related issues. In Sects. 4 and 5, we survey the various approaches for distinguishing photorealistic computer graphics and rephotographed images from photographs. It is common for a security system to face threat of attacks. In Sect. 6, we describe the potential attacks on a computer graphics or recaptured image detector and the corresponding counter-attack measures. In Sect. 7, we give a list of resources useful for researchers, including a few open benchmark datasets and a public evaluation system. Finally, we describe the open issues and future direction for this area of research in Sect. 8, before concluding in Sect. 9.

## 2 Challenges

Visual realism is no longer a hallmark exclusive for the common photographs. We will look into the level of visual realism achievable by computer graphics rendering and image rephotographing, and the challenges for human to discern them perceptually. We also explore the possibility of discerning these image formative processes using computer.

### *2.1 Visual Realism of Computer Graphics*

One of the important goals of computer graphics rendering is to produce photorealistic imageries that are perceptually close to real-scene images. Real-scene radiance induces visual stimuli that constantly impacts human visual system. Through biological evolution, human visual system is adapted to such visual stimuli [65] and hence develops a keen sensory for real-scene images. Ferwerda [18] defined three varieties of realism: *physical realism* which provides the same visual stimulation as the real-world scene, *photorealism* which produces the same visual response in human as the scene, and *functional realism* which allows human receives the same visual information, e.g., object shape and scene depth, as from the real scene. From the definition, achieving photorealism does not require faithful reproduction of real-scene radiance, although the physics-based computer graphics based on Kajiya's rendering equation [27] is capable of simulating real-scene radiance or achieving physical realism.

Studies of computer graphics photorealism and its perception are of interest to the computer graphics community, as the level of achievable photorealism represents a measure of success for computer graphics research. Knowing the visual elements of photorealism provide a guide for trading off rendering accuracy for efficient rendering

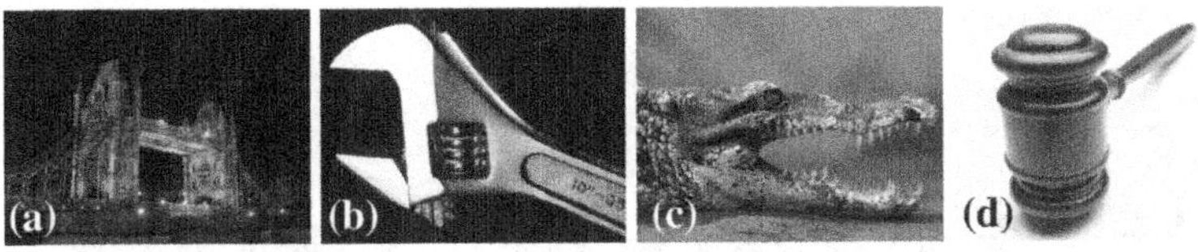


**Fig. 2** Visually challenging images from the Fake or Foto website. The true label of these images, photographic or computer generated, can be found in Appendix

without compromising photorealism. Such studies on perception of photorealism offer some clues about the perceptual differences between photographs and computer graphics.

Meyer et al. [43] asked human observers to label two images of the same scene displayed side-by-side on a monitor display. One of the images is photograph while the other is a computer graphics image rendered with a radiosity algorithm. The human observers found these images perceptually indistinguishable. McNamara [44] conducted a similar experiment with a diffuse scene that is more complex and found that computer graphics rendering that simulates up to second bounce reflection is sufficient to achieve photorealism for such type of scenes.

Rademacher et al. [59] performed a set of experiments on human perception to study what visual clues contribute to photorealism. They found that the softness of shadow and surface roughness correlate positively with photorealism perception, while scene complexity and the number of scene lightings do not show such correlation. This result implies that computer graphics can be made more believable to human by manipulating its soft shadow and rough surface despite the fact that hard shadow and smooth surface do exist in real scenes. Such trick has been employed in the *Fake or Foto*[3] visual quiz, where human observers are asked to label ten images as real or computer generated purely through visual inspection. Four images from the quiz are shown in Fig. 2.

Farid and Bravo [14] conducted a series of psychophysical experiments that used images of varying resolution, JPEG compression, and color to explore the ability of human observers to distinguish computer generated upper body images of people from photographic ones. The computer graphics images used in the experiments are downloaded from the Internet. The experiments provide a probability that an image that is judged to be a photograph is indeed a true photograph, which has 85% reliability for color images with medium resolution (between $218 \times 218$ and $436 \times 436$ pixels in size) and high JPEG quality. The reliability drops for lower resolution and grayscale images. This work shows that the computer graphics of human images in the Internet are quite distinguishable for human observers. This may indicate the level of difficulty for rendering highly photorealistic human images. However, this may change as computer graphics progresses while the ability of human observers remains unchanged.

[3] http://area.autodesk.com/fakeorfoto

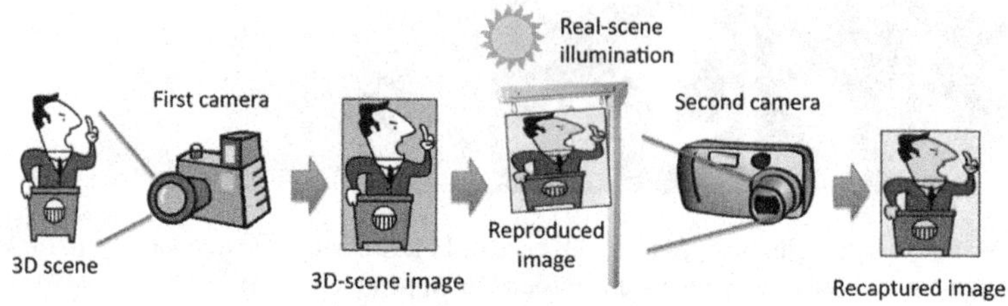


**Fig. 3** An image recapturing pipeline

**Fig. 4** One of the human face images (**a**) and (**b**) and one of the office scene images (**c**) and (**d**) were obtained by rephotographing a color laser printout on an ordinary office paper. The true label of these images, captured from true 3D scene or recaptured from 2D color laser printout, can be found in Appendix

## *2.2 Visual Realism of Recaptured Image*

The process of rephotographing in general involves steps as shown in Fig. 3. A 3D scene is first captured as an image and reproduced on a physical surface such as a printing paper or an LCD display before it is recaptured again under a different illumination. Under control environment, perspective distortion can be minimized by placing the image reproduction surface such that it is in parallel to the camera's sensor plane, in order to reduce the effective skew in the internal parameters of the recaptured image [24]. In general, rephotographing process is pure image-based and involves no graphics models or rendering, unless the first image is computer graphics. Recaptured images are also different from the common photographs in that what being captured is an image reproduction surface instead of a general scene.

Human observers may expect specific color tint associated with an image reproduction surface to be found in a recaptured image. However, such shades of color can exist in a real scene due to illumination. Figure 4 shows four images where two are produced by rephotographing a color laser printout on an ordinary office paper. Readers would notice how misleading is the color clue could be for distinguishing recaptured images.

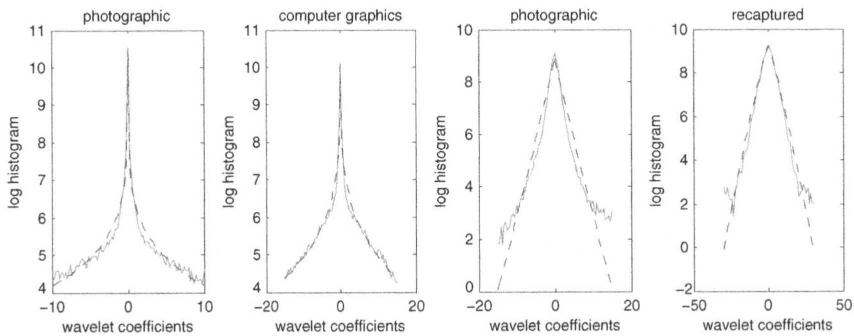


**Fig. 5** The log-histogram of the first-level detail wavelet coefficients computed using Daubechies 8 filters for the photographic image in Fig. 2b, computer graphics image in Fig. 2a, photographic face image in Fig. 4a, and recaptured face image in Fig. 4b (from *left* to *right*). The *dashed* is the least-squared fitted generalized Laplacian density

## *2.3 Statistics of Computer Graphics and Recaptured Images*

Natural images, as opposed to microscopic, astronomic, aerial or X-ray images, are images of the everyday scenes which serve as natural stimuli to human visual system. Natural images are believed to live in a very small subspace within the large image space. To characterize this subspace, researchers have proposed various statistics which demonstrate regularity over nature images [66]. One of the important natural image statistics is the sparse distribution of the wavelet coefficients of natural images that is aptly modeled by a generalized Laplacian density [40].

As a crude evaluation, we show in Fig. 5 the wavelet coefficient distributions of the second-level horizontal subband respectively for a photograph, a computer graphics, a non-recaptured photograph and its corresponding recaptured image. Their distributions appear to be visually similar and modeled well by a generalized Laplacian density. This experiment indicates the statistical similarity of these different types of images at a crude level. More detailed statistics has been considered for distinguishing different image types [47, 56].

## *2.4 Automatic Detection Using Computer*

By design, the current computer vision systems are rarely equipped with the capability to recognize an image beyond its appearance. The plain vulnerability of the laptop's face authenticator is a good example [54] and the various object recognition datasets [12, 16, 17, 39, 70] remain restricted to the pure appearance-based approach. The main reason is that the awareness about the potential security risks due to the inability to distinguish the various underlying image formation processes remains low.

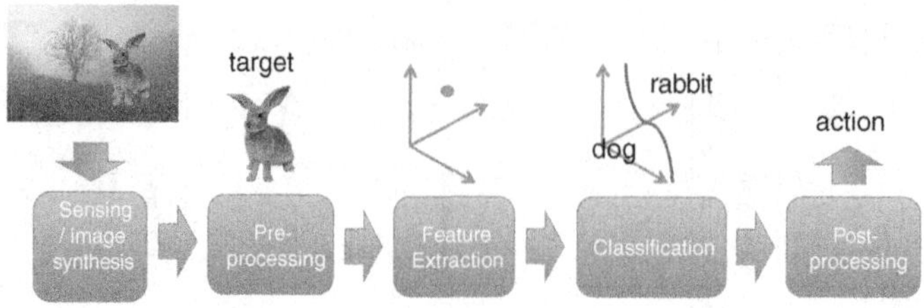


**Fig. 6** The typical process in a pattern recognition system

How computer performs as compared to human observers in recognizing an image beyond image appearance? When it comes to making visual judgement, human observers suffer from various forms of subjective bias and are insensitive to the low-frequency differences in visual signals. In contrast, computer is objective and particularly good at picking up fine features in signals. For example, a computer algorithm is capable of extracting the camera curve property from a single image [53] while this low-frequency and global signal is largely imperceptible to human. As shown in Sect. 4.6, computer-based detection meets with some success in classifying computer graphics and recaptured images from photographs. However, its performance will be much lower in an adversarial setting when intention to deceive is considered.

# 3 Pattern Recognition System and Design Issues

A pattern recognition system computes a pattern from an input and match it to a pattern model to form a decision [11]. This process in general involves *sensing*, *preprocessing*, *feature extraction*, *classification*, and *post-processing* as shown in Fig. 6. A recognition candidate is sensed and the input is preprocessed to remove the irrelevant information, e.g., segmenting the foreground objects from an input image. Pattern recognition is mainly based on distinguishing features of the targets, e.g., the distinctive color offer a good feature for distinguishing apples and oranges. The classifier then assigns the input pattern to one of the pattern models which are obtained from a set of previously observed data known as training samples before the decision is mapped into a recommended action.

Although computer graphics and recaptured image recognition can be approached through pattern recognition as in object recognition, these are new problems with unique characteristics as below:

*Recognizing Image Formation.* For typical pattern recognition such as object recognition, it is the sensor's output (e.g., image appearance) but not the sensor per se (e.g., image formation process) that is of interest. In contrast, for computer graphics and recaptured image recognition, the image formation process is the object of recognition and the image appearance per se is largely immaterial. Therefore, these problems calls for features that are beyond visual appearance such as those inspired by natural image statistics and steganalysis (see Sects. 4 and 5).

*Vague Definition of Classes.* In most pattern recognition problems, the target of recognition is well-defined. For example, gender recognition is a two-class recognition problem with well-defined male and female classes. However, the computer graphics images that we may encounter in real life may not be as well-defined in that the images may have been textured-mapped with photographic inputs or they may be images of a computer graphics foreground object composed with a photographic background scene. On the other hand, a recaptured image may be an image of a picture set against a natural-scene background. Ideally, instead of discrete classes, there should be a continuous measure for the level of computer-graphics-ness or recaptured-ness in an image.

*Dynamic Definition of Classes.* If there are distinctive features between photorealistic computer graphics, recaptured images and photographic images, these features are technology dependent. With the advancement in graphics rendering techniques and image reproduction devices, some of the distinctive features may disappear while new ones may surface. This points to the dynamic nature of the image class model that evolves with time. To maintain the effectiveness of the pattern recognition system, adaptiveness may be essential. Furthermore, in an adversarial setting, the system also needs to adapt to the attack patterns.

*Potential Application in the Court of Law.* For computer graphic detector to be useful for forensics, it has to comply with the forensics procedure and requirements. As forensics is an endeavor to use scientific methods to gain probative facts for criminal investigations, ensuring the objectivity and reliability of the forensics outcome is the primary requirement. The reliability and robustness of the decision also needs to be validated with rigorous and comprehensive test procedure to demonstrate its stability in the presence of noise. For admissibility to the court, the system should minimize the chance of falsely incriminating an innocent person by lowering the false position rate for instance. As the detection result will be debated in the court, any physical intuition on the detection result will make it more accessible and convincing to the legal professionals who may lack the technical background to grasp highly abstract technical details.

*Other Practical Issues.* Simple classification model and fast processing are also important. Simple classification model with small number of features will simplify the classification training procedure and requires a smaller number of training images. Fast processing is important especially for recaptured image detection so that a face authenticator secure against image or video replay attack can operate at an interactive rate.

## 3.1 Evaluation Metric

Current works mainly consider the problem of distinguishing photorealistic computer graphics or recaptured images from photographic images as a two-class classification problem. The performance of a binary classifier can be measured by a classification confusion matrix at an operating point of the classifier:

$$\begin{bmatrix} p(\mathrm{C} = \text{positive}|\ \mathrm{L} = \text{positive}) & p(\mathrm{C} = \text{negative}|\ \mathrm{L} = \text{positive}) \\ p(\mathrm{C} = \text{positive}|\ \mathrm{L} = \text{negative}) & p(\mathrm{C} = \text{negative}|\ \mathrm{L} = \text{negative}) \end{bmatrix}, \quad (1)$$

where the two classes are respectively identified as positive and negative while C and L respectively represent the assigned label (by the classifier) and the true label. The probability $p(\mathrm{C} = \text{positive}|\ \mathrm{L} = \text{negative})$ is known as false positive rate, and $p(\mathrm{C} = \text{negative}\ |\ \mathrm{L} = \text{positive})$ false negative rate. The averaged classification accuracy can be computed as

$$\frac{p(\mathrm{C} = \text{positive}|\ \mathrm{L} = \text{positive}) + p(\mathrm{C} = \text{negative}|\ \mathrm{L} = \text{negative})}{2}. \quad (2)$$

The operating point of a classifier can be adjusted by shifting the decision boundary or threshold values which in turn adjust the balance of the false positive and false negative rates. *Equal error rate* is often used for evaluating a biometric system. Equal error rate refers to the false positive rate or the false negative rate of a classifier when it functions at an operating point where the two rates are equal.

## 4 Approaches for Photorealistic Computer Graphics Detection

Computer graphics detection has been a problem of interest since the early days of content-based image retrieval [62]. The early work focuses on non-photorealistic computer graphics [1, 36, 72]. Only recently, digital image forensics [51] provides a strong motivation to study the problem of identifying photorealistic computer graphics.

Non-photorealistic computer graphics images such as cartoons, cliparts, logos and line drawings are abundant in the Internet. There are commercial incentives to separate such graphics from photographs. For web search companies, being able to identify non-photorealistic graphics offers a value-added service to web users who wish to find cliparts to enhance their presentation slides. On the other hand, this capability could improve the precision of image search by filtering out graphics images when users is interested in photographs. Some companies may be interested in scanning the logo images in the Internet to detect trademark infringement. In all the above-mentioned applications, computation speed is crucial to ensure interactive user experience.

In the Internet, graphics images are mainly kept in common image formats as photographs and metadata often offers no clue for the image type. Therefore, image content features are used for identifying computer graphics.

## 4.1 Methods Using Visual Descriptors

Visual descriptors refers to features motivated by visual appearance such as color, texture, edge properties and surface smoothness. Ideally, visual descriptors should not be effective in identifying photorealistic computer graphics aiming at simulating the appearance of photograph. However, if we look at photographic and photorealistic computer graphics images in the Internet, the *distribution of their visual properties* may be different. For instance, the level of difficulty in rendering a scene increases with its geometric and photometric complexity, hence computer graphics of lower complexity may be more common than the more complex ones in the Internet. However, there is no evidence that the photographic images in the Internet has such distributional property. Computer graphics of lower complexity here refers to images of simpler scene, less color variation or simpler textures.

*Color and Edges*. Simple visual descriptors have been proposed to identify non-photorealistic computer graphics [1, 6, 26, 36, 72]. For example, Ianeva et al. [26] observed that the cartoonist graphics has characteristics of saturated and uniform colors, strong and distinct lines, and limited number of colors. They devised a computer graphics detection method that used image features such as the average color saturation, the ratio of image pixels with brightness greater than a threshold, the Hue-Saturation-Value (HSV) color histogram, the edge orientation and strength histogram, the compression ratio and the distribution of image region sizes. Their work is motivated with the goal for improving the accuracy of video key-frame retrieval through graphics image pre-filtering. As computational efficiency is crucial for graphics detectors, Chen et al. [6] focused on this computational aspect of web graphics and photographs classification.

*Color and Texture*. Wu et al. [76, 77] used visual clues such as the number of unique colors, local spatial variation of color, ratio of saturated pixels and ratio of intensity edges to classify computer graphics and photographs. With these features, on their undisclosed dataset, a $k$-nearest neighbor ($k$-nn) classifier was able to achieve an average accuracy of 76%. They also considered Gabor texture descriptor which enabled a $k$-nn classifier to achieve an average classification accuracy of 95% while a *support vector machine* (SVM) classifier only achieved 75% with the same set of features.

*Fractal properties*. Pan et al. [56] considered that photorealistic computer graphics in the Internet is more surreal in color and smoother in texture as compared to photographic images. Fractal dimension, a self-similarity measure, was proposed to describe the mentioned characteristics. From a scalar image $I(x, y) \in [0, 1]$, a set of $N$ binary images are computed through thresholding:

$$I_k(x, y) = \mathbf{1}\left(\frac{k-1}{N} \leq I(x, y) \leq \frac{k}{N}\right), \qquad (3)$$

using an indicator function $\mathbf{1}(\cdot)$ that equals to 1 when the input argument is true, 0 otherwise. The fractal dimension of a binary image can be estimated with a simple

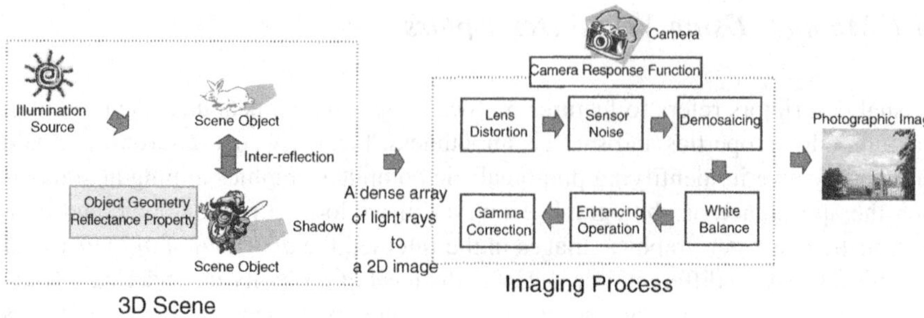


**Fig. 7** Photographic image formation process

method such as box counting [42]. They computed the simple fractal dimensions on the hue and saturation components of an image in the HSV color space. They also extracted a generalized fractal dimension [22] from an image to offer more detailed local information. On their undisclosed dataset, the SVM classifier achieved an averaged test accuracy of 91.2%, while the method by Lyu and Farid [38] achieved 92.7% on the same dataset.

## *4.2 Methods from Image Formation Process*

Incorporating knowledge from the problem domain can potentially lead to a simpler pattern model which requires less data for training. The problem of identifying photorealistic computer graphics and photographic images is essentially a problem of identifying the different image formation processes. Therefore, a detailed understanding on the image formation processes offers an inroad into the problem. As estimating the parameters of the generative models from a single image is mostly ill-posed, the existing methods mainly extract distinguishing features motivated by the generative models indirectly.

### 4.2.1 Formation of Photograph

Photographic images are in general snapshots of natural scenes with a camera. Technically, a camera samples the light radiance reflected from the scene with its optical sensor followed by a series of in-camera processing, as shown in Fig. 7. The scene radiance varies with the light sources, geometry and reflectance properties of the scene. Scene radiance could also be altered by the optical property of the participating medium such as fog and haze through which the light travels. Hence, the properties of real world scenes define the characteristics photographic images.

Common cameras are based on the pinhole camera model, where scene radiance is mapped perspectively onto the light sensors behind the pinhole. Modern digital cameras focus scene radiance for a better optical efficiency using a lens system and digitally processes the sensor measurements to produce visually pleasuring and storage-efficient images. As photographic images are produced by cameras, certain characteristics of the camera design and the in-camera processing are present in the images. For example, *vignetting*, a visual artifact of radial brightness fall-off that arises naturally from an uncompensated lens system, can sometimes be observed on photographs. *Chromatic aberration*, which manifests as color fringes at occlusion edges in an image due to lens with varying refractive index for different wavelengths, is also not uncommon. Apart from the mentioned *photometric distortions*, *geometric distortions* such as pincushion distortion can be visible in an image captured with a poorly designed lens.

Other imprints of camera on a photograph are related to the optical sensor and in-camera processing. Most image sensors used today including the charge-coupled device (CCD) and the complementary metal–oxide–semiconductor (CMOS) sensor are pixilated metal oxide semiconductor, which suffers from several forms of noise such as the pattern noise, dark current noise, shot noise, and thermal noise [25]. Although such camera noise is at a small degree, it can be estimated to certain extent [37].

Photographic images generally undergo *color filter array demosaicing* which is a form of image interpolation within and across color channels. Such interpolation is needed as most commercial cameras today sample the red, green and blue color components of the scene radiance with a single sensor array, instead of three separate ones. Hence, each sensor can only measure one of the color components at a snapshot and interpolation is employed to fill in the missing measurement. Others in-camera operations include white balancing that offsets the shade of the illumination color, edge sharpening, intensity contrast enhancement and gamma correction for dynamic range compression. The overall effect of all these operations can be modeled by a camera response function with a typical concave shape [23].

### 4.2.2 Formation of Photorealistic Computer Graphics

Physics-based graphics rendering described by the Kajiya's rendering equation [27] is the basis of photorealistic rendering. To achieve photorealism, graphics rendering needs to produce complex visual effects such as color blending from light interreflections between surfaces, complex outdoor illumination, and the appearance of some subtle reflectance properties of real-world objects.

Scene modeling refers to modeling of illumination, surface reflectance and object geometry. Although computer generated images of simple diffuse scenes can be visually indistinguishable from the photographs [43, 44], modeling and rendering complex scenes remains challenging. Image-based approach is an answer to the challenge. Image-based modeling incorporates photographs of real scene illumination and object appearance into the graphics pipeline and hence blurs the distinction between photographic and computer graphics images. For example, complex real-scene illumination can be modeled by an environment map derived from the photograph of a mirror sphere [46]. Similarly, spatially varying surface reflectance can be measured from multiple-view photographs [8].

An image-based model with high fidelity calls for elaborate measurement of the real illumination or objects. Hence, image-based measurement may require special devices or need to capture a large number of images which can seriously strain the storage and rendering efficiency in the graphics pipeline. To achieve efficiency, various forms of simplification are introduced in both graphics modeling and rendering. Representing color in three separate channels in the early stage of scene modeling and independent rendering of the color components are among the examples [59]. This creates differences between computer graphics and photographs.

At final stage, a synthesized image may be further touched up or color-adjusted using image editing software such as Adobe Photoshop. The post-processing step can be distinctively different from the in-camera processing in camera.

### 4.2.3 Prior Work Survey

Below we give detailed description of the methods and features motivated by the image formation processes.

*Contrasting Photographic and Computer Graphics Formation.* By contrasting the photographic and computer graphics formation processes, Ng et al. [49] identified three differences between them. First, photographic images are subject to the typical concave response function of cameras while computer graphics rendering pipeline may not have a standardized post-processing procedure that mimics the camera processing. Second, graphic objects may be modeled with simple and coarse polygon meshes. The coarseness of the polygons can give rise to the unnatural sharp edges and polygon-shaped silhouettes in computer graphics images. Third, the three color channels of graphics images are often rendered independently as graphics models are often in such color representation instead of the continuous color spectrum representation as in the real scenes.

The computational steps of the method in [49] are shown in Fig. 8. The three mentioned differences are described using image gradient, principal curvatures and Beltrami flow vectors. They also computed the local block-based fractal dimension and the local patch vectors. The local fractal dimension was meant to capture the texture complexity and self-similarity in photographs and the local patch vectors to model the local edge profile. A SVM classifier is trained with the features on the Columbia open dataset (more details in Sect. 7), they attained an average classification accuracy of 83.5%.

*Employing Device Noise Properties.* Dehnie et al. [9] demonstrated that noise pattern of photographic images, extracted by a wavelet denoising filter, is different from that of computer graphics images. Hence, with their respective reference noise patterns, a test image can be classified based on its correlation to the reference noise patterns. On their undisclosed dataset, the method achieved an averaged classification accuracy of about 72%.

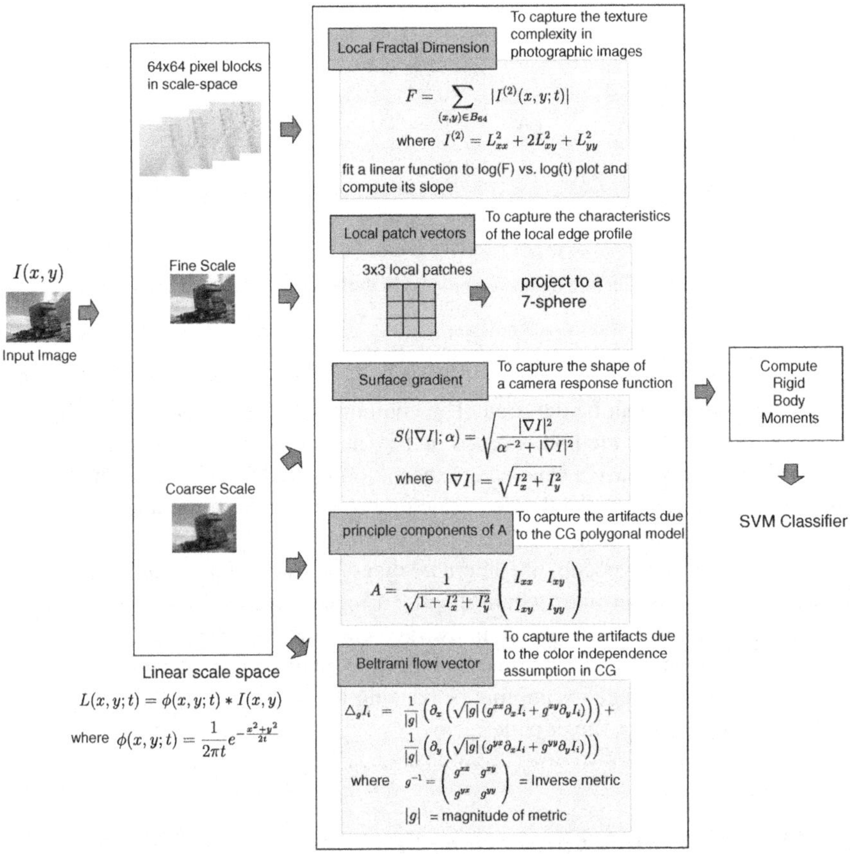


**Fig. 8** Illustration of the feature extraction process in the work by Ng et al. [49]

Inspired by the directional noise in scanners, Khanna et al. [29] employs the features for the device-dependent residual pattern noise computed row-wise and column-wise in an image to distinguish photographic, computer graphics and scanned images. On their undisclosed dataset, the method achieved an average accuracy of 85.9%.

*Employing Camera Demosaicing and Chromatic Abberation.* Dirik et al. [10] observed that an image from a camera with a Bayer color filter array experiences a smaller change if it is re-interpolated again according to the Bayer pattern as compared to other patterns. They also measured the misalignment among the color channels due to chromatic aberration where the camera lens diverges the incoming light of different wavelengths. With the two physical characteristics unique to photographs but often not present in computer graphics, the method achieved an average classification accuracy of about 90% on their undisclosed dataset.

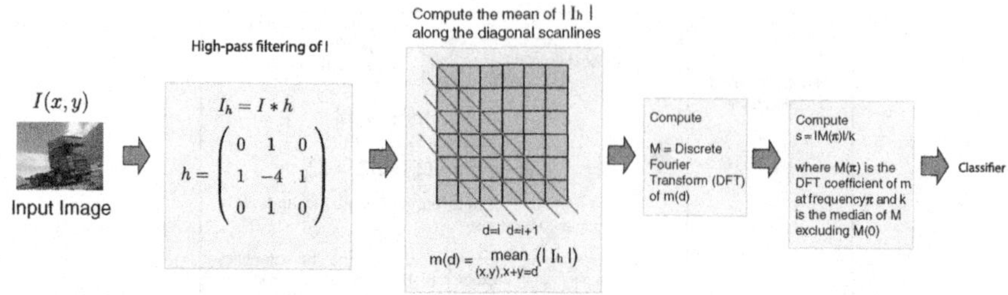


**Fig. 9** Illustration of the feature extraction process in the work by Gallagher and Chen [19]

Gallagher and Chen [19] showed that the Bayer-pattern demosaicing in original-size camera images can be detected. The computational steps for their method are shown in Fig. 9. Their method is based on two main observations; First, the demosaiced pixels always have a smaller variance as compared to the original pixels, and high-pass filtering can make the demosaicing property more prominent. Second, in the green-color component of a Bayer-pattern image, the interpolated and the original pixels respectively occupy the alternate diagonal lines. Hence, the variance plot from the diagonal scan lines would display a regular pattern with a frequency of two units. Their method achieved an average classification accuracy of 98.4% on the Columbia open dataset. Their method may be sensitive to image post-processing operations such as image resampling or resizing that may destroy the interpolation structure specific to the Bayer pattern.

## 4.3 Methods from Natural Image Statistics

The research for natural image statistics is motivated by efforts to observe, isolate, and explain the regularities inherent to natural images [66]. Due the high dimensionality of the image space, building a probability model directly on the space is intractable. Hence, statistics are instead derived from the lower dimensional subspaces in some transform domains such as the wavelet or Fourier domain. Some of the statistics that are motivated to explain the scale invariance properties in natural images which is in general only meaningful for image ensembles, while those motivated by applications such as image compression would be useful descriptors for single images. For example, the power law of the power spectra is statistically stable for an image ensemble while the sparse distribution of the marginal wavelet coefficients is stable for single images, as we have seen in Fig. 5. The methods motivated by natural image statistics for distinguishing computer graphics are based on the belief that computer graphics images are statistically different from the photographic ones.

*Wavelet Statistics*. Lyu and Farid considered that photographic images have different statistical characteristics in the wavelet domain as compared with photorealistic computer graphics images [15, 38]. The computation steps for their method are illus-

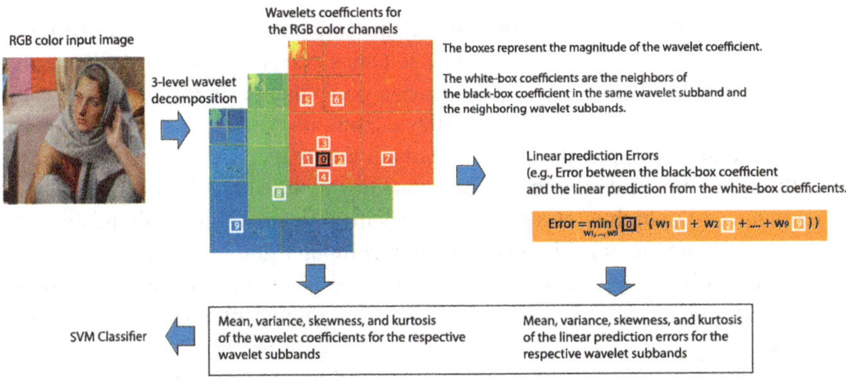


**Fig. 10** Illustration of the feature extraction process in the work by Lyu and Farid [38]

trated in Fig. 10. A RGB input image is first decomposed into three levels of wavelet subbands. For natural-scene images, the wavelet coefficients in a subband are modeled well with a generalized Laplacian distribution and correlation exists between wavelet coefficients of adjacent subbands [66]. Lyu and Farid modeled the former distribution with statistical moments of the wavelet coefficients within a subband, and the latter using the linear prediction error of the coefficients. Figure 10 illustrates how the prediction error is computed using an example where the black-box coefficient is linearly predicted using the neighboring white-box coefficients from subbands within a neighborhoud. Four moments (mean, variance, skewness, and kurtosis) of the wavelet coefficient distribution and the linear prediction error distribution are then computed for each subband as features. On an undisclosed Internet image set, a SVM classifier achieved a classification rate of 66.8% on the photographic images, with a false-negative rate of 1.2%.

Wang and Moulin [74] observed that the characteristic function of the coefficient histogram of a wavelet subband is different for photographic, photorealistic computer graphics and not-so-photorealistic computer graphics images. Such differences are distinct at the low and middle frequency regions. Hence, for each subband, they computed three simple features through low-passing and band-passing the characteristic functions. With the simple features, the computation speed is about four times faster than that of Lyu and Farid [38]. On their undisclosed dataset, the method has comparable performance as that of Lyu and Farid [38] on a simple Fisher linear discriminant classifier. They also tested their classifier (trained using their dataset) with the Columbia open dataset and an abnormally high false alarm was observed. This implied the statistical discrepancy between datasets. More comments on dataset differences are given in Sect. 4.6.

*Power Law for Fourier Power Spectrum and Local Patch Statistics*. Ng et al. [47] explored the usefulness of various natural image statistics for distinguishing photographic images from computer graphics. The study was based on the features related to the power law of the power spectrum of images, the wavelet statistics, and the local patch statistics. The study showed that the local patch statistics performed best in the classification, while the power law statistics performed the worst. This indicates a relationship between the classification performance and the spatial locality of these statistics. The power law statistics is computed on Fourier domain, hence it does not retain any spatial information about the image. The wavelet statistics is partially localized, while the local patch statistics with the smallest spatial support is computed on the high-contrast local patches in an image. This result indicates the importance of local features in distinguishing computer graphics images.

More recently, Zhang et al. [79] modeled local patch statistics as visual words and took an object recognition approach for recognizing photorealistic computer graphics.

*Color Compatibility*. Color composition of natural images is not random and some composition is more likely than the others. Lalonde and Efros [34] showed that the color compatibility between the foreground object and the background scene in a natural-scene image provides a statistical prior for identifying composite images. As computer graphics images may deviate from this statistical regularity, color compatibility can potentially be used to distinguish computer graphics from photographic images.

*Photorealism Measures*. Natural image statistics can essentially serve as a measure for photorealism. Wang and Doube [73] attempted to measure visual realism empirically. The measure consists of three visually perceivable characteristics of natural images which are surface roughness, shadow softness and color variance. As a valid photorealism measure should track the degree of photorealism in an image, this work brought up an interesting idea of using computer game images produced at different years for photorealism evaluation, with the assumption that newer computer games are more photorealistic. Such measure is still considered weak and hence a better measure is needed for real applications.

## *4.4 Methods from Steganalysis*

Steganography embeds confidential messages imperceptibly in a carrier (e.g., an image) in order to hide both the message and the act of message hiding. Whereas steganalysis aims at revealing the act of message hiding blindly without the help of the reference image. Some steganalysis methods detect the specific abnormal statistics in an image resulted from steganography. For example, the Chi-square test statistics on *pairs of values* that differ in the least significant bits (LSB) is good at detecting information hiding by EzStego [75]. Such technique is steganography-method-speccific. It is believed that there exists *universal steganalysis methods* [30] which can detect steganography regardless of its technique. These universal methods aim at extracting discriminative statistics or features which are highly sensitive to information hiding in general.

It is believed that the procedure for extracting the distinguishing features for hidden data can be applied for capturing the statistical characteristics of photorealistic computer graphics. For example, the wavelet statistics method by Lyu and Farid [15] was originally applied for steganalysis, although it is motivated by natural image statistics. Below we describe in detail several methods that are inspired by steganalysis methods.

*Moments of Characteristic Functions.* The method based on moments of characteristic functions on wavelet subbands and prediction error image originates from steganalysis [64]. Chen et al. [5] applied this method in hue, saturation and value (HSV) color space for distinguishing photorealistic computer graphics. The predication error image $I_e$ is the difference between an image $I$ and its predicted version $\hat{I}$, $I_e = |I - \hat{I}|$. The prediction $\hat{I}(x, y)$ with a threshold value $c$ can be computed as

$$\hat{I}(x,y)=\begin{cases} \max[I(x+1,y), I(x,y+1)] & c \leq \min[I(x+1,y), I(x,y+1)] \\ \min[I(x+1,y), I(x,y+1)] & c \geq \max[I(x+1,y), I(x,y+1)] \\ I(x+1,y)+I(x,y+1)-I(x+1,y+1) & \text{otherwise} \end{cases}. \tag{4}$$

Both the original and prediction images can be decomposed into wavelet and approximation subbands. The characteristic function $H(\omega)$ of a subband is the discrete fourier transform (DFT) of its coefficient histogram and the $n$-th order statistical moment of the characteristic function is given by

$$m_n = \frac{\sum_{\omega>0} \omega^n |H(\omega)|}{\sum_{\omega>0} |H(\omega)|}, \tag{5}$$

where only the positive half the characteristic function is considered in the computation. Chen et al. [5] computated three levels of wavelet decomposition on the original and the prediction images for each of the HSV color channels. The first three statistical moments were computed for each subband which gave 234 features in total. On a dataset expanded from the Columbia open dataset, they were able to achieve a classification accuracy of 82.1%.

Sutthiwan et al. [68] extended the work by Chen et al. [5] to include moments of 2D characteristic functions for the Y and Cb components of an image in YCbCr color space. A 2D characteristic function is the DFT of a 2D histogram. They computed the features on the original image, its JPEG coefficient magnitude image and their respective prediction error images, where wavelet decomposition were performed. With a total of 780 features, a SVM classifier was trained and tested on their undisclosed image dataset with an averaged test accuracy of 87.6%. With feature selection on a Adaboost classifier, the number of features was reduced to 450 while the averaged accuracy was improved to 92.7%.

In another work, Sutthiwan et al. [67] considered the JPEG horizontal and vertical difference images as first-order 2D markov processes and used transition probability matrices to model their statistical properties. A total of 324 features were extracted from Y and Cb channels. On their undisclosed dataset, the method achieved a classification accuracy of 94.0% with a SVM classifier. With the same feature reduction method with a Adaboost classifier, a classification accuracy of 94.2% can be achieved with 150 features.

*The ratio of Chi-squared Test Statistics*. Rocha and Goldenstein [61] considered the statistical response of an image to pixel perturbation as a property for distinguishing photographic and computer graphics images. Pixel perturbation is performed by replacing the least significant bits (LSB) of a randomly selected set of pixels with a random binary sequence generated from a uniform distribution of the binary symbols. This pixel perturbation process is similar to the bit embedding function of the EzStego steganography method [75].

Statistical test can be performed to measure the statistical similarity between the resulting distribution with the uniform distribution as reference. The chi-squared statistics $\chi^2$ and the Ueli Maurer Universal statistics $U_T$ of a LSB perturbed image $I_p$ were computed and the deviations from that of the original image $I$ were measured by

$$r_{\chi^2} = \frac{\chi^2(I_p)}{\chi^2(I)}, \quad r_{U_T} = \frac{U_T(I_p)}{U_T(I)}. \tag{6}$$

Six versions of perturbed images $I_p$ were generated for an image through perturbing a fixed percentage of randomly selected pixels, with the percentage corresponds to 1, 5, 10, 25, 50 and 75%. The authors validated their approach on an undisclosed image dataset with 12,000 photographs and 7,500 photorealistic computer graphics images. With a SVM classifier, the method achieved an averaged accuracy of 97.2% as opposed to 82.2% by the method of wavelet high order statistics [38].

## 4.5 Methods from Combining Features

The different types of features are meant to capture different characteristics of an image and they have different strengths and weaknesses. A set of features can be combined to improve performance. Sankar et al. [63] combined the general graphics features from Ianeva et al. [26] , the moments of characteristic function features from Chen et al. [5], the local patch statistics from Ng et al. [49], and the image resampling features from Popescu and Farid [58]. On the Columbia open dataset, a classifier with the aggregated set of features achieved an average classification accuracy of 90%.

## 4.6 Dataset and Performance Evaluation

Table 1 lists the performance of the various proposed methods. A direct comparison of their classification performances is not meaningful as some of the experiments were conducted on different datasets. The discrepancy between different datasets can be significant as observed by Wang and Moulin [74]. However, the methods evaluated on the Columbia open dataset may be compared with a caveat that the

**Table 1** Tabulation of the classification accuracy for various methods on distinguishing photographic and photorealistic computer graphics images

| Approach | Work | Feature dimension | Dataset | Highest classification accuracy |
|---|---|---|---|---|
| Image formation | Ng et al. [49] | 192 | Columbia open dataset | 83.5% |
| | Dehnie et al. [9] | 1 | Internet images | 72% |
| | Khanna et al. [29] | 15 | Internet images | 85.9% |
| | Dirik et al. [10] | 77 | Internet images | 90% |
| | Gallagher and Chen [19] | 1 | Columbia open dataset | 98.4% |
| | Lyu and Farid [38] | 216 | Internet images | 66.8% true-photo, 1.2% false-photo |
| Natural Image Statistics | Wang and Moulin [74] | 144 | Internet images | Comparable to Lyu and Farid [38] |
| | Ng et al. [47] | 24 | Internet images | 83% |
| | Chen et al. [5] | 234 | Columbia open dataset | 82.1% |
| Steganalysis | Sutthiwan et al. [68] | 450 | Internet images | 92.7% |
| | Sutthiwan et al. [67] | 150 | Internet images | 94.2% |
| | Rocha and Goldenstein [61] | 96 | Internet images | 97.2% |
| Visual cues | Wu et al. [76] | 38 | Internet images | 95% with $k$-NN, 75% with SVM |
| | Pan et al. [56] | 30 | Internet images | 91.2% |
| Combining features | Sankar et al. [63] | 557 | Columbia open dataset | 90% |

quoted classification performance merely corresponds to a single operating point on the performance curve. More comments on performance evaluation are given below. *Properties of Internet Image Sets*. Fundamentally, our aim is to build a system for recognizing photorealism and the inherent properties of camera, which are independent of image scenes. Hence, an ideal dataset would be one composed of pairs of photographic and computer graphic images with identical image scenes. Such image pairs have been used for subjective experiments [43, 44, 59]. However, how to synthesize such a dataset efficiently remains an open problem.

The evaluation of current works are mainly based on datasets of Internet images, hoping that the image content is diverse enough to qualify as random samples in the image space or at least the two sets of images have similar distribution in the image space. Unfortunately, the photographic images and the photorealistic computer graphics in the Internet may form different distributions in the image space. For example, computer graphics that are harder to render can be less common in the Internet than the easier ones, but such statistics may not apply to photographic images in the Internet. Although filtering has been in place for the Columbia open dataset to reduce the number of simplistic computer graphics, the dataset is still considered far from the ideal dataset. Therefore, the classification performance in Table 1 can only serve as a proxy for the capability in recognizing photorealism or the properties of camera.

*Usefulness for Real Applications*. The classification accuracy based on the Columbia open dataset ranges from 82.1% of Chen et al. [5] to 98.4% of Gallagher and Chen [19]. The quoted performance may not be a sufficient indicator for the usefulness of the methods when it comes to real applications. For example, the method by Gallagher and Chen exploits the interpolation clues related to the Bayer color filter array and works well on recognizing the original-size photographic images in the dataset. However, the images may be resized in real applications. The Columbia group has built a website for detecting computer graphics images submitted by web users [48]. This exercise offers a realistic evaluation scenario for the detectors with adversarial users. It was observed that the true performance of the detectors under such scenario is in general lower than the performance numbers given in Table 1.

## 5 Approaches for Recaptured Image Detection

Technically, a recaptured image is a photograph of an image reproduction medium. There are many ways to recreate an image on a physical surface. For example, an image on paper reproduced by an inkjet printer is represented as ink dots modulated by half-toning to recreate the appearance of the image. For color laser print, the color toner particles, in a combination of cyan, magenta, yellow and black color, are deposited and fused on a paper through heat treatment in multiple scans. Photo print recreates an image on a photo paper with finite grain size which is generally about 300 dots per inch (DPI). When an image is displayed on LCD screen, each pixel is represented by the emitted light modulated by a liquid crystal of finite size arranged in a 2D array. Each of these physical methods has different color reproduction capability and prefered viewing conditions. Hence, the reproduced color and image appearance depends on the color reproduction technique, the physical medium and the ambient light. Below we described in detail the proposed recaptured image methods.

A recaptured image detection system can be used as a counter-measure for image replay attack on a face authentication system. Liveness detection [7, 32, 35, 57] generally identifies image replay through the specific motion of the human subject that is captured in video. Unlike the image recaptured detection approach, none of the liveness detection methods can resist video play attack, e.g., playing back a face

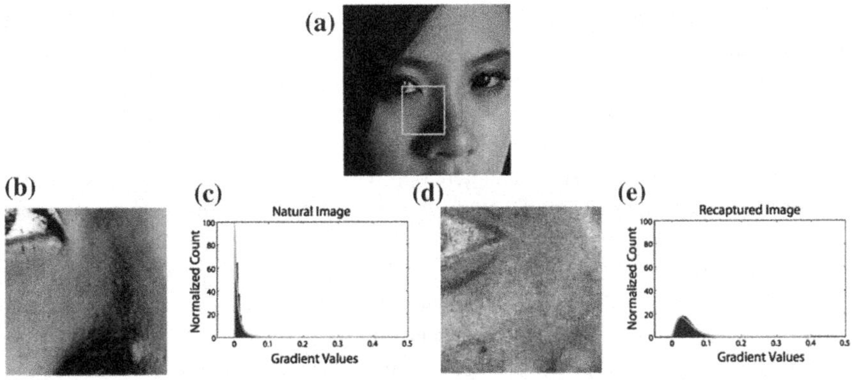


**Fig. 11** The normalized specular components of a 3D face image and its recaptured image, shown together with their corresponding histograms of the gradient of the specular images. **a** Photographic face image, showing the zoom-in region. **b** Specular ratio image (zoom-in) computed from the photographic image. **c** Specular Gradient Histogram for the photographic image. **d** Specular ratio image (zoom-in) computed from the corresponding recaptured image. **e** Specular Gradient Histogram for the recaptured image

video clip on a tablet computer. Furthermore, unlike image recaptured detection, liveness detection would not be effective in detecting image replay for static objects.

*Reproduction Medium Property*. Yu et al. [78] studied the ambient light reflected off the paper as recaptured by a high resolution camera. Part of the light is reflected as specularity. Such reflectance carries a spatial pattern similar to the fine texture of a paper which is more pronounced in the specular component of the recaptured image. There are various methods to decompose an image $I$ into its diffuse $D$ and specular $S$ components, where $I = D + S$ [41, 69, 71]. Figures 11b and d shows the normalized specular component $\hat{S} = S/(S + D)$ for a cropped out image of a real face as shown in Fig. 11a and its corresponding recaptured image. The fine texture is a characteristic of recaptured images from paper printouts.

Bai et al. [2] modeled this texture pattern with a histogram of gradient magnitude on the specular component. The texture results in a heavy tail for the histogram as shown for the recaptured face in Fig. 11e. The shape of the histogram after being normalized to a unity area is modeled by a generalized Rayleigh distribution [28]

$$f(x) = kxe^{(\frac{x}{\alpha})^{\beta}} \tag{7}$$

parametrized by two parameters $\alpha$ and $\beta$. The parameters can be used as features for distinguishing recaptured images from non-recaptured 3D scene images. A linear SVM classifier is trained with a set of 45 real face images and 45 recaptured face images. The classifier achieved 2.2% false acceptance rate and 13% false rejection rate with 6.7% equal error rate.

*Color Reproduction and Recapture Scene Properties*. To resolve the fine texture of a reproduction medium such as paper or computer screen requires high-resolution camera for image recapturing. Such limitation will preclude recaptured image detection on mobile cameras which have relatively lower pixel resolution. To enable recaptured image detection on mobile devices, Gao et al. [20] proposed a set of distinguishing features related to the photometric and scene-related properties due to the image recapturing process, which is an extension of the common photographing process as shown in Fig. 12. The extension is related to the color reproduction response function $f_m$, the reflected radiance from the recaptured scene $R$ and the photometric property of the second camera $f_2$.

A color reproduction technique may involve specific color profile, which could be a smaller portion of the full color space or has limited color resolution. This can result in specific color shade on the reproduced image such as the blue tint on some LCD display. Therefore, the covariance of the color distribution of an image can be as a distinguishing feature.

The photometric response for the reproduction process and the second recapture are in general non-linear. Hence, the cascaded photometric response of $f_1$, $f_m$, and $f_2$ in the image recapturing pipeline could be different from $f_1$ of common photographs. The difference in photometric characteristics can be captured by image gradient [49].

Apart from the reflected specularity, the light may transmit through a reproduction medium which is not entirely opaque such as a paper. Similar to specularity, the light transmitted from the back can significantly reduce the contrast and saturation of a recaptured image. Therefore, image contrast and the histogram of the chromatic components for an image can be computed to capture these phenomena.

One may place a photograph or a printout at a specific distance from the camera to control its size when appearing in the recaptured image. Doing so can potentially place the photograph outside of the camera depth of field and result in image blur. Therefore, image blur can be a tell-tale sign of recapture attack. Finally, it is possible that the natural-scene background is visible in the recaptured image, besides the image reproduction medium.

Using the above-mentioned features, Gao et al. [20] devised a recaptured image detector for mobile devices using a SVM classifier trained on a recaptured image dataset [21]. They compared the performance of the proposed features with that of the wavelet features by Lyu and Farid [38]. When the natural-scene background is visible, the detector achieved an averaged detection accuracy of 93% as compared to 86% for the wavelet features. Without the background information, the detector achieved 78% detection accuracy as compared to 68% for the wavelet features.

*Statistical Property for LCD Screen Recaptured Images.* Cao and Kot [3] studied the issue of image recapturing from LCD screens. They performed a subjective study which showed that recaptured images from LCD screen are largely perceptually indistinguishable to human. However, there are fine differences between LCD screen recaptured images and the non-recaptured ones. For instance, there is a fine grid pattern on LCD screen recaptured images which can be described statistically with multi-scale local binary pattern features [55]. They also considered the loss of details due to recapturing as a distinguishing feature and model it with the statistics of wavelet coefficients. To capture the chromatic property of LCD scene recaptured images, color features in both RGB and HSV color spaces were computed.

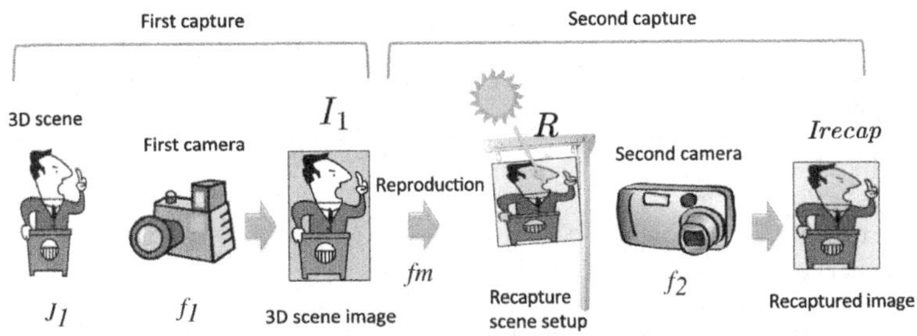


**Fig. 12** An image recapturing pipeline where image recapturing (*second capture*) as an extension of the common photographing pipeline (*first capture*)

The features were evaluated on an image dataset with 2,700 LCD scene recaptured images and 2,000 non-recaptured images with a SVM classifier that achieved an equal error rate of 0.5% as compared to that of 3.4% for the wavelet features by Lyu and Farid [38].

# 6 Possible Attacks and Counter Attacks

In digital image forensics setting, a computer graphics detector can potentially face attacks as described below. Although attacks on recaptured image detection have not been studied, we can imagine that similar attack strategies are equally applicable to recaptured images.

1. **Recapture attack:** Ng et al. [49] showed that recapturing computer graphics is a convenient way to turn a computer graphics into a photograph with little perceptible changes on the image content. To reduce the risk of such attack, the computer graphics image set used for classifier training can be expanded to include the recaptured images.
2. **Histogram manipulation attack:** Sankar et al. [63] showed that a computer graphics detector with color histogram features is vulnerable as the histogram of a computer graphics image can be warped to mimic that of a photograph. Such attack can be prevented by detecting histogram manipulation with a local pixel correlation measure, as histogram manipulation naturally alters the correlation of local pixels.
3. **Hybrid image attack:** Sankar et al. [63] showed that the performance of a computer graphics detector drops significantly for hybrid images, a composite of photographic and computer graphics image regions. They found that local patch statistics [49] are effective in distinguishing hybrid images from non-hybrid ones.

Attacks on a detector succeed as the attack images deviate from the pattern model of the detector. Therefore, an approach with a system of classifiers that builds in a

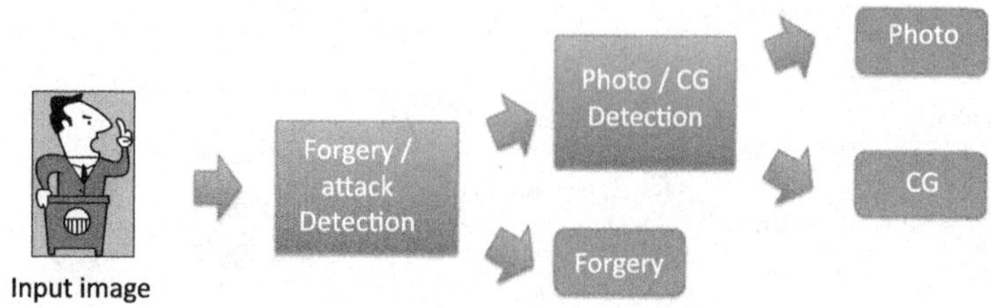


**Fig. 13** A system design for computer graphics detector with a forgery or attack detection feature

mechanism for anticipating and handling attacks can be helpful. Figure 13 shows an example of such system proposed by Sankar et al. [63]. The system begins with a classifier detecting the anticipated types of attack or forgery images. Although such an open-ended system is more secure than a single non-adaptive classifier, it is not capable of handling unseen types of attacks. An ideal system would adapt to the inputs and minimize misclassification through online learning with some forms of supervision.

Design of a digital image forensics system secure against attacks is still an open problem. For learning-based system, a potential issue would be how to gather enough training data for constructing attack models. Furthermore, if the attackers have easy and unlimited access to the forensics system, an exhaustive search can be performed to arrive at the best attack parameters. Even when system access is limited to each user, multiple attackers can still collude to search for the best attack parameters.

## 7 Resources

With photorealistic computer graphics and recaptured image detection formulated as a pattern recognition problem, dataset becomes an essential component of the research. Open benchmark dataset does not only provide data for experiments, it also serves as a basis for comparing various detection methods. For classification of photographic and photorealistic computer graphics images, Ng et al. [50] constructed the Columbia open dataset. Whereas the classification of recaptured and non-recaptured images, Gao et al. [21] constructed the $I^2R$ open dataset.

These datasets are limited in diversity and could not cover all types of images one may encounter in real applications. For example, when Ng and Chang [48] deployed their computer graphics detector online, images with different characteristics from those in the open Columbia dataset were observed. These images include composite images, recaptured images and images with graphics posters. Hence, the online classifier presents a useful case study for a real-world application [52] and help to address the limitation of the datasets. Samples from such online studies may be used to grow the datasets, whenever their ground truth labels are reliable.

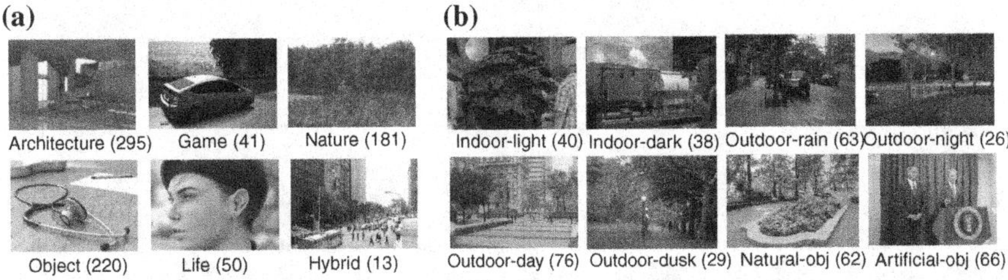


**Fig. 14** The image categories in the Columbia open dataset

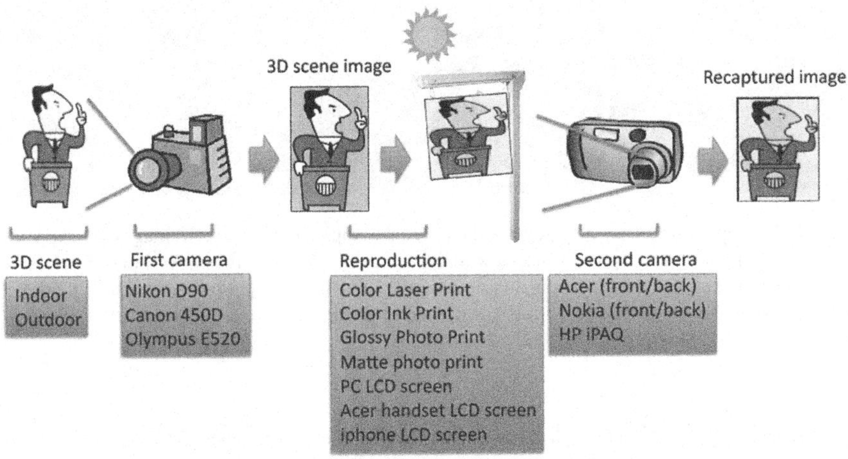


**Fig. 15** The diversity in image recapturing pipeline considered for constructing the I$^2$R open dataset

## 7.1 Benchmark Datasets

*Columbia open dataset.* The Columbia benchmark dataset was constructed and made accessible to the research community [50]. The dataset consists of 800 personal photographic images, 800 photographic images obtained through Google Image Search, 800 photorealistic computer graphics images from 3D artist websites, and 800 recaptured computer graphics images. To ensure diversity in the image content and lighting, there are various subcategories in the dataset, as shown in Fig. 14. As the goal for the dataset was to support studies on photorealism instead of just as a sample set of Internet computer graphics per se, the downloaded computer graphics are perceptually filtered with majority votes from three human observers by assessing the photorealism level of the images through visual inspection.

I$^2$R *open dataset for Smart Phone Recaptured Images.* Gao et al. [21] constructed a smart phone recaptured and non-recaptured image dataset, which considered the varations in the image recapturing pipeline, as shown in Fig. 15. There are variety in the first camera, the reproduction device and the second camera. The first cameras

**Fig. 16** Example images in the $I^2R$ open dataset. **a** Examples of recaptured images. The top row shows the examples of recaptured images with the real environment as the background. The bottom row shows the corresponding cropped images of the first row. **b** Example geometrically aligned images in the $I^2R$ open dataset. The top row of images are the real-scene images, while the bottom row of images are the corresponding recaptured ones

are mainly high-quality single-lens reflex (SLR) cameras. The second cameras are smart-phone cameras with lower sensor quality, which includes the front and back cameras of a smart phone. The back camera which is meant for photo-taking in general has a higher resolution than the front camera which is meant for video conferencing or facetime applications. Various forms of reproduction medium were considered including laser print, ink print, photo print, computer LCD screen and smart-phone LCD screen. An important characteristic of this dataset is the matched-content pairs for the recaptured and non-recaptured images as shown in Fig. 16a. Each pair of images can be geometric aligned as shown in Fig. 16b to facilitate a unbiased study on their photometric difference, independent of the image content. In fact, the matched-content property is equally desirable for the open Columbia dataset as it would make the dataset ideal for studying photorealism in a content-independent manner.

## 7.2 Online Evaluation System

The Columbia online demo system [48] offers a platform for users to try out the fully automatic computer graphics detection function on any test images of their choices. Users are free to try out any attack strategy on the online system. In [52],

Fig. 17 The images with vague classes submitted to the Columbia online computer graphics detector. The classification results for four types of classifier are shown under each image. The shorthands G, W, C and F respectively represent the geometry classifier [49], wavelet classifier [38], cartoon classifier [26] and the fusion of the former three classifiers. **a** G=cg, W=cg, C=cg, F=cg. **b** G=photo, W=cg, C=cg, F=photo. **c** G=photo, W=photo, C=photo, F=photo. **d** G=photo, W=photo, C=photo, F=photo

an evaluation on the Columbia online system was done considering its performance and the specifics of the images submitted by users. The system also allows comparison of different detection algorithms and features (geometry, wavelet, and cartoon features). Among the images submitted by users, there are unconventional images with vague classes. These images include photographs with graphic content embedded in the 3D real-world scenes, and the images composed of both photograph and computer graphics as shown in Fig. 17.[4]

[4] Figure 17a was from http://www.latimes.com/media/alternatethumbnails/photo/2006-06/24010006.jpg, Fig. 17b http://www.iht.com/images/2006/06/25/web.0626city9ss4.jpg, Fig. 17c http://www.spiegel.de/img/0,1020,681928,00.jpg and Fig. 17d http://www.spiegel.de/img/0,1020,681938,00.jpg

## 8 Open Issues and Future Research Directions

Computational measure for photorealism or photograph-ness remains an open problem despite advances in multimedia forensics and perceptual studies for computer graphics. While perception of photorealism can be studied through subjective experiments, the computational model of photorealism requires detailed modeling of the photographic image pipeline and the non-photographic ones such as computer graphics rendering and image recapturing. Hence, estimating the related distinguishing features from the image models is a research challenge for computer vision and graphics.

The definition of photograph evolves with time, so are that for computer graphics and recaptured images. The current camera model is largely based on the pinhole model. The camera model will eventually evolve for enabling new functionalities for cameras, as has been actively pursued by researchers in computational photography [60]. While disruptive changes may not be imminent, the current camera imaging pipeline has always been changing in a gradual manner with small advances in hardware. For example, Foveon introduced a new sensor called the Foveon X3 sensor (a CMOS sensor) which can mimic the color negative film by stacking the RGB color sensitive elements on top of each other, in layers, at each pixel site. As a result, it can measure three RGB colors at each site and hence demosaicing is no longer needed. The dynamic definition of photography poses challenges to machine learning. Adaptive and online learning methods [4, 31] are required to keep track of the changes in the photography model and update its pattern model without relearning from scratch.

A good dataset is important for a pattern recognition problem. As pointed out in Sect. 4.6, the current photographic and computer graphics dataset can be bettered by have content matching pairs, a feature similar to that of the $I^2R$ open dataset for smart phone recaptured images [21]. The photographic and computer graphics pairs with matched content enables a better investigation of computational photorealism in a content-independent manner. Establishing this dataset would require significant efforts in computer graphics modeling and rendering.

The current approach for distinguishing computer graphics and recaptured images from photographs has been mainly through discriminative learning. This approach lacks flexibility in adding new classes of images. Generative model for various types of images probably is more efficient for future-proof system design.

Eventually, the computer graphics or recaptured image detectors are meant for real world scenarios. One of the applications is in the court of law. As described in Sect. 3, such detector needs to be robust, rigorously evaluated, interpretable in physical terms and capable of handling attacks. The current works have been lacking in addressing these practical issues and more efforts are needed to bring the research closer to real-world applications.

## 9 Conclusions

Distinguishing computer graphics and recaptured images from photographic images is important for image classification and countering security risks. Despite a number of works in these areas, there are still many issues remain open. The security risk due to not being able to distinguish photographic and non-photographic images is real. This risk greatly reduces the value of photograph and its application in computer vision. The alternative image synthesis will become more sophisticated with advances in computer graphics and computer vision. Therefore, the security risk will certainly become more prominent in coming years. The body of work presented in this chapter represents an initial step in containing the mentioned security risks and more exciting results are anticipated in this area of research as we move forward into the future.

## Appendix: True Image Labels for Images in Fig. 10.2 and 10.4

Figure 2a and c are computer generated, while Fig. 2b and d are photographic images. Figure 4a and b are photographic images of 3D face, while Fig. 4c and d are recaptured images.

## References

1. Athitsos V, Swain MJ, Frankel C (1997) Distinguishing photographs and graphics on the world wide web. In: IEEE Workshop on Content-Based Access of Image and Video Libraries, pp 10–17
2. Bai J, Ng T-T, Gao X, Shi Y-Q (2010) Is physics-based liveness detection truly possible with a single image? In: Proceedings of the IEEE International Symposium on Circuits and Systems
3. Cao H, Kot AC (2010) Identification of recaptured photographs on LCD screens. In: Proceedings of the IEEE International Conference on Acoustics, Speech and Signal Processing
4. Caruana R (1997) Multitask learning. Mach Learn 28(1):41–75
5. Chen W, Shi Y-Q, Xuan G (2007) Identifying computer graphics using HSV color model and statistical moments of characteristic functions. In: Proceedings of the IEEE International Conference on Multimedia and Expo, pp 1123–1126
6. Chen Y, Li Z, Li M, Ma WY (2006) Automatic classification of photographs and graphics. In: Proceedings of the IEEE International Conference on Multimedia and Expo, pp 973–976
7. Choudhury T, Clarkson B, Jebara T, Pentland A (1999) Multimodal person recognition using unconstrained audio and video. In: International Conference on Audio- and Video-Based Person Authentication, pp 176–181
8. Dana KJ, Van Ginneken B, Nayar SK, Koenderink JJ (1999) Reflectance and texture of real-world surfaces. ACM Trans Graph 18(1):34
9. Dehnie S, Sencar T, Memon N. Digital image forensics for identifying computer generated and digital camera images. In: Proceedings of the IEEE International Conference on Image Processing, pp 2313–2316

10. Dirik AE, Bayram S, Sencar HT, Memon N (2007) New features to identify computer generated images. In: Proceedings of the IEEE International Conference on Image Processing, vol 4, pp 433–436
11. Duda RO, Hart PE, Stork DG (2001) Pattern classification. Wiley-Interscience, Hoboken
12. Everingham M, Zisserman A, Williams C, Van Gool L, Allan M, Bishop C, Chapelle O, Dalal N, Deselaers T, Dorko G (2006) The 2005 PASCAL visual object classes challenge. Mach Learn Chall, 117–176
13. Farid H (2004) Creating and detecting doctored and virtual images: Implications to the child pornography prevention act. Technical report, TR2004-518, Dartmouth College, Computer Science
14. Farid H, Bravo MJ (2011) Perceptual discrimination of computer generated and photographic faces. Digital Investigation
15. Farid H, Lyu S (2003) Higher-order wavelet statistics and their application to digital forensics. In: IEEE Workshop on Statistical Analysis in Computer Vision
16. Fei-Fei L, Fergus R, Perona P (2004) Learning generative visual models from few training examples: An incremental bayesian approach tested on 101 object categories. In: Workshop on Generative-Model Based Vision
17. Fergus R, Perona P, Zisserman PA (2003) Object class recognition by unsupervised scale-invariant learning. In: Proceedings of the Computer Vision and Pattern Recognition
18. Ferwerda JA (2003) Three varieties of realism in computer graphics. In: Proceedings of the SPIE Human Vision and Electronic, Imaging, vol 3
19. Gallagher AC, Chen T (2008) Image authentication by detecting traces of demosaicing. In: Proceedings of the Computer Vision and, Pattern Recognition
20. Gao X, Ng T-T, Qiu B, Chang S-F (2010) Single-view recaptured image detection based on physics-based features. In Proceedings of the IEEE International Conference on Multimedia and Expo
21. Gao X, Qiu B, Shen J, Ng T-T, Shi Y-Q (2010) A smart phone image database for single image recapture detection. In Proc. of International Workshop on Digital Watermarking
22. Grassberger P (1983) Generalized dimensions of strange attractors. Phys Lett A 97(6):227–230
23. Grossberg M, Nayar SK (2003) What is the space of camera response functions? In: Proceedings of the Computer Vision and, Pattern Recognition
24. Hartley R, Zisserman A (2000) Multiple view geometry, chapter 6, p 164. Cambridge university press, Cambridge
25. Healey G, Kondepudy R (1994) Radiometric CCD camera calibration and noise estimation. IEEE Trans Pattern Anal Mach Intell 16(3):267–276
26. Ianeva TI, de Vries AP, Rohrig H (2003) Detecting cartoons: A case study in automatic video-genre classification. In: Proceedings of the IEEE International Conference on Multimedia and Expo, vol 1
27. Kajiya JT (1986) The rendering equation. In: Proceedings of the ACM SIGGRAPH, pp 143–150
28. Kam AH, Ng T-T, Kingsbury NG, Fitzgerald WJ (2000) Content based image retrieval through object extraction and querying, In: IEEE Workshop on Content-based Access of Image and Video Libraries
29. Khanna N, Chiu GTC, Allebach JP, Delp EJ (2008) Forensic techniques for classifying scanner, computer generated and digital camera images. In: Proceedings of IEEE International Conference on Acoustics, Speech and Signal Processing, pp 1653–1656
30. Kharrazi M, Sencar HT, Memon N (2005) Benchmarking steganographic and steganalysis techniques, In: Proceedings of the SPIE Electronic Imaging
31. Kivinen J, Smola AJ, Williamson RC (2004) Online learning with kernels. IEEE Trans Signal Process 52(8):2165–2176
32. Kollreider K, Fronthaler H, Bigun J (2009) Non-intrusive liveness detection by face images. Image Vis Comput 27(3):233–244
33. Kovach S (2011) This guy just exposed a major security flaw in ice cream sandwich. http://www.businessinsider.com/ice-cream-sandwich-face-unlock-2011-11

34. Lalonde JF, Efros AA (2007) Using color compatibility for assessing image realism. In: Proceedings of the International Conference on Computer Vision
35. Li J, Wang Y, Tan T, Jain AK (2004) Live face detection based on the analysis of fourier spectra. In: Proceedings of the Society of Photo-Optical Instrumentation Engineers (SPIE) Conference Series, vol 5404. pp 296–303
36. Lienhart R, Hartmann A (2002) Classifying images on the web automatically. J Electron Imaging 11(4):445–454
37. Lukáš J, Fridrich J, Goljan M (2006) Digital camera identification from sensor pattern noise. IEEE Trans Inf Secur Forensics 16(3):205–214
38. Lyu S, Farid H (2005) How realistic is photorealistic? IEEE Trans Signal Process 53(2): 845–850
39. Magee DR, Boyle RD (2002) Detecting lameness using re-sampling condensation and multistream cyclic hidden markov models. Image Vis Comput 20(8):581–594
40. Mallat SG (1989) A theory for multiresolution signal decomposition: The wavelet representation. IEEE trans pattern anal mach intell 11(7):674–693
41. Mallick S, Zickler T, Belhumeur P, Kriegman D (2006) Specularity removal in images and videos: A pde approach. In: Proceedings of European Conference on Computer Vision, 550–563
42. Mandelbrot BB (1982) The fractal geometry of nature. W. H. Freeman, San Francisco
43. Mayer GW, Rushmeier HE, Cohen MF, Greenberg DP, Torrance KE (1986) An experimental evaluation of computer graphics imagery. In: Proceedings of the ACM SIGGRAPH, pp 30–50
44. McNamara A (2005) Exploring perceptual equivalence between real and simulated imagery. In: Proceedings of the ACM symposium on Applied perception in graphics and visualization, p 128
45. Michels J, Saxena A, Ng AY (2005) High speed obstacle avoidance using monocular vision and reinforcement learning. In: Proceedings of the International Conference on Machine Learning, pp 593–600
46. Miller G, Hoffman CR (1984) Illumination and reflection maps: Simulated objects in simulated and real environments. In: SIGGRAPH 84 Advanced Computer Graphics Animation seminar notes, vol 190
47. Ng T-T, Chang S-F (2004) Classifying photographic and photorealistic computer graphic images using natural image statistics. Technical report, ADVENT Technical Report, 220-2006-6, Columbia University
48. Ng T-T, Chang S-F (2006) An online system for classifying computer graphics images from natural photographs. In: Proceedings of SPIE, 6072, pp. 397–405, 2006. http://apollo.ee.columbia.edu/trustfoto/trustfoto/natcgV4.html
49. Ng T-T, Chang S-F, Hsu J, Xie L, Tsui M-P (2005) Physics-motivated features for distinguishing photographic images and computer graphics. In: Proceedings of the ACM International Conference on Multimedia, pp 239–248
50. Ng T-T, Chang S-F, Hsu Y-F, Pepeljugoski M (2004) Columbia photographic images and photorealistic computer graphics dataset. Technical report, ADVENT Technical Report, 203-2004-3, Columbia University
51. Ng T-T, Chang S-F, Lin C-Y, Sun Q (2006) Passive-blind image forensics. In: Multimedia Security Technologies for Digital Rights, pp 111–137. Elsevier
52. Ng T-T, Chang S-F, Tsui M-P (2007) Lessons learned from online classification of photorealistic computer graphics and photographs. In: IEEE Workshop on Signal Processing Applications for Public Security and Forensics
53. Ng T-T, Tsui M-P (2009) Camera response function signature for digital forensics—part I: Theory and data selection. In: IEEE Workshop on Information Forensics and Security (WIFS)
54. Ngo D (2008) Vietnamese security firm: Your face is easy to fake. http://news.cnet.com/8301-17938_105-10110987-1.html
55. Ojala T, Pietikainen M, Maenpaa T (2002) Multiresolution gray-scale and rotation invariant texture classification with local binary patterns. IEEE Trans Pattern Anal Mach Intell 24(7):971–987

56. Pan F, Chen JB, Huang JW (2009) Discriminating between photorealistic computer graphics and natural images using fractal geometry. Science in China Series F: Information Sciences 52(2):329–337
57. Pan G, Sun L, Wu ZH, Lao SH (2007) Eyeblink-based anti-spoofing in face recognition from a generic webcamera. In: Proceedings of the International Conference on Computer Vision, pp 1–8
58. Popescu AC, Farid H (2005) Exposing digital forgeries by detecting traces of resampling. IEEE Trans Signal Process 53(2):758–767
59. Rademacher P, Lengyel J, Cutrell E, Whitted T (2001) Measuring the perception of visual realism in images. In: Proceedings of the Eurographics Workshop on Rendering, Techniques, pp 235–248
60. Raskar R, Tumblin J, Mohan A, Agrawal A, Li AY (2006) Computational photography. Proceedings of the Eurographics State of the Art Report
61. Rocha A, Goldenstein S (2006) Is it fake or real?. In: XIX Brazilian Symposium on Computer Graphics and Image Processing, pp 1–2
62. Rui Y, Huang TS, Chang S-F (1999) Image retrieval: Current techniques, promising directions, and open issues. J vis commun image represent 10(1):39–62
63. Sankar G, Zhao V, Yang YH (2009) Feature based classification of computer graphics and real images. In: Proceedings of the IEEE International Conference on Acoustics, Speech and Signal Processing, pp 1513–1516
64. Shi YQ, Xuan G, Zou D, Gao J, Yang C, Zhang Z, Chai P, Chen W, Chen C (2005) Image steganalysis based on moments of characteristic functions using wavelet decomposition, prediction-error image, and neural network. In: Proceedings of the IEEE International Conference on Multimedia and Expo
65. Simoncelli EP, Olshausen BA (2001) Natural image statistics and neural representation. Annu review neurosci 24(1):1193–1216
66. Srivastava A, Lee AB, Simoncelli EP, Zhu SC (2003) On advances in statistical modeling of natural images. J Math Imaging vis 18(1):17–33
67. Sutthiwan P, Cai X, Shi Y-Q, Zhang H (2009) Computer graphics classification based on markov process model and boosting feature selection technique. In: Proceedings of the IEEE International Conference on Image Processing, pp 2877–2880
68. Sutthiwan P, Ye J, Shi Y-Q (2009) An enhanced statistical approach to identifying photorealistic images. Digit Watermark, 323–335
69. Tan RT, Ikeuchi K (2005) Separating reflection components of textured surfaces using a single image. IEEE Trans Pattern Anal Mach Intell 27(2):178–193
70. Torralba A, Murphy KP, Freeman WT (2007) Sharing visual features for multiclass and multiview object detection. IEEE Trans Pattern Anal Mach Intell, 854–869
71. Umeyama S, Godin G (2004) Separation of diffuse and specular components of surface reflection by use of polarization and statistical analysis of images. IEEE Trans Pattern Anal Mach Intell 26(5):639–647
72. Wang F, Kan MY (2006) NPIC: Hierarchical synthetic image classification using image search and generic features. Image Video Retr, 473–482
73. Wang N, Doube W (2011) How real is really? a perceptually motivated system for quantifying visual realism in digital images. In: Proceedings of the IEEE International Conference on Multimedia and Signal Processing 2:141–149
74. Wang Y, Moulin P (2006) On discrimination between photorealistic and photographic images. In: Proceedings of the IEEE International Conference on Acoustics, Speech and Signal Processing 2:161–164
75. Westfeld A, Pfitzmann A (2000) Attacks on steganographic systems. In: Information Hiding, pp 61–76. Springer
76. Wu J, Kamath MV, Poehlman S (2006) Color texture analysis in distinguishing photos with computer generated images. In: Proceedings of the UW and IEEE Kitchener-Waterloo Section Joint Workshop on Knowledge and Data Mining

77. Wu J, Kamath MV, Poehlman S (2006) Detecting differences between photographs and computer generated images. In: Proceedings of the 24th IASTED international conference on Signal processing, pattern recognition, and applications, pp 268–273
78. Yu H, Ng T-T, Sun Q (2008) Recaptured photo detection using specularity distribution. In: Proceedings of the IEEE International Conference on Image Processing, pp 3140–3143
79. Zhang R, Wang R, Ng T-T (2011) Distinguishing photographic images and photorealistic computer graphics using visual vocabulary on local image edges. In: Proceedings of the International Workshop on Digital-forensics and Watermarking

# Part IV
# Digital Image Forensics in Practice

# Courtroom Considerations in Digital Image Forensics

**Rebecca Mercuri**

**Abstract** The manner in which results of digital image forensic investigations are presented in the courtroom can be positively or adversely affected by case law, federal rules, legislative acts, availability of discovery materials, tools, and the use of exhibits. Issues pertinent to effective and successful testimony are addressed, along with suggestions for improvement of methods.

## 1 Computer Forensics

Computer forensic investigations, including those applied to all types of digital imagery, primarily deal with the analysis and reporting of evidence that is typically collected after (or possibly also during) an incident, with the intended goal of producing legally acceptable evidence for courtroom purposes. The forensic process thus necessarily differs from and involves considerably more effort than that which is required to perform simple data recovery and image restoration tasks. This is because one cannot merely present digital evidence materials to the court without also providing extensive accompanying documentation, detailing numerous salient aspects, especially those pertinent to establishing chain of custody, knowledgeable possession, and control. The methods used by the forensic examiner must be well documented, as they may likely be subjected to scrutiny by attorneys, judges, and other investigators. Forensic preparations, therefore, place a high emphasis on data authentication and report writing, and involve the use of tools that enhance the examiner's ability to perform these tasks.

The field of computer forensics, having evolved predominantly over the last decade, is relatively young, especially as compared to other forensic endeavors (such as those used by coroners or in ballistics) that have been performed for a century

R. Mercuri (✉)
Notable Software, Inc., PO Box 1166, Philadelphia, PA 19105, USA
e-mail: notable@notablesoftware.com

H. T. Sencar and N. Memon (eds.), *Digital Image Forensics*,
DOI: 10.1007/978-1-4614-0757-7_11,

or more. Presently, there is no uniform certification process, training, degree, or license required in order to become a computer forensic or digital imaging examiner, and it is rare to find such an expert with a decade or more of court-related experience. There is no particular laboratory configuration specified for the performance of this work, although best practice recommendations are now beginning to appear. This lack of standardization is due, at least in part, to the rapid growth and ever-changing challenges of the computer field. New products are emerging constantly and commonly used forensic tools may not be able to handle certain items properly, or even at all, at the time when an examination is needed. For image forensics, software tools must be able to identify and display the various different types of digital picture and video formats. Hardware has to safely accommodate the changing array of media storage devices, such as SIM and other memory chips and cards, as well as drives and disks. To fill in the gaps, when needed, experts often supplement the commercially and publicly available tools with ones they have personally created or commissioned. Operating systems, programming languages, interfaces, and other aspects of the computational environment are also evolving and the examiner must stay informed in order to be able to properly assess and address the various and unique situations presented in each case.

The method of examination should not be unduly constrained by the forensic tools. Often it happens, though, that magnetic media (drives, disks, etc.) have been archived by an impounding facility using a process that results in a proprietary format (EnCase is one of these) that cannot be decoded without access to the specific software or a particular version thereof. This may impose additional and unnecessary time and cost on an examination facility, placing their client and legal team at a practical disadvantage. It is important that a court recognize the fact that each side should be allowed the same access to the data, preferably in the form that it was originally found, or at least a common and openly available one.

Of course, the volatility of electronic recording devices (especially those that can be written to or altered) comes into play. Individuals charged with the responsibility of collecting original evidence materials must ensure that their process does not damage or alter data in any way. Occasionally, evidence has been contaminated by overzealous information technology staff who may scan a drive without first placing it into a write-protected bay, or file recovery tools may be erroneously employed that overwrite other data. The use of original digital media evidence must be limited to the minimal steps necessary for archiving and authentication. These archived materials should then be used to generate other forensic clones or derivative evidence to be provided for examination. Here again, there have been little or no controls or standards imposed or required, either for the collection, copying, or storage of these materials, and quality assurance has been observed to vary widely depending upon the type of facility involved. As an example, one would think it prudent to perform a hash value (authentication) calculation at the initial time of impounding, or minimally at the time the first copy is made, in order to ensure that no data has been altered, but this is rarely done.

## 2 Knowledge, Possession, Control

It is especially important to keep track of who has had access to the original evidence and establish timelines for possession. Once items have been impounded, they should be safeguarded under lock and key, such as in storage lockers, at all times when these materials are not in use. Chain of custody documents, where the individuals who have been responsible for the items are identified, along with access dates and times, should be maintained and made available for review as appropriate. This is true for derivative materials (data that has been copied from the originals) as well as the evidence itself. It is always preferable to hand-courier these materials, especially originals, rather than risk damage, loss, or interception by using shipping services. If one must ship items, signatures, tracking, and insurance (if available) should be required.

Where child pornography digital imagery investigations are concerned, most laws refer to "knowing possession and control," which must be proven in order to convict. Federal law 18 U.S.C. § 2252A(a) prohibits certain activities relating to any visual depictions involving the use of a minor engaging in sexually explicit conduct, and makes it a felony to knowingly transport or ship in interstate or foreign commerce by any means including by computer or mail, or knowingly possess, receive or distribute such materials. Unfortunately, courts have quite often been persuaded by prosecutors to turn a blind eye to the aspects involving "knowledge" and "control" with the result being that mere possession is tantamount to guilt. Even though the law specifically refers to "sexually explicit conduct," the definition of this varies widely, even between counties in the same state. The photos of baby's first diaper change or cousins skinny-dipping in the lake, especially if genitalia are evident, could even be deemed explicit and used to constitute prurient interest. Some prosecutors and judges still use the "you know it when you see it" test for pornography, which can be flawed or deceiving when older but youthful looking people are depicted.

Many computer users have no idea that when they try to delete files that have been accidentally downloaded, items may still persist in the "recycling bin" (or desktop "trash can") until perhaps the computer is turned off or rebooted, or after the trash is manually emptied via a file menu selection. This is also true of electronic mail, where deletion may only put the message into a trash folder that requires another action to empty it. As well, with electronic mail, attachments may be saved in a completely different folder in the computer's file system that users may not know about or is hard to find. Misleading and even preposterous comments have been made in grand jury testimony, and when obtaining search warrants, to the effect that "collectors" of contraband materials will deliberately place these items in the trash to conceal them from other users of the computer, when clearly the accused recipient was actually trying to dispose of unwanted material.

Many computer users do not realize that deletion, even from the "trash," typically only removes the name of the file from the active directory tree, and the file itself nevertheless continues to persist on the drive until the operating system overwrites it with other data. Given the proliferation of larger sized storage devices, it is common to find and recover a substantial percentage of the files that had ever existed on the system, in the unallocated space of the hard drives, some days, months, or even years after the user(s) attempted and intended to delete them. Recovery tools have become more sophisticated too, such that even partial files for images and videos can be reconstructed out of existing fragments.

Certainly, it is not illegal to possess or use programs that will effectively overwrite (with random characters or 1s or 0s) the unused (unallocated and slack) space of one's own magnetic media. This is often necessary for legitimate purposes (such as HIPAA compliance for health care records [1]), but prosecutors are fond of making it seem that ownership of such tools constitutes intention to conceal contraband files. Nevertheless, many of these tools do not work properly, or are installed incorrectly, and individuals are often lulled into a false sense of confidence that their drives are being "wiped" of any evidence of possible misuse, when actually considerable data continues to persist.

In fact, whether a defendant actually opened and/or viewed a downloaded file that they deliberately downloaded may be considered irrelevant in terms of possession. An analogous situation is that one can certainly be charged with drug possession whether or not they are a user. The problem, especially with digital imagery, is that when multiple files are transmitted together in a group (such as in a zip file), the person requesting the file may be wholly unaware that contraband exists within the archive. When an archive is expanded, the contents typically are placed into a folder, which may never be opened for review. Here, a rapid-fire sequence of "last accessed" timestamps, within seconds of each other, is generally indicative of automatic file extraction without viewing.

File names can be highly misleading, since nondescriptive names (such as sequence numbers) are often used to aid in concealing contents from unsuspecting downloaders. A GAO study [2] that performed searches on peer-to-peer networks using benign topics of interest to children or specific keywords, underscored the broad availability of such contraband materials while also noting problems with respect to the file names. Although their keyword search using words related to child pornography yielded file names indicating 42% child porn, 34% adult porn, and 24% non porn, the actual images revealed the percentages to be 44% child porn, 13% child erotica, 29% adult porn, and 14% non porn, thus showing that a significant amount of material is miscategorized based on naming. The statistical likelihood that someone using peer-to-peer networks for other purposes would unknowingly download felonious pictures or videos is therefore rather high.

Content identification is an area of imagery research that has seen more common use in recent years. Photo archives can allow keying of particular faces and objects to names, with automatic generation of labels as the database is developed. The rigor of such algorithms is not yet well understood, but it will certainly become another method of ferreting out contraband and pinpointing possible victims, likely with considerable false positives, in future investigations. With this, issues of privacy and censorship will need to be considered. Although it might be more reasonable for Internet Service Providers to monitor for content in imagery transmissions, these entities were provided with safe harbor in the United States of America under the

Digital Millennium Copyright Act of 1998, with liability for misuse shifted to the end users, networks (such as peer-to-peer) and archivers (like YouTube). Other nations (Australia is a good example) do require the ISPs to perform some filtering. Such requirements (or lack thereof) may shift in the future as new laws and rulings are developed and instituted.

When attorneys and experts are able to successfully present the court with rationale that mere possession is not sufficient to demonstrate knowledge and control of contraband digital imagery, they must also be prepared to deal with evidence that can potentially corroborate allegations that the particular files were intentionally downloaded or actually seen. Such evidence typically involves the recorded timestamps on the files at issue, as well as data that can be extracted from system logs.

The trio of file timestamps, often referred to as "created," "modified," and "last accessed," can be especially difficult to deal with. Recovery efforts, if not performed properly, may alter this data, and when overwritten on the original media, the earlier information is essentially unrecoverable. It should be noted that deleted files (after the trash can is emptied, but still existing in the unallocated space of the drive) will not show this associated data, since it is maintained by the file system and lost on deletion. The terminology referring to these three timestamps is rather misleading, since "created" does not necessarily (or even at all) pertain to when the file was actually generated, but rather it might be when it was placed into or taken out of an archive, or copied, or when operations are performed. The other two timestamps have the same properties, with "last accessed" being particularly unreliable because it often is routinely affected/updated by system operations (such as periodic virus scans). System time clocks may be altered or may not be reliable (especially if the computer is old and the battery has run down). There may also be confusion over the time zone of this information, especially when tools are used that reflect universal (Greenwich Mean) time, and do not account for Daylight Savings differences. Essentially, these dates and times are not really reliable proof of particular user activities, although they are often held up as incontrovertible corroboratory evidence. This is not to say that the timestamps cannot also be used to provide exculpatory evidence, especially when there may be time or date intervals showing temporally when a defendant can demonstrate that he/she had no access whatsoever to the computer system.

System files are another area from which information can be harvested that appears to corroborate allegations. With respect to digital imagery, the Windows Media Player logs are often used to provide "proof" of viewing. For peer-to-peer situations, the Web browser software (such as Limewire) may use a slightly modified file name if the item is previewed during downloading, which can also be deemed to demonstrate proof of "knowing possession" especially if the modified name appears in a viewing log. That these logs occur in the background without the user's explicit authorization may seem to imply that their accuracy is unquestionable, but in fact, there is no true way to tie this information to any particular user, and it could also have resulted from malware (virus activity) or access by an unauthorized individual or distant computer. It is important to scan the system in order to determine if it may have been operating as a remotely controlled "bot" or was infected in a manner that may have enabled adverse activity to occur. Peer-to-peer situations are especially dangerous, in that if a contraband file is discovered in the file share folder, where it would normally reside following downloading, the charges against a defendant can include distribution as well as possession if the sharing feature has been turned on. Vigilante groups and special police task forces have been known to harvest the IP address information from file share browsers in an effort to identify possible contraband holders who will later be asked to provide their computers or will be served with search warrants in sting operations. The best way to prevent such problems from occurring is to avoid using peer-to-peer services, but many feel that this is an unfair constraint.

The thumbnail images that can be viewed when a directory is opened are also highly dubious as evidence. Some of these have been used to convict on the grounds of knowing possession and control, even when the original material has not been found to be present on the drive. The contention is that such thumbnails are only created when the related file has existed on the user's computer, but this is flatly untrue. When a folder is viewed in thumbnail mode, even after particular files are deleted, the thumbnail file (such as thumbs.db on the Windows operating system) still retains all of the individual thumbnails that previously were generated. Copying or moving an entire folder typically results in copying of the contained thumbnail file, which may not accurately reflect what pictures actually exist at the destination location. Hence, a folder copied from one user to another may contain concealed contraband that the recipient never knew about and could not see. This retention effect is difficult to demonstrate in court such that a jury or judge will understand that the appearance of a particular thumbnail in a system file does not at all constitute knowing possession or control, since the file is hidden by default and specialized tools (such as the ones experts use in drive examinations) are required in order to extract and view any contents that are not also affiliated with the larger versions of pictures in the folder. Often the thumbnails are provided as discrete evidence elements in discovery without identifying them as being part of hidden system files, something that the defense expert needs to be careful to sort out in preparing their report. Although this tactic is becoming less common, individuals have certainly received convictions, with rejected appeals, on the basis of such thumbnails.

Digital image files will likely also contain header data that pertains to the time of creation and the equipment and settings used. This information can be readily modified, spoofed, or deleted, using publicly available tools. Recent scientific studies, though, have revealed subtle content within the image itself, allowing the particular digital camera that created the photo or video to be identified from others of the same model and type. Such analysis is still so new as to not have led to significant reported case law, but will likely become a factor in the future.

## 3 Expert Testimony

The court system is hierarchical in nature and, in formulating rulings, judges often look to the decisions of others, especially precedential decisions of higher courts within the same jurisdiction. It is thus in the best interest of all experts to try to avoid creating "bad law" where cases are "won" but the underlying expert theories are false. Image analysis could be one area where we may look back and see some matters whose outcome was based on expert testimony that was contrary to later-known facts. It is difficult to balance the need to move ahead with a demonstration that has not yet been fully vetted in the scientific community, with the desire to help a client as much as reasonably possible. Similar questions have been raised even with regard to the reliability of some common forensic tools, such that NIST has begun qualifying these via extensive examinations conducted in their Computer Forensics Tool Testing Project. We will likely see more of such tool validation in image analysis.

Differences in the manner in which expert testimony is handled by different courts, along with verification of scientific and forensic methods, is most evident when considering admissibility of evidence. The case of *Daubert v. Merrell Dow Pharmaceuticals, Inc.*, 509 U.S. 579 (1993) dramatically altered the manner in which experts could present testimony. *Daubert* involved the claim that birth defects of children were related to the mothers' prenatal use of Bendectin, a prescription drug. Although testimony from eight experts concluded that Bendectin was responsible for birth defects on the basis of animal studies, chemical analyses, and an unpublished analysis of human statistical studies, the District Court in *Daubert* deemed that these methods of testing or analysis did not meet the "general acceptance" judicial test for expert opinion. The Court of Appeals for the 9th Circuit agreed with this decision of the District Court, citing *Frye v. United States*, 293 F. 1013, 1014 (D.C. Cir. 1923).

At particular issue in *Daubert* were the Federal Rules of Evidence, which had been adopted by the United States Supreme Court on November 20, 1972 and enacted by the U.S. Congress on January 2, 1975. Confusingly, even though the Supreme Court reversed the lower courts' decisions in *Daubert* by trumping the earlier *Frye* standard with the newer Rules, this was only mandated for testimony in federal cases. Under Rule 702, scientific, technical or other specialized knowledge is only admissible if "(1) the testimony is based on sufficient facts or data, (2) the testimony is the product of reliable principles and methods, and (3) the witness has applied the principles and methods reliably to the facts of the case" [3].

Where previously, under *Frye*, a level of "general acceptance" of the proffered testimony in the scientific community in which one claimed to have expertise was all that was necessary for admissibility, *Daubert* requires an independent judicial assessment of reliability of the testimony, even though a court may often be hard-pressed to provide such affirmations and often relies on the witness to say so themself. The phrase "I attest to this assertion with a high degree of scientific certainty" or "this opinion is stated within a reasonable degree of scientific probability or certainty" is used by experts in reports or during hearings in order to underscore the reliability of an opinion, but there is really no substantive proof demanded or required in order to support such claims. Although a blatant falsehood about an underlying fact relied upon for opinion might be considered perjurious, if such misinformation is pointed out an expert may still be able to claim "I believed this at the time, based on then-current methods and data."

For state court matters, some states have decided to use *Daubert* and adopted versions modeled on the federal rules, others are still using the earlier *Frye* test, some have developed their own tests of admissibility, and a number of states have not chosen to instantiate any particular method. Results in computational areas have occasionally failed to satisfy *Daubert* criteria because experts are often incapable of explaining the algorithms they have used to a nontechnical audience [4]. So, when delivering opinion, it is important for the expert to be familiar with and apply the appropriate testimony standards for the court one is appearing before. If there are any questions during pretrial preparation, typically the attorneys working with the expert should advise.

The Federal Rules of Evidence are quite comprehensive and cut across diverse fields of forensics, such that computer experts have been able to adapt them to their investigations even though these were not even necessarily being considered (and computational devices barely existed) when Congress created this body of law. Certain particular rules are especially important to consider when dealing with digital image forensics.

For example, Rule 1003 (Admissibility of Duplicates) states that "a duplicate is admissible to the same extent as an original unless (1) a genuine question is raised as to the authenticity of the original or (2) in the circumstances it would be unfair to admit the duplicate in lieu of the original." Here, authentication for digital imagery is certainly a subject of contention. Law enforcement and prosecution is fond of referring to MD5 and SHA1 hash values as "digital fingerprints" while these are actually highly lossy and certainly not unique. Given that both of these hashes have been broken using collision search attack methods [5], one would think that more rigorous algorithms (such as SHA-256) would be encouraged, but the earlier methods are still commonplace, likely due to the toolsets being used and the time it takes to create the larger hash values.

That the smaller hashes have been compromised certainly comes into play when the values for contraband image files (such as those depicting child pornography) are used to identify such materials, possibly from remote locations via file sharing searches. Cracking has enabled files to be constructed such that hashes match those of common system files, so that when a hash-based search is done, numerous false positives are yielded.

Rule 1004 (Admissibility of Other Evidence of Contents) goes further with regard to duplicates, stating that "the original is not required" if it has been lost or destroyed, not obtainable, in possession of an opponent who has failed to produce it, or is not related to issues pertaining to control. Whereas in the law enforcement community, the MD5 and SHA1 hashes are still "generally accepted" as indicative of authenticity, under *Daubert* it might be possible to raise questions about the material (although proving this may be difficult) in an effort to have derivative evidence or copies excluded from trial use. As well, in police sting or honeypot (luring) operations, the full set of data collected during the investigation is rarely provided on the basis that such detail could potentially compromise other active cases. The introduction by prosecution of partial, out-of-context evidence (such as image or video fragments) may place defendants at risk, and attorneys may move a court to suppress such evidence.

Although Rule 601 states that "every person is competent to be a witness except as otherwise provided in these rules" the Rules have been used to provide a customary framework by which experts are validated by the court. Typically at the beginning of an expert's testimony, there is a review (voir dire) of the individual's qualifications by the attorney who has called them to the stand. The credentials, in the form of a curriculum vitae (CV) or résumé that includes publications as well as lists of cases in which the expert has previously testified or their opinion was made public, are given to the opposing side and the judge beforehand, providing fuel for cross-examination. The questioning may be perfunctory if there are no objections, but is often rather extensive (for example, after publication, this chapter and even the entire book will likely be brought up in the qualification part of some hearings for this author when she provides expert testimony), with minute details used to try to impress (or malign) the level of expertise, or to point out potential areas of bias or prejudice, before the judge and jurors. Ultimately, it is the judge who determines whether the expert is deemed qualified and only then are they allowed to proceed to offer an opinion to the judge and jury on the case material. Rarely is an expert entirely disqualified, although an attorney may try to obtain numerous qualifications (such as for computer systems, computer forensics, and image forensics) some of which may be declined on the basis of the documentation and responses provided. Although part of a witness' CV may include a list of areas where judges have previously issued qualification, experts are only qualified for each particular case, requiring that this process occur in each litigation.

A problematic aspect of Rule 612 involves the use of notes when an expert is testifying. Basically, anything in writing can potentially be subject to subpoena by the opposition, so there may be instances where an expert should not create any documented record of discussions. Electronic messages and reports that are intended as drafts (not for distribution) are typically marked "ATTORNEY WORK PRODUCT PROTECTED" (or some variation thereof) because certain privileges recognized in the Rules of Evidence attach to preparation and collaboration with experts for reports and testimony. An expert's billing statements may often be perfunctory, leaving out details of what topics or individuals were involved, for similar reasons (although the hourly rate of remuneration may also be asked about during the expert's qualification). For the busy expert, subject to human frailty, who may not have a 100% precise memory of each case, many of which may be very similar in nature, the way around not bringing notes to the stand that could be required to be revealed, is that the attorney will mark the expert report as an exhibit at trial and then this document may be referred to and read from as a prompting vehicle for responses to questioning. As with the CV, typically the expert's written reports are disclosed in the discovery phase of litigation so that there will be no surprise at trial. Of course, all aspects of the report are subject to being picked apart by opposing counsel on cross-examination, so it is generally advisable to write just enough to make your points clear and go no further beyond that. This is generally a good rule of thumb for oral testimony as well.

## 4 Discovery

If the expert is retained early enough prior to trial, they may have an opportunity to assist with the discovery process. Although a novice might just rely on the hiring attorney or client to provide access to materials for review, the true value of an experienced expert, especially in the area of computer forensics, is often first seen in their ability to ascertain what additional discovery materials might be available in the case. The expert should advise the attorney and client as to what they may be able to request, and will work with the impounding authority or opposition to obtain discovery materials in a format that best facilitates investigation. This is typically an iterative process, and may take months or even years to conclude. The good thing about being on the receiving end of discovery is that if you make your requests promptly, and the opposition is tardy in their responses, this can be called to the court's attention and should be helpful in obtaining continuances (extensions) on trial dates if additional time is required in order to perform the investigation and satisfy "due process."

When dealing with contraband digital imagery, the defense team is often held at a severe disadvantage by the customary belief that their expert cannot be allowed to examine a copy of the materials in their own laboratories for fear that they will be making and distributing copies of the alleged illicit data, despite the fact that doing so would likely constitute a felony and may even result in the loss of their business or occupation. Yet, even when court orders have been obtained requiring impounding agencies to provide contraband data to an examining laboratory on behalf of the accused defendant, requests have been denied on such grounds (especially since the enactment of the Adam Walsh Act in 2006). Defense has argued in numerous cases (sometimes successfully) that such discovery and investigation constraints provide an uneven playing field. One way of dealing with this is to use just the materials that have been released for off-site investigation, but occasionally it is not possible to perform an adequate job with only this data (particularly if photographs are provided in printed form and not as digital files). This type of discovery obstruction can also occur in civil cases where sides are unable to agree to the release of proprietary (trade secret) information.

If the forensic expert needs to be present at an opposition site to review data, typically they will be required to use (often substandard) equipment at that location, they may not have access to the same tool set that they are accustomed to using in their laboratory, their activities in performing the investigation may be monitored or even recorded (thus potentially revealing strategy), and subtle or overt harassment may even occur. In contrast, the holder of these materials essentially has 24/7 access to all of the evidence data, and can run exhaustive searches using whatever equipment they choose. Especially in public defender matters, where fee ceilings are typically imposed and funding is limited, being required to repeatedly visit an unfamiliar lab to conduct investigations can unfairly increase litigation costs. The defense team will often not have access to the impounded discovery materials during the actual conduct of trial, so they cannot easily check items on a break or overnight when issues are raised in the immediacy of testimony. Prosecution investigators have been known to take advantage of their unlimited and easy access to the evidence (their office may even be located in another part of the building where the trial is taking place), and if they tip their hand to having performed additional mid-trial searches, the defense attorney must object to the introduction of any new results, and request suppression of these, on the grounds of unfairness or surprise in discovery.

Since this type of restrictive investigation process often occurs with contraband cases, and since child pornography is increasingly at issue, especially with file sharing services, one would think that the defense lawyers and public defender offices would have joined together long before now (as prosecution did years ago with the FBI and state police in forming the various Regional Computer Forensic Laboratories around the country), in creating authorized settings where evidence examinations can be conducted without opposition calling the shots. Perhaps someday the gross inequities of this type of investigation will be challenged as a civil rights matter, and the situation will be rectified, but in the meanwhile it continues to hamper and prejudice the process.

The original intention of federal and state laws (as mentioned earlier) pertaining to possession and distribution of child pornography were to protect actual individuals from harm. This is why cartoons (such as South Park) depicting children engaged in sexual acts are not presently considered illegal, and also why the National Center for Missing and Exploited Children (NCMEC) maintains a database that identifies particular images and series with the people involved. (One should keep in mind that NCMEC is working for the prosecution, so defense access to their data and reports is typically prohibited or highly limited.)

With the subsequent enactment of Megan's Laws in all states, there has been an increased perception that individuals who happen to have obtained alleged child pornography, even if they came across it accidentally and are certainly not producing it, are actual pedophiles rather than just curious or foolish Web surfers. This has caused the pendulum to swing in the direction of considering some fabricated (i.e. computer generated or Photoshopped) images as contraband, even when no child was ever involved. States may also have bestiality laws, and imagery of this nature can be considered contraband, although it does not necessarily carry the same penalties as does a child porngraphy conviction, where lifelong residence registration is typically required even after all jail time and probation has been served. Note that the "sexting" phenomenon (where underage youth send nude or erotic photos of themselves to other youngsters) was not considered when this law was being devised. Megan's mother, Maureen Kanka, has indicated that prosecution of juveniles for such offenses was "harming the children more than helping them [6]."

For these many reasons and others (such as the authenticity and even the camera angles of the monitoring videos from police cars or shopping malls), it is important to enable digital image files to be analyzed for source and content. A vivid example is the air show footage of a jet that appears to be flying directly over the Golden Gate Bridge, actually considerably further away, with the FCC later reporting that objects were foreshortened through the camera angle and telephoto lens [7]. As tools become more sophisticated in their ability to discern modifications (editing, morphing, etc.)

or image distortions, defense will increasingly need the same type of 24/7 access to the data that prosecution now has, for analysis purposes. Experimentation and algorithm improvements can be done with benign files, but this also needs to be demonstrated with the actual charged items in order for results to be confirmed and for effective statements to be made, that will stand up to challenges, in reports and at trial.

The charges themselves also motivate the type of forensic analysis that should be used. Since time and funds are always a factor, a more focused investigation will typically yield stronger results than a scattershot approach. For example, if the charges do not include distribution or creation of images, then results pertaining to those issues may not need to be addressed. Especially in civil matters, it is helpful to understand the actual dispute in order to best select the methods that will potentially yield beneficial data.

## 5 Exhibits

The manner in which digital images are displayed as exhibits at trial can also be problematic. Many courtrooms are now equipped with enormous display screens, where a movie or photograph can be shown some ten or more times larger than it would have appeared on a laptop computer. The shocking nature of this is often exploited with contraband imagery, whose deleterious effects defense attorneys should be cautious in ensuring are suppressed. Displays that are inappropriately oversized must not be allowed to unduly influence the judge or jury.

At the other extreme, the perceived prurient nature of an image can be used to provide a "peep show" at trial. In one instance, prosecution provided copies of two thumbnails of naked children (not engaged in sexual acts) to each juror in a manila folder. The jurors were informed that they should not open the folder until told to do so, and that what they would see they might find "shocking." Unfortunately, the defense attorney did not feel it appropriate to object to this presentation and the jury subsequently came back with a felony conviction, largely on the basis of these tiny, blurry images.

## 6 Concluding Thoughts

Courtroom issues involving the presentation of testimony related to digital image forensics create a balancing act for forensic experts, where the development and use of novel techniques must be weighed against the merits of the case and the ability to successfully introduce results into evidence. Digital image forensics is in its infancy, as is the field of computer forensics. As such, it is helpful to continue to gather insight from other forensic endeavors in an effort to establish good science and good law. As well, the pace at which technology advances occur, far outstrips the ability of courts and legislatures to establish law that can reasonably address new situations. Judges and jurors may often have limited knowledge of how computers and the Internet work. The development of evidence and discovery rules have tended to be reactive rather than proactive, and these delays can adversely affect rulings. There are no simple answers to these problems, other than to attempt to create a level playing field where all parties can fairly review the evidence and present expert opinions in such fashion that justice will be able to be reasonably served.

**Acknowledgments** The author would like to recognize Attorneys Anna M. Durbin and Michael W. Hoffman who provided salient feedback during the writing of this chapter. Clients and attorneys whom the author has worked with (or against!) on cases involving digital imagery, whose names have not been instantiated for reasons of privacy, are also thanked for the numerous insights they have provided. Dr. Norman Badler, director of the Center for Human Modeling and Simulation at the School of Engineering and Applied Science of the University of Pennsylvania is also acknowledged for his long-standing encouragement and support of research in computer graphics in general and this author in particular.

## References

1. Mercuri RT (2004) The HIPAA-potamus in Health Care Data Security. Security Watch, Communications of the Association for Computing Machinery, 47(7), July 2004
2. United States General Accounting Office (2003) Child pornography is readily accessible over peer-to-peer networks. GAO-03-537T, 13 Mar, 2003
3. Morse MA, Gaugler AC (2007) Daubert challenges to experts in federal criminal cases: an overlooked defense. The National Association of Criminal Defense Lawyers Champion, July 2007
4. Stevens M (2010) Admissibility of scientific evidence under Daubert. http://faculty.ncwc.edu/mstevens/425/lecture03.htm. Accessed 22 Oct 2010
5. Wang X, Yin YL, Yu H (2005) Finding collisions in the full SHA-1. In: Shoup V (ed) Crypto 2005, LNCS vol. 3621. Springer, Heidelberg, pp. 17?36
6. National Public Radio (2009) 'Megan's Law' Mom Criticizes 'Sexting' Charges, 26 Mar, 2009
7. Myers C, Doft D (2010) Nothing wrong with jet's air show maneuver, FAA says. CNN, 19 Oct, 2010

# Counter-Forensics: Attacking Image Forensics

**Rainer Böhme and Matthias Kirchner**

**Abstract** This chapter discusses counter-forensics, the art and science of impeding or misleading forensic analyses of digital images. Research on counter-forensics is motivated by the need to assess and improve the reliability of forensic methods in situations where intelligent adversaries make efforts to induce a certain outcome of forensic analyses. Counter-forensics is first defined in a formal decision-theoretic framework. This framework is then interpreted and extended to encompass the requirements to forensic analyses in practice, including a discussion of the notion of authenticity in the presence of legitimate processing, and the role of image models with regard to the epistemic underpinning of the forensic decision problem. A terminology is developed that distinguishes security from robustness properties, integrated from post-processing attacks, and targeted from universal attacks. This terminology is directly applied in a self-contained technical survey of counter-forensics against image forensics, notably techniques that suppress traces of image processing and techniques that synthesize traces of authenticity, including examples and brief evaluations. A discussion of relations to other domains of multimedia security and an overview of open research questions concludes the chapter.

## 1 Definition of Counter-Forensics

This final chapter changes the perspective. It is devoted to digital image *counter*-forensics, the art and science of impeding and misleading forensic analyses of digital images.

R. Böhme (✉)
Department of Information Systems, Westfälische Wilhelms-Universität Münster, Leonardo-Campus 3, 48149 Münster, Germany
e-mail: rainer.boehme@wi.uni-muenster.de

M. Kirchner
International Computer Science Institute, 1947 Center Street, Berkeley, CA 94704, USA
e-mail: kirchner@icsi.berkeley.edu

H. T. Sencar and N. Memon (eds.), *Digital Image Forensics*,
DOI: 10.1007/978-1-4614-0757-7_12,

Digital image forensics has become a hot topic over the past couple of years. Initially, academic schemes found applications in domains like law enforcement, intelligence, private investigations, and media. Passive image forensics has promised to reestablish trust in digital images, which otherwise were deemed too easy to manipulate. But what stops perpetrators, spies, and swindlers, who make efforts to manipulate images for their own profit anyway, from finding out forensic investigators' latest tricks and techniques? Then they can use this knowledge to cover up traces or—even worse—plant misleading traces.

Many forensic algorithms were not designed with such behavior in mind, hence they are easy to deceive. To justify the additional trust we place in digital images through forensics, it is important that the limits of forensics are known and will eventually be overcome. The only alternative would be a closed infrastructure of trustworthy acquisition devices that authenticate images actively [15]. This vision is certainly costly and most likely politically unviable. The other apparent solution of keeping passive forensic techniques secret is determined to fail. Its advantages are only temporary and highly uncertain [23]. Not to mention that the chain of causality leading to a conviction in a public court must withstand scrutiny of independent experts, who would learn the techniques and potentially compromise their secrecy.

The research field that challenges digital forensics and systematically explores its limitations against intelligent counterfeiters is called *counter-forensics*, or anti-forensics. Both terms are used synonymously in the literature. We prefer the former because it better reflects the pragmatic *reaction to* forensics, as opposed to a normative *disapproval of* forensics. Counter-forensics stands in a equally productive relation to forensics like cryptanalysis to cryptography. Therefore, we borrow from the cryptanalysis terminology and call a counter-forensic scheme *attack* (against forensics). This reflects the strategic intention of the counterfeiter's action.

Besides the need to assess and improve the reliability of forensic methods, two more reasons motivate research on counter-forensics. First, many forensic techniques link images to the circumstances of their acquisition, e. g., the acquisition device or time. This indirectly reveals information about identities of the author or depicted subjects. This is not always desired. Researchers have studied systems providing *unlinkability* and *anonymity* in digital communications for some time [37]. All these efforts are useless if the link to the subject can be reestablished by forensic analysis of the message. Hence counter-forensic techniques to suppress traces of origin in digital images are a relevant building block for anonymous image communication, which can be useful in practice to protect the identity of sources, e. g., in the case of legitimate whistle-blowing.

Second, some authors argue that counter-forensics, if implemented in image acquisition devices, can be useful to hide details of the internal imaging pipeline and thus *discourage reverse engineering* [43]. We mention this motivation for completeness, but remain reserved on whether current counter-forensics is ripe enough for this purpose. Our main concern is that counter-forensics often imply a loss of image quality. Camera manufacturers, for instance, compete on quality. It is questionable if they would sacrifice a competitive edge for making reverse engineering a little harder. Yet we are curious to see promising applications in the future.

The outline of this chapter is as follows. In the next section, we define counter-forensics formally in a general decision-theoretic framework. This framework is interpreted and extended to encompass the requirements to forensic analyses in practice in Sect. 3, including a discussion of the notion of authenticity in the presence of legitimate processing, and the role of image models with regard to the epistemic underpinning of the forensic decision problem. The Sect. 4 develops a terminology that distinguishes security from robustness properties, integrated from post-processing attacks, and targeted from universal attacks. The Sect. 5 contains a technical survey of counter-forensics against image forensics. The survey is structured by techniques that suppress traces of image processing and techniques that synthesize traces of authenticity. It includes examples of each technique as well as a brief evaluation. The Sect. 6 adds a discussion of relations to other domains of multimedia security, notably steganography and robust digital watermarking. We conclude this chapter with an overview of open research questions in the final Sect. 7.

## 2 Theory

Digital image forensics refers to the collection of scientific methods to systematically infer particulars of an unknown image generation process from a given (set of) digital image(s). For a formal treatment of image forensics and counter-forensics, we have to define this generation process (Sect. 2.1), then describe forensic inference as a decision problem (Sect. 2.2), before counter-forensics can be introduced as means to influence the outcome of such decisions (Sect. 2.3).

### *2.1 Image Generation*

Generation of natural images links the real world with digital representations thereof.

**Definition 1** The *image generation function* $\mathsf{generate} : \mathscr{N} \times \Theta \to \mathscr{I}$ maps observations of the infinite set of all conceivable natural phenomena $N \in \mathscr{N}$ to the finite set of digitized images $\mathbf{I} \in \mathscr{I}$. This mapping is parameterized with a collection of parameters $\theta \in \Theta$.

The parameters include, inter alia, the perspective, the time of the acquisition, the choice of the acquisition device, and its configuration (e.g., settings, lenses). It is convenient to understand the image generation process as a combination of both the image acquisition with a digital imaging device and subsequent post-processing.

**Definition 2** Function $\mathsf{generate}$ is composed of a concatenation of an *image acquisition function* $\mathsf{acquire} \in \mathscr{A} : \mathscr{N} \to \mathscr{I}$ and an *image processing function* $\mathsf{process} \in \mathscr{P} : \mathscr{I}^+ \to \mathscr{I}$, where $\mathsf{acquire}$ and $\mathsf{process}$ are elements of the respective families of functions $\mathscr{A}$ of all possible image acquisition methods and $\mathscr{P}$

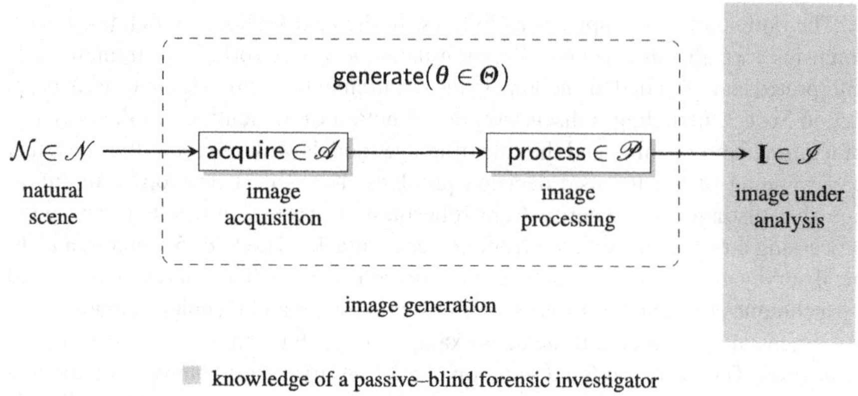


**Fig. 1** General image generation function in the context of passive–blind image forensics

of all possible image processing operations. The exact composition is defined by the parameters $\theta$ of generate.

In the above definition, operator $^+$ is the 'Kleene plus', which for a given set $\mathscr{I}$ is defined as $\mathscr{I}^+ = \bigcup_{n=1}^{\infty} \mathscr{I}^n$. Hence, function process may take an arbitrary positive number of digital images as input. The block diagram in Fig. 1 illustrates the concept of image generation suggested by Definitions 1 and 2 (ignoring the possibility of multiple inputs to function process for the sake of simplicity). The case of *passive–blind* image forensics implies that the forensic investigator has no influence on generate (passive) and her knowledge about generate is limited to the information given in the image under analysis (blind).

There exist two parameter settings that deserve a special note. First, digital images not necessarily undergo a processing step after initial acquisition with a digital imaging device. Set $\mathscr{P}$ thus explicitly includes the identity function, $\perp_{\mathscr{P}}: \mathbf{I} \mapsto \mathbf{I}$, i. e., no post-processing.

**Definition 3** All image generation functions (acquire, $\perp_{\mathscr{P}}$) $\in \mathscr{A} \times \mathscr{P}$ produce *original images* as opposed to *processed images* that result from generation functions (acquire, process) $\in \mathscr{A} \times \mathscr{P} \setminus \{\perp_{\mathscr{P}}\}$.

Similarly, set $\mathscr{A}$ includes a pathologic function, $\perp_{\mathscr{A}}$, which refers to no acquisition with an imaging device. This is particularly useful to differentiate between natural images and computer-generated images.

**Definition 4** All image generation functions (acquire, process) $\in \mathscr{A} \setminus \{\perp_{\mathscr{A}}\} \times \mathscr{P}$ produce *natural images* as opposed to *computer-generated images* that result from generation functions ($\perp_{\mathscr{A}}$, process) $\in \mathscr{A} \times \mathscr{P}$. By definition, computer-generated images are processed images.

A further important attribute with regard to the image generation process is the notion of authenticity. A digital image $\mathbf{I}$ is called authentic if it is a valid projection of the natural phenomenon $\mathcal{N}$. Instances of $\mathsf{process}$ may impair authenticity.

To give a formal definition of authenticity, we note that the projection of one particular natural phenomenon $\mathcal{N}$ to an authentic image is not necessarily unique. There may exist many different mappings that yield *semantically equivalent* images. This means each element in a set of many different images $\mathbf{I}_1 \neq \mathbf{I}_2 \neq \cdots \neq \mathbf{I}_n$ is a valid representation of the same realization of nature $\mathcal{N}$. For example, in many cases it makes no difference with which digital camera a given event is captured, but each camera will produce a slightly different image. Within certain limits also the change of resolution or lossy compression may retain an image's authenticity. In this sense, authenticity is an attribute of the tuple $(\mathbf{I}, \theta, \mathcal{N})$ where $\mathcal{N}$ must be the realization of $\mathscr{N}$ under parameters $\theta$.

Intuitively, also the *semantic meaning* of an image refers to the link between a depicted scene and the corresponding natural phenomenon. Yet this is more difficult to formalize, as the association of semantic meaning requires interpretation and it is highly context-dependent in general. We work around this difficulty and assume that semantic equivalence is measurable between images.

**Definition 5** Two images $\mathbf{I}$ and $\mathbf{J} \in \mathscr{I}$ are *semantically equivalent* if there exists $\mathcal{N} \in \mathscr{N}$ such that

$$|\mathsf{dist}(\mathbf{I}, \mathcal{N}) - \mathsf{dist}(\mathbf{J}, \mathcal{N})| < d,$$

where $\mathsf{dist} : \mathscr{I} \times \mathscr{N} \to \mathbb{R}_+$ is a measure of the *semantic distance* between an image and a—real or imaginary—natural phenomenon, and $d$ is a given threshold.

The *semantic resolution* is the ability of function $\mathsf{dist}$ to differentiate between very similar natural phenomena for a fixed image $\mathbf{I}$. This resolution depends on the *quality* of an image, or, more precisely, on the information conveyed in an image $\mathbf{I}$ about $\mathcal{N}$. Threshold $d$ has to be chosen commensurate with the semantic resolution of the image with the lowest quality.

Equipped with the notion of semantic equivalence, we can finally define what qualifies an image as authentic.

**Definition 6** All original natural images are authentic. Furthermore, for a given authentic image $\mathbf{I} = \mathsf{generate}(\mathcal{N}, \theta)$, a processed version $\mathbf{J} = \mathsf{process}(\mathbf{I})$ is called authentic if $\mathbf{I}$ and $\mathbf{J}$ are semantically equivalent with respect to $\mathcal{N}$.

Definitions 3 and 6 reflect the subtle yet significant difference between processed and counterfeit images. While each non-trivial instance of $\mathsf{process}$ destroys the originality of an image, it does not necessarily impair its authenticity. Whether or not a processed image will be considered as counterfeit ultimately depends on a given context and established habits. We further point out that computer-generated images are not counterfeits by definition, because function $\mathsf{process}$ can always be defined to replace a natural image with a computer-generated version (or parts thereof).

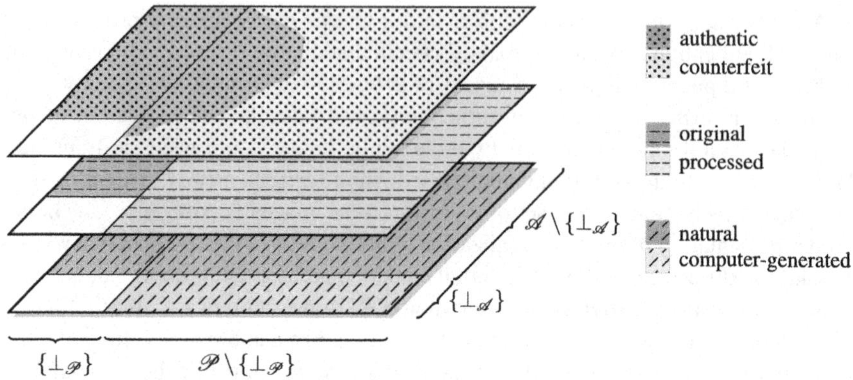


**Fig. 2** Function space $\mathscr{A} \times \mathscr{P}$ of conceivable combinations of image acquisition and processing functions along with different classifications of the resulting digital images

This is viable as long as synthesis algorithms are sophisticated enough to generate semantically equivalent images.

Figure 2 gives a schematic overview of the different classifications of digital images discussed so far. Each of the three layers depicts a different partition of the same function space $\mathscr{A} \times \mathscr{P}$. The square in the lower left corner is left blank intentionally, as processing functions $(\perp_{\mathscr{A}}, \perp_{\mathscr{P}})$ have no practical and meaningful equivalent.

## 2.2 Digital Image Forensics as a Classification Problem

The ultimate objective of passive–blind image forensics is to infer the authenticity of a given image without knowledge about the inputs of generate, notably the scene $\mathcal{N}$ and the parameters $\theta$. Yet authenticity is often too hard to prove, so that forensic investigations resort to the inference on the inputs to generate. These serve as indicators which can be combined with side-information from other sources to make statements about the authenticity. Forensic analyses are possible if the functions acquire and process leave *identifying traces* in the resulting images. These traces can be used to distinguish between samples from different generation processes. Hence, digital image forensics is best described as a classification problem.

### 2.2.1 Classes in Digital Image Forensics

Forensic investigators define classes $C_0, \ldots, C_k$ to encapsulate parameter ranges of the generation function. The choice of the class space, denoted by $\mathscr{C}$, depends on the concrete application. For example, manipulation detection is usually stated as a binary classification problem, $|\mathscr{C}| = 2$, with one class $C_0$ for original images and another class $C_1$ for processed images. In the case of source identification, the classes represent different imaging sources, e. g., different digital camera models or individual devices (typically $|\mathscr{C}| \gg 2$).

**Definition 7** A class $C \in \mathscr{C}$ partitions the function space $\mathscr{A} \times \mathscr{P}$ into two subspaces, $(\mathscr{A} \times \mathscr{P})_{(C)}$ and $(\mathscr{A} \times \mathscr{P})_{(\not{C})}$ so that all images $\mathbf{I}_{(C)}$ generated by $(\mathsf{acquire}, \mathsf{process}) \in (\mathscr{A} \times \mathscr{P})_{(C)}$ share common identifying traces.

*Convention* To keep notations simple, we use $\mathbf{I}_{(k)}$ equivalent for $\mathbf{I}_{(C_k)}$ when referring to instances of images of a particular class $C_k \in \mathscr{C}$. Moreover, we write $\mathbf{I}_{(0)}$ for authentic images and $\mathbf{I}_{(1)}$ for counterfeits whenever the context prevents ambiguities and the class space contains only these two classes.

Definitions 1–7 allow us to express various kinds of image forensics in a unified formal framework, as illustrated by the following examples.

*Example 1 Natural versus computer-generated images:* Class $C_0$ of natural images contains all instance of images $\mathbf{I}_{(0)}$ generated by functions in the subspace $(\mathscr{A} \setminus \{\perp_{\mathscr{A}}\} \times \mathscr{P})$. Class $C_1$ of computer-generated images entails all instance of images $\mathbf{I}_{(1)}$ generated by functions in the subspace $(\{\perp_{\mathscr{A}}\} \times \mathscr{P})$.

*Example 2 Manipulation detection:* Class $C_0$ of original images contains all instances of images $\mathbf{I}_{(0)}$ generated by functions in the subspace $(\mathscr{A} \times \{\perp_{\mathscr{P}}\})$. Class $C_1$ of processed images entails all instances of images generated by functions in the subspace $(\mathscr{A} \times \mathscr{P} \setminus \{\perp_{\mathscr{P}}\})$.

*Example 3 Source identification via sensor noise:* Class $C_k$ of acquisition with sensor $k = 1, \dots$ contains all instances of images $\mathbf{I}_{(k)}$ generated by functions in the subspace $(\mathscr{A}_k \times \mathscr{P})$ where $\mathscr{A}_k \subset \mathscr{A}$ is the set of all image acquisition functions of sensor $k$ and $\bigcap_k \mathscr{A}_k = \emptyset$.

*Example 4 (Ideal) temporal forensics:* Class $C_{t_1,t_2}$ of images acquired in the time interval $t_1 < t < t_2$ contains all instances of images $\mathbf{I}_{(C_{t_1,t_2})}$ generated by functions in the subspace $(\mathscr{A}_{t_1,t_2} \times \mathscr{P})$ where $\mathscr{A}_{t_1,t_2} \subset \mathscr{A}$ is the set of all image acquisition functions invoked between time $t_1$ and $t_2$. In practice, temporal forensics today requires prior knowledge of the acquisition device so that more specific partitions have to be defined.

Note that classes are intentionally defined by partitioning the image generation process, and not the image space $\mathscr{I}$. This is why in the examples above, we refer to *instances* of images, i. e., outputs of specific invocations of $\mathsf{generate}$. A given image $\mathbf{I} \in \mathscr{I}$ with unknown provenance may be the result of different generation functions spanning more than one class. To resolve this ambiguity in the possibilistic framework, it is useful to take a probabilistic perspective.

**Definition 8** Function $\mathcal{P}_C : \mathscr{I} \rightarrow [0, 1]$ is the likelihood function returning the conditional probability $\Pr(\mathbf{I} \mid C)$ of observing image $\mathbf{I}$ if the generation process falls in the partition of class $C$. The probability reflects the empirical distributions of $\mathcal{N} \sim \mathscr{N}$ and $(\mathsf{acquire}, \mathsf{process}) \sim (\mathscr{A} \times \mathscr{P})_{(C)}$.

This probabilistic perspective allows us to quantify the ambiguity and derive decision rules, which on average minimize forensic decision errors.

### 2.2.2 Decision Rules

Given a class space $\mathscr{C}$, $|\mathscr{C}| \geq 2$, and an observed digital image $\mathbf{I}$ with unknown class, the forensic investigator needs a decision rule to assign $\mathbf{I}$ to a class $C_*$.

**Definition 9** A *digital image forensics algorithm* is given by a function $\mathsf{decide} : \mathscr{I} \rightarrow \mathscr{C}$ that assigns an image $\mathbf{I} \in \mathscr{I}$ to a class $C \in \mathscr{C}$.

This decision rule now partitions the image space into disjoint classes, $\mathscr{I} = \bigcup_k \mathscr{R}_k$, such that all elements within a *decision region* $\mathscr{R}_k$ are assigned to class $C_k$,

$$\mathscr{R}_k = \{\mathbf{I} \in \mathscr{I} \mid \mathsf{decide}(\mathbf{I}) = C_k\}. \tag{1}$$

It is reasonable to assume that decisions are based on the class probabilities conditional to the observed image, $\Pr(C_k \mid \mathbf{I})$, which can be calculated from the likelihood functions in Definition 8 by using Bayes' theorem,

$$\Pr(C_k \mid \mathbf{I}) = \frac{\Pr(\mathbf{I} \mid C_k) \cdot \Pr(C_k)}{\sum_i \Pr(\mathbf{I} \mid C_i) \cdot \Pr(C_i)} = \frac{\mathcal{P}_{C_k}(\mathbf{I}) \cdot \Pr(C_k)}{\sum_i \mathcal{P}_{C_i}(\mathbf{I}) \cdot \Pr(C_i)}. \tag{2}$$

In general, the larger the $\Pr(C_k \mid \mathbf{I})$, the more evidence exists that $\mathbf{I}$ was generated by a function $(\mathsf{acquire}, \mathsf{process}) \in (\mathscr{A} \times \mathscr{P})_{(C_k)}$. The concrete transformation of posterior probabilities into decisions depends on the algorithm $\mathsf{decide}$ and its decision rule. Many image forensics algorithms adhere to the minimum probability of error principle and decide for the class that maximizes $\Pr(C_k \mid \mathbf{I})$ [2],

$$\mathsf{decide}(\mathbf{I}) = C_* \quad \Leftrightarrow \quad C_* = \arg\max_{C_i \in \mathscr{C}} \Pr(C_i \mid \mathbf{I}). \tag{3}$$

It is possible to impose additional constraints to ensure a reliable decision, for example by requiring a minimum a-posteriori probability,

$$\Pr(C_* \mid \mathbf{I}) \geq p_{\min}, \tag{4}$$

or a minimum separability from the second-most probable class,

$$\Pr(C_* \mid \mathbf{I}) - \max_{C_i \in \mathscr{C} \setminus C^*} \Pr(C_i \mid \mathbf{I}) \geq p_{\text{sep}}. \tag{5}$$

Here, function $\mathsf{decide}$ has to return a special value for undecidable cases.

In many forensic problems, for which the class space is defined to comprise only two classes $C_0$ and $C_1$ (see for instance Examples 1 and 2), the decision problem can be expressed as a simple hypothesis test between

- $H_0$: the image under analysis $\mathbf{I}_{(k)}$ is generated by a process belonging to class $k = \mathcal{C}_0$, and
- $H_1$: the image under analysis $\mathbf{I}_{(k)}$ is generated by another process $k = \mathcal{C}_1$.

Because $\Pr(\mathcal{C}_0 \mid \mathbf{I}) + \Pr(\mathcal{C}_1 \mid \mathbf{I}) = 1$, and according to the theory of hypothesis tests, the optimal decision rule is given by the likelihood-ratio test,

$$\mathsf{decide}(\mathbf{I}) = \mathcal{C}_k \quad \Leftrightarrow \quad \frac{\mathcal{P}_{\mathcal{C}_k}(\mathbf{I})}{\mathcal{P}_{\mathcal{C}_{|k-1|}}(\mathbf{I})} > \tau\,, \qquad k \in \{0, 1\}\,. \tag{6}$$

## 2.3 *Counter-Forensics*

For a given image $\mathbf{I} = \mathbf{I}_{(k)}$, counter-forensics aims at preventing the assignment to the image's class $\mathcal{C}_k$. By suppressing or counterfeiting identifying traces, the counterfeiter creates a *counterfeit* $\mathbf{J} = \mathbf{J}_{(\hat{l})}$ with the intention to let it appear like an authentic member of an alternative class $\mathcal{C}_l \in \mathscr{C}$, $l \neq k$, when presented to the forensic investigator's function $\mathsf{decide}$.

*Convention* We use the subscript notation $_{(\hat{l})}$ to denote the intended class change of the counterfeit.

**Definition 10** A digital image forensics algorithm $\mathsf{decide}$ is *vulnerable* to a *counter-forensic attack* if for a given image $\mathbf{I} = \mathsf{generate}(\theta)$

$$\exists\, \mathsf{attack} \in \mathscr{P},\ \mathbf{J} = \mathsf{attack}(\mathbf{I}) \quad \text{so that} \quad \mathsf{decide}(\mathbf{J}) \neq \mathsf{decide}(\mathbf{I})$$

subject to the constraints

1. $\mathbf{I}$ and $\mathbf{J}$ are semantically equivalent (*semantic constraint*), and
2. the probability of finding $\mathsf{attack}$ for a given $\mathbf{I}$ is not negligible within a given complexity bound (*computational constraint*).

The following examples illustrate how this definition matches existing counter-forensic strategies.

*Example 5* A typical counter-forensic image manipulation of an authentic image $\mathbf{I}_{(0)}$ will involve two steps, first a transformation $\mathbf{I}_{(0)} \mapsto \mathbf{I}'_{(1)}$ which changes the semantic meaning according to the counterfeiter's intention, and second a counter-forensic attack $\mathbf{I}'_{(1)} \mapsto \mathbf{I}'_{(\hat{0})}$ to pretend authenticity of the counterfeit, i. e., $\mathsf{decide}\left(\mathbf{I}'_{(\hat{0})}\right) = \mathcal{C}_0$. Images $\mathbf{I}'_{(1)}$ and $\mathbf{I}'_{(\hat{0})}$ are semantically equivalent.

*Example 6* Counterfeiting the source of an authentic image $\mathbf{I}$ involves a single application of a counter-forensic attack $\mathbf{I} \mapsto \mathbf{I}'$, possibly with the additional requirement that a specific target class $\mathcal{C}_{\text{target}} \overset{!}{=} \mathsf{decide}(\mathbf{I}') \neq \mathsf{decide}(\mathbf{I})$ is pretended.

Note that counterfeiting a particular target class generally needs to address both the suppression of identifying traces of the original class and synthesis of artificial traces of the target class. For example in PRNU-based digital camera identification [13], inserting the reference noise pattern of a target camera may lead to decisions for the new class $C_{\text{target}}$. But the (distorted) fingerprint of the true camera is still present. A thorough forensic investigator may find abnormally high likelihood values $\mathcal{P}_{C_k}\left(\mathbf{I}'_{(k)}\right)$, $k \neq C_{\text{target}}$, suspicious.

This leads us to the notion of *reliability* of a counter-forensic attack $\mathbf{I}_{(k)} \mapsto \mathbf{J}_{(\hat{l})}$ against all possible decision rules on a given class space $\mathscr{C}$. (Unlike Definition 10, which is formulated for a specific choice of function $\mathsf{decide}$.) From the counterfeiter's viewpoint, every forensic analysis can be reduced to a two-class decision problem on a class space $\mathscr{C}' = \{C'_0, C'_1\}$ by defining classes $C'_0$ and $C'_1$ to represent the set of target (i.e., admissible) generation functions and attacks, respectively:

$$(\mathscr{A} \times \mathscr{P})_{(C'_0)} \subseteq (\mathscr{A} \times \mathscr{P})_{(\mathscr{C}_k)} \tag{7}$$

$$(\mathscr{A} \times \mathscr{P})_{(C'_1)} = (\mathscr{A} \times \mathscr{P})_{(C_k)} \times \{\mathsf{attack}\}\,. \tag{8}$$

The concrete definition of class $C'_0$ depends on the counterfeiter's agenda. It may correspond to a combination of several classes (if the goal is only to suppress identifying traces of class $C_k$) or to a particular class $C_{\text{target}}$ (for instance to pretend a specific source device, cf. Example 6).

Careful counterfeiters in general strive to design their attacks so that samples of both classes $C'_0$ and $C'_1$ are indistinguishable. The decidability of the hypothesis test in Eq. (6) for all realizations of $\mathbf{I} \in \mathscr{I}$ can be measured by the Kullback–Leibler divergence between the two conditional probability distributions,

$$\mathsf{D}_{\mathrm{KL}}\left(\mathcal{P}_{C'_0}, \mathcal{P}_{C'_1}\right) = \sum_{\mathbf{I} \in \mathscr{I}} \mathcal{P}_{C'_0}(\mathbf{I}) \log \frac{\mathcal{P}_{C'_0}(\mathbf{I})}{\mathcal{P}_{C'_1}(\mathbf{I})}. \tag{9}$$

**Definition 11** A counter-forensic attack is $\varepsilon$-*reliable* against all digital image forensics algorithms on a class space $\mathscr{C}$ if for each pair $(C'_0, C'_1) \in \mathscr{C}' \times \mathscr{C}'$, $\mathbf{I}_{(C'_k)} \sim \mathcal{P}_{C'_k}$,

$$\mathsf{D}_{\mathrm{KL}}\left(\mathcal{P}_{C'_0}, \mathcal{P}_{C'_1}\right) \leq \varepsilon.$$

The unit of $\varepsilon$ is bits or nats, depending on the base of the logarithm in Eq. (9). For the special case $\varepsilon = 0$, authentic and counterfeit images are drawn from the same distribution and the forensic investigator cannot gain any information from the analysis of $\mathbf{I}$. Hence, the counter-forensic attack is called *perfectly reliable*.

Note the similarity between Definition 11 and the notion of $\varepsilon$-*secure*, respectively *perfect steganography* in [5, 14]. Also in image forensics, $\varepsilon$ bounds the error rates of the forensic investigator from below by the binary entropy function via the deterministic processing theorem. Further similarities between image forensics, counter-forensics, and other fields in information hiding are discussed in more detail in the Sect. 6 below.

## 3 Practical Considerations

The theory in the Sect. 2 provides a framework general enough to discuss a wide range of questions regarding digital image forensics and counter-forensics. However, only a few of the definitions are directly applicable in practice.

### 3.1 Epistemic Bounds

The main difficulty in applying the above equations is the lack of knowledge about the conditional probability distributions $\mathcal{P}_C(\mathbf{I})$, which are given only empirically (cf. Def. 8). According to widely accepted epistemological paradigms, natural phenomena in the real world can never be fully known but merely approximated by consequent falsification and refinement of theories about the real world [3, 4]. This is also reflected in Definition 1, which states that the support of $\mathscr{N}$ is infinite. Even after the transformation to the finite space $\mathscr{I}$, in general the support of $\mathcal{P}_C(\mathbf{I})$ is too large and too heterogeneous to efficiently estimate distributions by sampling.

Even if we ignore this for a moment and assume that authentic images can be efficiently sampled, there remains the difficulty of sampling counterfeit images. Generating good counterfeits is a time-consuming manual task that largely depends on the counterfeiters' creativity. And it highly depends on the original image. This process is very hard to automate. The high cost of sampling is also reflected by the size and quality of available data sets that have been compiled in controlled environments for the purpose of image forensics research. Typical counterfeits are obtained by copying patches from within the same or other images, without sophisticated post-processing and without adaptivity to the depicted scene [35, 19]. Only few databases provide more realistic forgeries [8, 9]; however, without resolving the general trade-off between quality and quantity.

### 3.2 Image Models

To reduce complexity and avoid the epistemic obstacles, all practical digital image forensics algorithms make use of *models* of digital images. Such models may be stated explicitly or—more often—implicitly. Models can be seen as a dimensionality reduction by projecting the high-dimensional image space $\mathscr{I}$ to a much smaller and more tractable subspace, e. g., a scalar in $\mathbb{R}$, on which decide is defined as a *discriminant function*.

Modeling images in low-dimensional model domains is effective as long as the mismatch with the real world is not substantial. By accepting the general need for image models, it is clear that a forensic algorithm can only be as good as the model it employs. The better the underlying model can explain and predict observed samples of a particular class, the more confident a forensic investigator can base her decisions on it. Conversely, this also implies that the more restricted a forensic investigator's model of digital images is, the easier a counterfeiter can find ways to construct successful counter-forensic techniques. In the light of this important observation, counter-forensic techniques clearly benefit from the modus operandi of using low-dimensional projections when assigning digital images to particular classes.

Counterfeiters are subject to the same limitations. They can never gain ultimate knowledge whether their image model is good enough so that no decision function can discriminate between two classes of authentic and counterfeit images. The theoretic reliability in Definition 11 cannot be calculated in the absence of knowledge of $\mathcal{P}_{\mathcal{C}_0}$ and $\mathcal{P}_{\mathcal{C}_1}$. A counter-forensic attack will only be successful as long as image forensics algorithms have not been refined accordingly. The interaction of forensics and counter-forensics can therefore be framed as competition for the best image model.

## *3.3 Blurred Notion of Authenticity*

Another obstacle is that authenticity is hard to evaluate in practice. Although Definition 6 is convenient for a formal approach, many shades of gray may exist in practical situations (and need to be reflected in appropriate definitions of the class space $\mathscr{C}$).

### 3.3.1 Device-Internal Processing

The definition of authenticity is deeply entangled with the question of what constitutes an acquisition device and thus the separation between functions $\mathsf{acquire}$ and $\mathsf{process}$. Common sense suggests to equate the function $\mathsf{acquire}$ with imaging devices and then strictly apply Definitions 3 and 6: all original images captured with these devices are authentic. However, considering the sophistication of modern imaging devices, the situation is not as easy. Increasingly, post-processing and image enhancement have become parts integral to the *internal* imaging pipeline. A forensic investigator has to accept that such processing inevitably raises the uncertainty of forensic decisions because device-internal processing and post-processing can generate very similar results. For example, many forensic techniques are sensitive to heavy device-internal quantization. While quantization in most cases preserves the semantic meaning of an image, it causes information loss by definition. This can make it harder to distinguish authentic from counterfeit images. The situation is even more complex when consumers are in the position to actively modify and extend the firmware of their devices.[1]

### 3.3.2 Analog Hole

The convention in Definition 6 to assume authenticity for every image generated with function $\mathsf{process}$ equal to the identity function does not reflect the possibility of image manipulations in or via the analog domain. While one can argue that detecting posed scenes in the real world is beyond the scope of digital image forensics, this limitation also excludes attacks which involve the reproduction and reacquisition of digitally processed images. In these attacks, and all iterations thereof, elements of $\mathscr{A}$ become part of $\mathscr{P}$. This blurs the distinction between the empirical functions $\mathsf{acquire} \in \mathscr{A}$ and deterministic functions $\mathsf{process} \in \mathscr{P}$ further. To be very precise, one would have to extend our theory and define a new set of functions for digital-to-analog transformations. Then, pairs of this functions and $\mathsf{acquire}$ would have to be included into the transitive hull of $\mathscr{P}$. For the sake of brevity and clarity, we refrain from introducing new terminology and leave it by drawing the reader's attention on this blind spot of our theory.

### 3.3.3 Legitimate Post-Processing

While it is tempting to deny authenticity to every processed image *per se*, this simplification is too narrow for many realistic applications. Instead, there may be a subset $\mathscr{P}_{\text{legitimate}} \subset \mathscr{P}$ of legitimate processing operations which do not impair the authenticity of an image (cf. Fig. 2).

This subset certainly depends on the context. For instance, it is common practice to downscale and compress digital images with JPEG prior to publication on the Internet. Glossy magazines, by contrast, may not permit any quality reduction when acquiring images from photographers. Instead they may allow some basic color enhancement operations. There exist special codes of conduct, which specify what is considered legitimate post-processing for scientific journals [36, 11].

So a realistic model of authenticity will contain at least three categories, namely

1. *original images*, where $\mathsf{process}$ can be nothing but the identity function $\perp_{\mathsf{process}}$,
2. *plausible images*, which have been subject to legitimate post-processing $\mathsf{process} \in \mathscr{P}_{\text{legitimate}}$, and
3. *manipulated images*, for processing with all other elements of $\mathscr{P}$.

The context in a practical situation defines whether the first two categories or just the first category shall be considered as authentic.

[1] The Canon Hack Development Kit, for instance, allows to run virtually arbitrary image processing routines inside most modern Canon digital cameras, http://chdk.wikia.com/wiki/CHDK.

*Example 7* Imagine a case where a judge who has to rule on a traffic accident may consider JPEG-compressed images as authentic if they have been mailed to the insurance company via e-mail. Since the authenticity (and in particular the semantic integrity) of JPEG-compressed images is more difficult to prove than of never-compressed images, a party in doubt may present (or demand) the original raw files. The claim "these are the original raw files" alters the notion of authenticity. JPEG artifacts in the presumably never-compressed files would be an indication of inauthenticity and raise suspicion that the images are counterfeits.

Remark that technically, this claim imposes an exogenous condition on the likelihood function $\mathcal{P}_{\mathcal{C}}(\mathbf{I}\,|\,\text{claim})$. In this way, contextual knowledge can be incorporated in the formal framework and sharpen the notion of plausibility with probability distributions. Again, for brevity we refrain from extending our terminology in this chapter.

# 4 Classification of Counter-Forensic Techniques

Counter-forensic techniques against passive–blind image forensics can be classified along three dimensions. First, counterfeiters can generally exploit robustness or security weaknesses to mislead forensic analyses. Second, integrated and post-processing attacks vary in their position in the image generation process. Third, targeted and universal attacks differ in the (range of) attacked forensic algorithms. Figure 3 illustrates our classification and names as representative examples for each relevant category. The following sections discuss each of the dimensions in more detail.

## *4.1 Robustness Versus Security*

The distinction between legitimate and illegitimate post-processing becomes relevant for the distinction of robustness and security properties of forensic algorithms (see Sect. 3.3.3).

**Definition 12** The *robustness* of a digital image forensics algorithms is defined by its reliability under legitimate post-processing.

In terms of counter-forensics, the lack of a clear separation between original and manipulated images increases the strategic options of a counterfeiter. If quality reduction, such as lossy compression or downscaling, is considered plausible and thus inconspicuous, a counterfeiter can eliminate subtle traces of illegitimate processing by subsequent quality reduction. Many known forensic algorithms are sensitive to strong quantization and fail to identify subtle traces in low-quality images. Yet some exceptions exist. For example, scans of printed and dithered images in newspapers are coarse digital representations of the real world, but traces of inconsistent lighting may still be detectable [21].

As a counter-forensic technique, legitimate post-processing does not require much knowledge of the image generation process. Its sole objective is to generate plausible counterfeits. It is sufficient if the counterfeit is moved *somewhere* outside the decision region $\mathscr{R}_{C_1}$ entailing all manipulated images (subject to the constraints in Def. 10).

The experimental literature is mainly concerned about the robustness of novel forensic algorithms. Most authors measure and report the performance loss as a function of JPEG compression quality. While this is a good indicator of the average reliability, it does not permit conclusions on the overall reliability. To complete the picture, also worst-case scenarios with sophisticated and intentional counterfeiters have to be considered. Resistance against *attacks* is directly associated with the *security* of forensic algorithms.

**Definition 13** The *security* of a digital image forensics algorithm is defined by its reliability to detect intentionally concealed illegitimate post-processing.

In other words, security is the ability to withstand counter-forensics.

Counterfeiters attacking security properties exploit specific knowledge about and shortcomings of the image model used by forensic investigators. The counterfeits are therefore purposely moved *in a particular direction* toward the decision boundary of the primary class (and just beyond). Such attacks are more powerful because their success does not depend on adjustable definitions of plausibility, but rather on weaknesses of forensic algorithms.

As has been pointed out in the context of digital watermarking, robustness is necessary but not sufficient for security. If counter-forensic attacks against a forensic algorithm with $\mathsf{attack} \in \mathscr{P}_{\text{legitimate}}$ exist, this algorithm cannot be considered secure. However, truly secure algorithms need to be reliable under all possible countermeasures $\mathsf{attack} \in \mathscr{P}$.

## *4.2 Integrated and Post-Processing Attacks*

*Post-processing attacks* modify an image $\mathbf{I}_{(k)}$ such that the resulting counterfeit $\mathbf{I}_{(\hat{l})}$ does not reveal traces of the original class $C_k$ anymore (cf. Example 5 in Sect. 2.3). Figure 4 illustrates that such attacks can be thought of as an additional processing step $\mathsf{attack} \in \mathscr{P}$ that supplements the original generation process $(\mathsf{acquire}, \mathsf{process}) \in (\mathscr{A} \times \mathscr{P})_{(C_k)}$. Particular examples include lossy compression to take advantage of robustness issues, but also inverse flatfielding as a means to exploit specific weaknesses of digital camera identification based on sensor noise (cf. Sect. 5.2.2).

*Integrated attacks*, on the other hand, interact with or replace parts of the image generation process such that, instead of $\mathbf{I}_{(k)}$, the counterfeit is generated directly by a tuple $(\mathsf{acquire}', \mathsf{process}') \in (\mathscr{A} \times \mathscr{P})$. The modified functions $\mathsf{acquire}'$

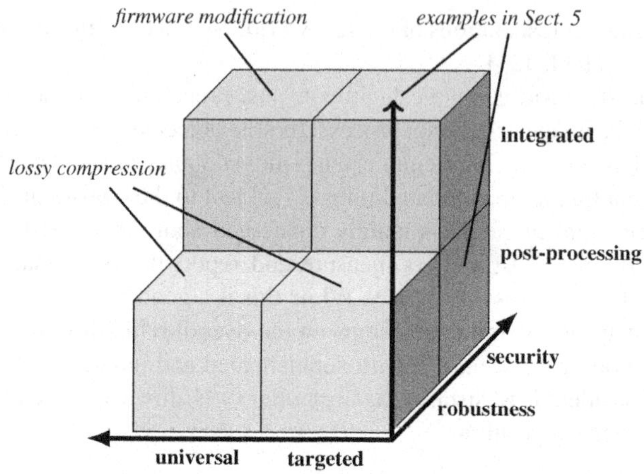


**Fig. 3** Design space for counter-forensic techniques

and/or $\mathsf{process}'$ are specifically designed to avoid the formation of identifying traces or to mimic characteristics of the target class (see also Fig. 4). In the aforementioned Example 5, an integrated attack would directly transform the original authentic image $\mathbf{I}_{(0)}$ into a semantically different counterfeit $\mathbf{I}'_{(\hat{0})}$ without ever releasing the detectable manipulation $\mathbf{I}'_{(1)}$. Note that this procedure is also covered by our formal description of counter-forensic attacks in Definition 10. It is always possible to express the (imaginary) map $\mathbf{I}'_{(1)} \mapsto \mathbf{I}'_{(\hat{0})}$ in terms of a post-processing function $\mathsf{attack}$. Because integrated methods obviously require deep knowledge of the image generation process, they do not address robustness issues of forensic algorithms by definition. This is indicated in Fig. 3, where the corresponding regions are left blank.

Integrated methods are relevant for manipulation detectors, where counterfeiters are hardly restricted in the choice of image processing primitives (or variants thereof). We will encounter several examples in the Sect. 5. Integrated attacks to impede source identification are less obvious (apart from the pathological case of capturing a scene with a completely different device). Nevertheless, it is conceivable to modify software for raw image processing for counter-forensic purposes. With freely available open-source firmware modifications, device-internal counter-forensics may become very powerful because device-specific traces need not leave the device at all.

## *4.3 Targeted and Universal Attacks*

A further classification is borrowed from the context of steganalysis [14] and digital watermarking [10]. We call an attack *targeted*, if it exploits particulars and weaknesses of one specific forensic algorithm $\mathsf{decide}$, which the counterfeiter usually

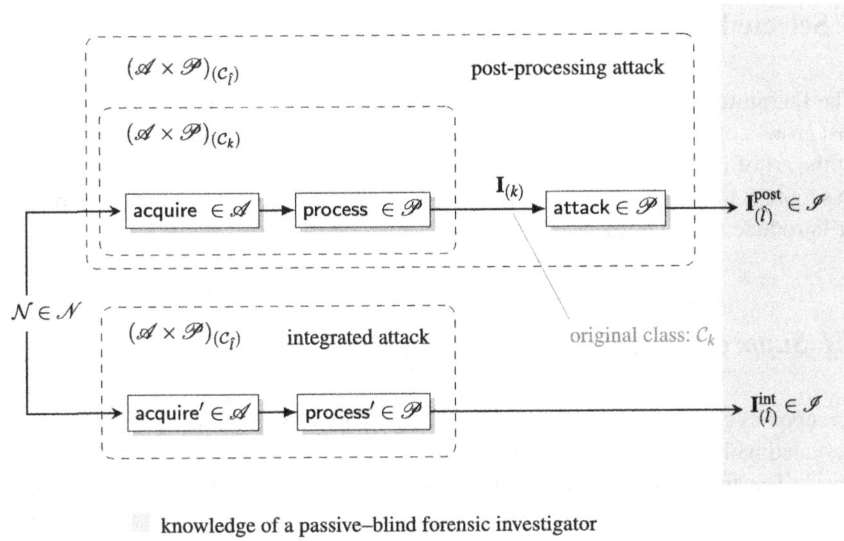


**Fig. 4** Post-processing and integrated counter-forensic attacks. Post-processing attacks suppress and/or synthesize identifying traces subsequent to the original image generation process. Integrated attacks directly replace the original process with a counter-forensic variant

knows. Such vulnerabilities directly relate to the image model implemented in the forensic algorithm, which leads to the notion of *ε-reliability with respect to a specific model*—yet another term that has is roots in steganography [14]. Clearly, it is possible (and likely) that other forensic algorithms using alternative or improved image models can detect such counterfeits.

Conversely, *universal attacks* try to maintain or correct as many statistical properties of the image in order to conceal manipulations even when presented to unknown forensic tools. This is the more difficult task, and it is an open research question whether image models can be found good enough to sustain analysis with combinations of forensic algorithms. In general, a mere combination of all known targeted counter-forensic techniques does not yield a universal attack; at least when typical low-dimensional image models are employed, which ignore the interference and interdependence of different attacks.

In the meantime, counterfeiters can exploit weak robustness and use lossy but legitimate processing whenever plausible. Recall that this variant of universal attacks is always a tradeoff between originality and plausibility of the counterfeit. Even if strong quantization removes identifying traces of the true class, it very likely precludes claims about the originality of the corresponding image (cf. Sect. 3.3).

## 5 Selected Targeted Attacks

The literature on counter-forensic techniques is still very limited compared to the fast growth of publications on forensic techniques. In this section we survey the state of the art of image counter-forensics. The presentation is structured into techniques to suppress traces of possibly malicious image processing and techniques to restore or introduce artificial traces of seemingly authentic images.

### *5.1 Suppressing Traces of Image Processing*

To suppress characteristic traces, variants of typical image processing operators were designed which destroy detectable structure by modulating the process with random noise. The methods differ in where in the process the randomness is introduced and how it is controlled to find a good tradeoff between visual imperceptibility and undetectability by known forensic detectors.

#### 5.1.1 Resampling

Resampling with interpolation is a very common image processing primitive which takes place during scaling, rotating, and shearing of images. For such affine coordinate transformations, each pixel of the resampled image $\mathbf{I}'_{(1)}$ is a weighted sum of one or more pixels of the source image $\mathbf{I}_{(0)}$,

$$\mathbf{I}'_{(1)}(x, y) = \text{round}\left(\sum_i \sum_j \phi\left(\Delta\left(\mathbf{A}^{-1}\begin{pmatrix}x\\y\end{pmatrix}, \begin{pmatrix}i\\j\end{pmatrix}\right)\right)\mathbf{I}_{(0)}(i, j)\right), \tag{10}$$

where $\mathbf{A}$ is a $2 \times 2$ transformation matrix, $\Delta : \mathbb{R}^2 \times \mathbb{Z}^2 \rightarrow \mathbb{R}^+$ is a distance function that returns the distance between two coordinate vectors, and $\phi : \mathbb{R}^+ \rightarrow \mathbb{R}$ is the interpolation kernel.

The interpolation step ensures visually appealing outcomes when pixel intensities are mapped from discrete positions of a source grid to a discrete destination grid without a one-to-one correspondence between source and destination positions. Depending on the relative position of source and destination grid, systematic dependencies between pixels and their local neighborhood are introduced. In many cases, the strength of this dependence alternates periodically in space, which leads to identifiable resampling artifacts. This is so because function $\Delta$ takes only a few different values for many combinations of input coordinates. See Chap. 9 in this volume for a review of methods to measure and interpret these traces.

Most automatic resampling detectors exploit the periodicity of dependencies between pixels, largely because those traces can be added up over multiple

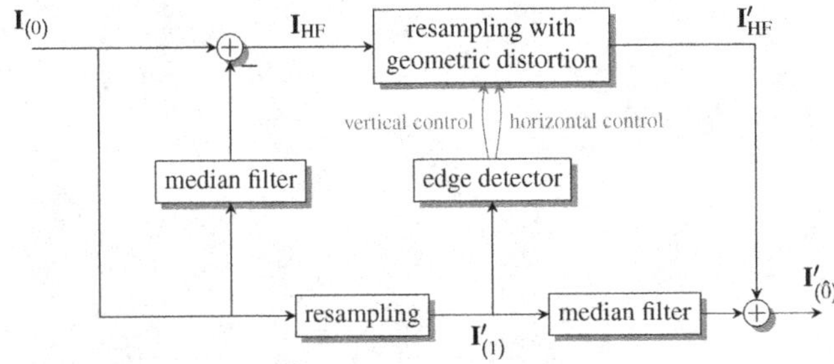


**Fig. 5** Block diagram of undetectable resampling [25]

periods, thereby reducing the influence of noise and other interfering factors like the image content. Therefore, to effectively suppress traces of resampling, any periodic structure must be avoided. Basic signal processing theory suggests interpolation with a sinc-kernel, which is theoretically optimal for downscaling [47]. However, this kernel requires infinite support. Apart from computational demands, it is also impractical because of boundary effects in finite images.

Another approach is to perturb the interpolation process with nonlinear filters and noise [24, 25]. Both ideas complement each other in the so-called *dual-path attack*. This integrated attack decomposes the image into a low-frequency component and a high-frequency component as depicted in Fig. 5. The high-frequency component $\mathbf{I}_{\mathrm{HF}}$ of a source image $\mathbf{I}_{(0)}$ is obtained by subtracting the output of a median filter, a nonlinear low-pass filter with windows size $2s + 1$, from the source image:

$$\mathbf{I}_{\mathrm{HF}}(x, y) = \mathbf{I}_{(0)}(x, y) - \mathrm{median}\left\{\mathbf{I}_{(0)}(i, j) \mid \sup\left(|x - i|, |y - j|\right) \le s\right\}. \quad (11)$$

This component is resampled by a modified function which adds noise to the real-valued destination positions before interpolation,

$$\mathbf{I}'_{\mathrm{HF}}(x, y) = \sum_i \sum_j \phi\left(\Delta\left(\mathbf{A}^{-1}\begin{pmatrix} x \\ y \end{pmatrix} + \begin{pmatrix} n_{\mathrm{hor},x,y} \\ n_{\mathrm{ver},x,y} \end{pmatrix}, \begin{pmatrix} i \\ j \end{pmatrix}\right)\right) \mathbf{I}_{\mathrm{HF}}(i, j). \quad (12)$$

This procedure is called *geometric distortion*. Scalars $n_{\mathrm{hor}}$ and $n_{\mathrm{ver}}$ are real-valued displacements for horizontal, respectively vertical geometric distortion drawn independently for each pair $(x, y)$ from a zero-mean Gaussian random source $N(0, \sigma)$. Moderate geometric distortion effectively reduces detectable periodic artifacts, but it suffers from the side-effect that visible jitter appears at sharp edges in the image. To prevent this, the degree of the geometric distortion, i. e., the variance of the random source, is attenuated by the output of an edge detector. For better results, this control can be implemented for horizontal and vertical edges independently. This ensures that a horizontal edge is not affected by visible vertical geometric distortion while at the same time avoiding that measurable periodicities appear along the horizontal direction, and vice versa.

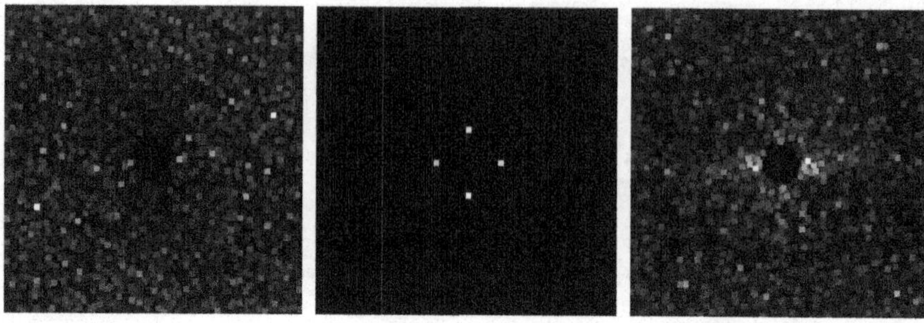

**(a)** original **(b)** scaled to 105% **(c)** scaled to 105% with dual-path attack

**Fig. 6** Resampling peaks in the FFT-transformed linear predictor residue of a grayscale image. The spectral images were enhanced with a maximum filter and scaled to the maximum contrast range for better printing quality (*source* [25]). **a** Original, **b** scaled to 105%, **c** scaled to 105% with dual-path attack

The low-frequency component is first resampled with conventional methods. This intermediate result serves as input to the edge detector. In a second step, a median filter is applied. This nonlinear filter effectively removes detectable periodicities from resampled images. However, by its very nature, it removes high frequencies with a nonlinear and non-trivial cutoff function. To minimize the visual impact, both frequency components are added together after independent resampling,

$$\mathbf{I}'_{(\hat{0})}(x, y) = \mathrm{round}\left(\mathbf{I}'_{\mathrm{HF}}(x, y) + \mathrm{median}\left\{\mathbf{I}'_{(1)}(i, j) \mid \sup\left(|x - i|, |y - j|\right) \leq s\right\}\right). \tag{13}$$

Figure 6 demonstrates the effectivity of this method for a single example. All three images are Fourier transforms of linear predictor residue. Figure 6 (a) is the reference for an untampered image. Figure 6 (b) shows the result for the same image after scaling to 105 % of the original size using conventional bilinear interpolation. The characteristic peaks are clearly visible. Finally, Fig. 6 (c) displays the same analysis results after resampling with the described dual-path attack. The characteristic peaks have disappeared and the spectrum is dominated by scattered local maxima, similar (but not identical) to the spectrum of the original image.

Larger experiments on the detectability of this method are documented in [25]. The quantitative results from up to 200 images are summarized in Figs. 7 and 8 for scaling and rotation, respectively. While rotation, upscaling and moderate downscaling is very well detectable when done in the conventional way, the dual-path attack can lower the detection rates substantially. Rotations of more than 45° are equally (un)detectable by symmetry. Nevertheless, there remains a residual risk for the counterfeiter since in 10–20% of the cases, the processing is still detectable. We also suspect that the dual-path attack, while suppressing known traces of resampling, may leave new kinds of traces. Besides some specific works on the detectability of

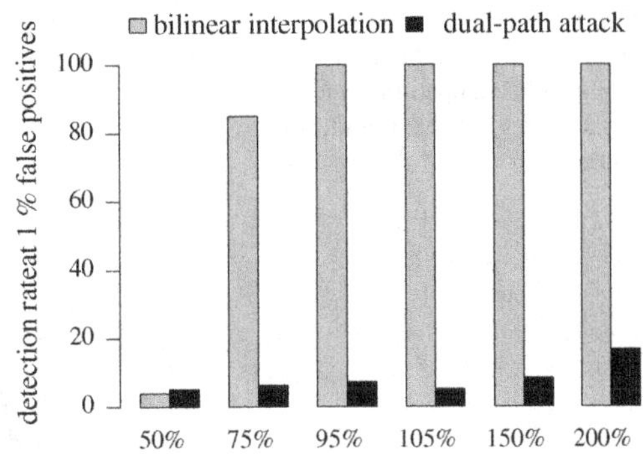


**Fig. 7** Quantitative results for undetectable scaling: the dual-path attack reduces detectable traces of upscaling and moderate downscaling (experimental results of [25])

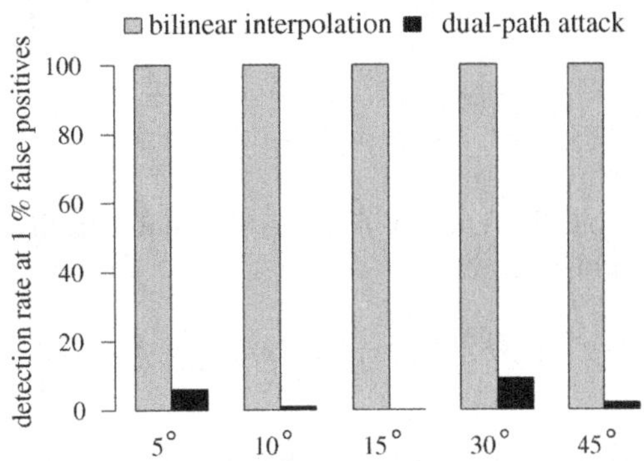


**Fig. 8** Quantitative results for undetectable rotation: the dual-path attack reduces detectable traces of resampling after rotation (experimental results of [25])

sole median filtering [27, 7, 51], we are not aware on any research on how detectable these traces are.

The authors of [25] also evaluated the quality loss of the dual-path attack and report peak signal-to-noise ratios (PSNR) between 30 and 40 dB for scaling (more distortion for downscaling) and about 40 dB for rotation independent of the angle. The distortion is measured between the attacked resampled image $\mathbf{I}'_{(\hat{0})}$ and the conventionally resampled version of the image $\mathbf{I}'_{(1)}$ for each image and then aggregated by taking the average distortion over 100 images.

### 5.1.2 Contrast Enhancement

Contrast enhancement is prone to leave characteristic traces in an image's histogram due to the nonlinear mapping $f : \mathbb{Z} \to \mathbb{R}$ of integer values and subsequent rounding to integers,

$$\mathbf{I}'_{(1)}(i, j) = \text{round}\big(f\left(\mathbf{I}_{(0)}(i, j)\right)\big). \tag{14}$$

**Fig. 9** Formation of gaps and peaks after rounding of nonlinearly mapped integers; dashed lines indicate rounding margins

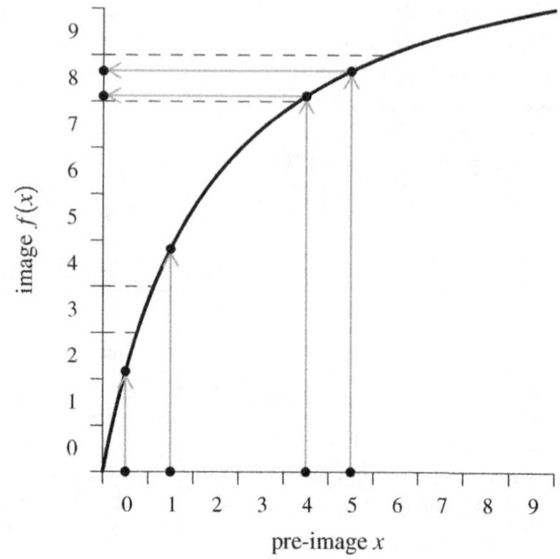


Typically function $f$ increases monotonically. For gamma correction, it takes the form

$$f(x) = (2^\ell - 1)\left(\frac{x}{2^\ell - 1}\right)^\gamma, \tag{15}$$

where $\ell$ is the bit depth in bits and $\gamma$ is the correction parameter. Values $\gamma < 1$ imply contrast reduction and values $\gamma > 1$ imply contrast enhancement.

Figure 9 visualizes the effect of Eq. (14) on the histogram of the modified image $\mathbf{I}'_{(1)}$. We distinguish two situations. First, not a single discrete value of the pre-image $x$ is mapped after rounding to the value $f(x) = 3$ in the image. The corresponding histogram bin in $\mathbf{I}'_{(1)}$ remains empty, thus leading to a *gap* in the histogram. Second, multiple discrete values of the pre-image $x$ are mapped after rounding to the value $f(x) = 8$ in the image. For typical pre-images with smooth histograms, this bin of the image's histogram receives a multiple of the hits of its adjacent bins, thus leading to a *peak* in the histogram. Forensic detectors can use the presence of gaps and peaks in a histogram as indications of prior processing with contrast enhancement [41].

An integrated attack against detectors which exploit these characteristics has been proposed in [6]. The idea is to identify histogram bins which are suspect of becoming a gap or peak. Then the result of the nonlinear mapping is perturbed by adding noise to all pixels in the affected bins and their directly adjacent bins,

$$f^*(x) = f(x) + n \quad \text{with} \quad n \sim N(0, \sigma_x). \tag{16}$$

In principle, $\sigma_x$ can be set depending on the proximity of a histogram bin to the next peak or gap. In practice, the authors of [6] report good results for a fixed dispersion parameter $\sigma = 1$ for all values of $x$. Figure 10 shows that this procedure effectively smoothes out the histogram of gamma-corrected versions of an example

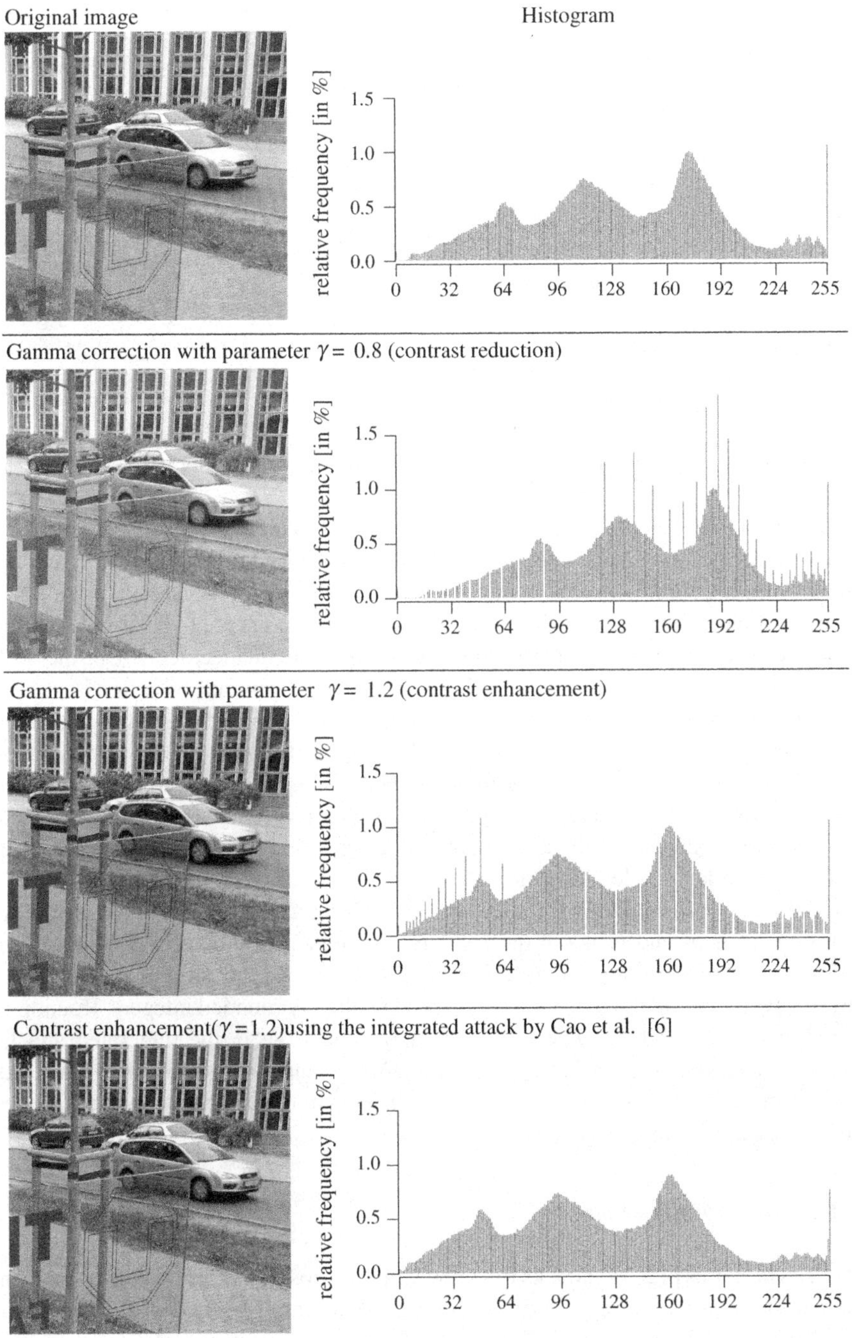


**Fig. 10** Suppressing traces of double-quantization after nonlinear mappings

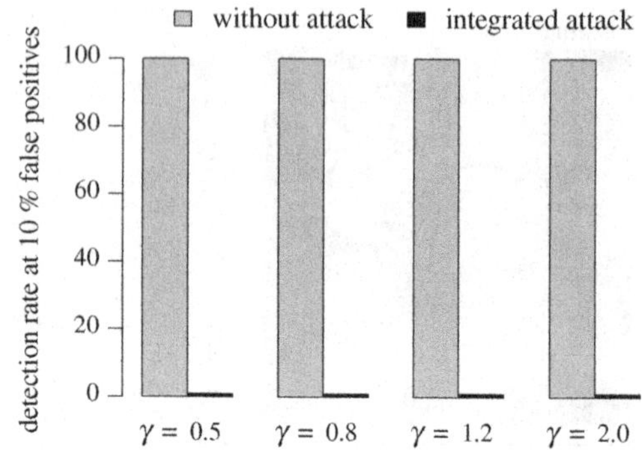


**Fig. 11** Quantitative results for undetectable contrast enhancement: the integrated attack suppresses detectable traces in the histogram (aggregation of experimental results of [6])

image (bottom and top rows), unlike conventional contrast enhancement (middle rows).

Figure 11 reports the results of a quantitative evaluation of this attack. The bars show detection rates of a detector which measures histogram peaks and gaps by identifying high-frequency components in the Fourier-transformed histogram [41]. The threshold has been set to allow 10% false positives in a set of 693 natural images. Observe that the detector correctly detects all contrast enhancements regardless of the parameter choice. This is effectively prevented if the gamma correction is carried out with the integrated attack. The detection rates of the attacked images drop further to empirically zero if the number of tolerable false positives is set to 5%. In practice, thresholds of less than 1% might be required. For this experiment, Cao et al. [6] report PSNRs between the gamma-corrected image with attack and the gamma-corrected image without attack of about 47.5 dB. Some images deviate toward even higher values (i.e., relatively less distortion).

The attack can be extended to a post-processing attack by first estimating the positions of gaps and peaks in the histogram's frequency domain and then adding noise. This variant produces slightly inferior PSNR because the unrounded real values are not available and the noise has to be added to already rounded integers. Rounding errors and noise add up, whereas they partly cancel out in the integrated attack. Another potential problem is incorrect estimation of gaps and peaks, which can lead to inappropriate values of $\sigma_x$ if $\sigma$ is not fixed. We are not aware of any research on the practical relevance of this issue.

### 5.1.3 Lossy Compression

A series of works [42–44] attacks to remove traces of lossy compression from a given image. This involves fixing two kinds of features which are exploited by current forensic detectors of the compression history: traces of dequantization and discontinuities at block boundaries. Specific counter-forensics have been proposed against detectors of each feature.

### Hiding Traces of Dequantization

JPEG compression involves cutting a source image into blocks of $8 \times 8$ pixels. Each block is transformed into the frequency domain with the discrete cosine transformation (DCT). The resulting 64 real-valued coefficients are quantized with frequency-depedent quantization factors $q \geq 1$ before they are rounded to integers and finally stored with lossless Huffman encoding. Let $y$ be the unquantized DCT coefficient, then the stored value $\tilde{y}$ is given by

$$\tilde{y} = \text{round}\left(\frac{y}{q}\right). \tag{17}$$

For a fixed frequency band, say DCT (2, 2) coefficients, the quantization factor $q$ depends on the desired compression ratio and thus the retained quality. Higher values of $q$ imply lower quality (and smaller file size) because the error between recovered coefficient value $\hat{y}$ and the original value $y$ on average increases with $q$,

$$\begin{aligned} \text{E}\left(\left|\hat{y} - y\right|\right) &= \text{E}\left(\left|q\tilde{y} - y\right|\right) &(18)\\ &= \text{E}\left(\left|q \cdot \text{round}\left(\frac{y}{q}\right) - y\right|\right) &(19)\\ &= \text{E}\left(\left|q \cdot e\right|\right) = q \cdot \text{E}\left(\left|e\right|\right) \quad \text{and} \quad \text{E}\left(\left|e\right|\right)\rangle 0. &(20) \end{aligned}$$

Now if a JPEG image is decompressed, the recovered coefficients $\hat{y}$ only take multiples of $q$. It is easy to detect a previous JPEG compression of any given image in spatial domain by applying block-wise DCT and checking the histograms of the DCT coefficients for gaps between multiples of a previous quantization factor. Figure 12 illustrates this and contrasts the smooth DCT histogram of a never-compressed image to the histogram of the same image after JPEG decompression with quality setting $Q = 60\%$. Observe the obvious peaks at multiples of the quantizationfactor factor $q = 10$. Note that unlike in the case of contrast enhancement, the gaps are not necessarily perfect in the sense that the actual frequency is zero. We rather observe "elephant feet" artifacts at the bottom of each histogram peak. This is so because during decompression, the real-valued intensity values from the inverse DCT are rounded to integers and truncated to the eligible value range. The rounding and truncation errors from all pixels in the block add up and may perturb the recalculated DCT coefficient beyond one rounding margin. Yet the errors are typically too small to smooth out the gaps entirely, even for $q = 2$.

A post-processing attack to smooth out and restore such histograms is described in [43]. It uses the fact that DCT coefficients of natural images can be modeled reasonably well with a Laplace distribution [29]—except frequency band (1, 1), which requires special treatment. The good fit can be confirmed in the left histogram of Fig. 12. The Laplace distribution is a symmetric one-parameter distribution with density function

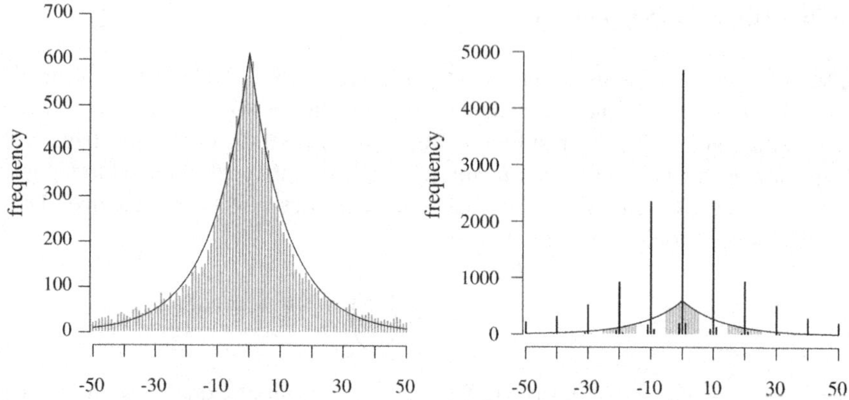


**Fig. 12** Histograms of DCT (2, 2) coefficients for never-compressed (*left*) and JPEG-decompressed (*right*) image with fitted Laplace distribution superimposed. The JPEG quality is $Q = 60\%$, corresponding to a quantization factor of $q = 10$. Small pertubations from the theoretical values in the *right* histogram are from rounding errors in the spatial domain and subsequent transformation back into the DCT domain. *Shaded* areas in the *right* histogram indicate the value range over which the *middle* peak is distributed according to the Laplace model. Note the different scales

**Fig. 13** Histogram of DCT (2, 2) coefficients after JPEG compression with quality $Q = 60\%$ and subsequent histogram restoration as proposed in [43]. The *solid line* is the fitted Laplace image model and the *shaded* bars indicate values which were expanded from the same quantized coefficient value. Compare to Fig. 12

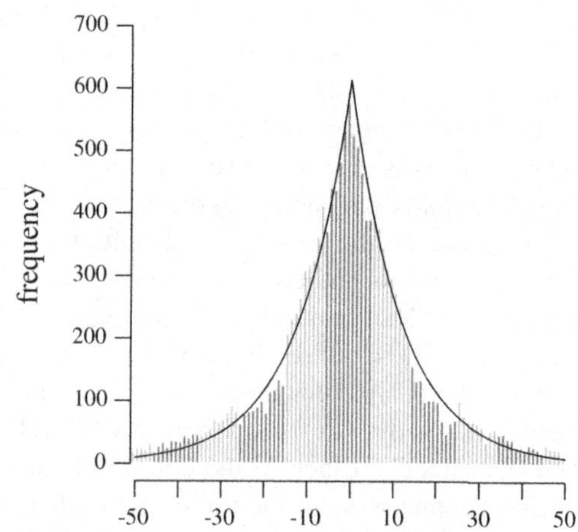


$$f(x) = \frac{\lambda}{2} \exp(-\lambda \cdot |x|). \tag{21}$$

The authors of [43] employ a maximum likelihood estimator of parameter $\lambda > 0$ for the discretized case of the Laplace distribution. This model can be estimated from the peaks of a JPEG-decompressed histogram. In a second step, all coefficients are redistributed by adding noise according to conditional probability functions for each peak and each DCT frequency band. In this way, the shape of a never-compressed DCT histogram is restored pretty well, as confirmed in the example of Fig. 13.

Since no parametric model is known for the DCT (1, 1) coefficients, noise following a uniform prior between two rounding margins is introduced. This is very similar to the post-processing variant of the attack against contrast enhancement detectors (see Sect. 5.1.2 above).

The effectiveness of this attack has been evaluated with quantitative experiments on 244 images, which were compressed using JPEG qualities $Q = 90$, 70, and 50%. The restored images were presented to a state-of-the-art JPEG quantization matrix estimator [12]. An image was classified as pre-compressed if one or more quantization factors were estimated greater than one, i.e., $\hat{q} > 1$. After applying the post-processing attack, the compression history could be detected for only 4% of the images pre-compressed with quality $Q = 90$, 7% of the images pre-compressed with quality $Q = 70$, and 18% of the images pre-compressed with quality $Q = 50\%$ [43]. The original publication does not report comparable detection rates for unattacked images, nor does it state the false positive rate. While the former are likely to reach almost 100% for the chosen quality settings, we lack a good prior for the false positive rate. With regard to the retained image quality, only a single PSNR value is given at 41.6 dB. This example image has been pre-compressed with JPEG quality $Q = 60\%$ before restoration. We are not aware of more generalizable quality measurements. Note that recent publications call into question the perceptual quality [49] and the claimed statistical undetectability [28, 50] of this attack.

This attack is adapted to wavelet-based image compression in [42].

### Suppression of Block Boundaries

Discontinuities at the boundaries of the $8 \times 8$ blocks are another indicative artifact of previous JPEG compression. Several researchers have investigated ways to remove blockiness artifacts from decompressed images, however, mainly with the aim to increase the perceptual quality [31, 52]. The requirements in counter-forensics differ in that statistical undetectability has higher priority than visual experience. Therefore a post-processing attack against the blockiness detector of [12] is proposed in [44]. The attack uses a combination of a median filter and additive Gaussian noise,

$$\mathbf{I}'_{(\hat{0})}(x, y) = \text{round}\left(\text{median}\left\{\mathbf{I}'_{(1)}(i, j) \mid \sup\left(|x - i|, |y - i|\right) \leq s\right\} + n_{x,y}\right), \tag{22}$$

where $n \sim N(0, \sigma)$ is drawn from a zero-mean Gaussian distribution independently for each pixel. The method is surprisingly effective, as can be seen in Fig. 14. With moderately invasive parameter settings, $s = 1$ and $\sigma^2 = 2$, the attack outperforms the best performing of the two visual blockiness reduction methods [31, 52] for JPEG qualities $Q = 70\%$ and above (hatched black bars in the figure). Stronger compression creates more pronounced block artifacts, which can be suppressed by adjusting the strength of the attack to $s = 2$ and $\sigma^2 = 3$. This way, even substantial and definitely noticeable JPEG compression can be effectively hidden from detectors

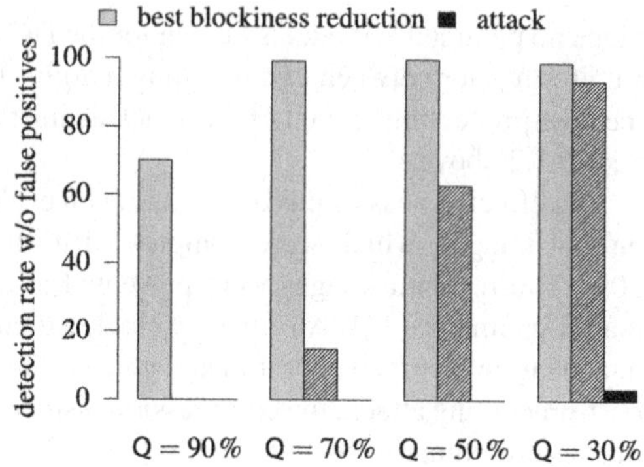


**Fig. 14** Quantitative results for the suppression of JPEG blockiness artifacts: comparison of moderate ($s = 1, \sigma^2 = 2$, hatched bars) and strong ($s = 2, \sigma^2 = 3$, solid) counter-forensics against the best performing visual blockiness reduction (results from [44])

which evaluate just a single blockiness criterion. For extreme settings like $Q = 10\%$, even the strong attack is not effective anymore. The original publication [44] does not provide any measures of retained image quality.

While the results against the specific detector in [12] are very clear, a note of caution is warranted. The attack has not been tested against a detector of median filtering [27]. While it is uncertain if this detector still works in the presence of additive noise and thus poses a serious threat, another concern arises on the choice of the blockiness detector. Prior efforts of suppressing even subtle JPEG blockiness after steganographic embedding in the DCT domain [40] has proven effective only against the very specific blockiness criterion used in the objective function of the blockiness reduction algorithm. However, it turned out to be easily detectable as soon as the statistical criterion is replaced by an alternative one [48]. More research is needed to understand the relaibility of this attack against detectors based on different or a combination of several blockiness indicators.

## 5.2 *Adding Traces of Authentic Images*

Certain forensic techniques test the authenticity of images by verifying the integrity of traces of the common image acquisition or processing chain. Tampering very likely destroys these traces. In this case, the task of counter-forensics is to restore or synthesize the traces of authentic images. All attacks presented in this section are post-processing attacks.

### 5.2.1 Synthesis of Color Filter Array Pattern

Local dependencies between pixel values in different color channels of color images have attracted the interest of forensic investigators pretty early on [39]. These dependencies emerge from the way color images are captured by different sensors using color filter arrays (CFA pattern, see also Chap. 1 in this volume). More precisely, typical sensors capture a $n \times m$ matrix $\mathbf{I}$ of intensity values. The light received at each cell (i.e., pixel) went through one of three physical color filters: red, green, and blue. To obtain a full color image of the dimension $n \times m$, the missing color information of the two remaining color channels is filled in by interpolation,

$$\mathbf{I}_k = \mathsf{interpolate}(\mathsf{select}(\mathbf{I}, k)) \quad \text{for} \quad k \in \{\text{red, green, blue}\}. \tag{23}$$

Function $\mathsf{select} : \mathbb{Z}^{n \times m} \times \{\text{red, green, blue}\} \rightarrow \mathbb{Z}^{n' \times m'}$ isolates all pixels captured with a filter of a specific color. Function $\mathsf{interpolate} : \mathbb{Z}^{n' \times m'} \rightarrow \mathbb{Z}^{n \times m}$ expands the color channel to the original size by filling the missing values with interpolated values from adjacent pixels. (Note that $m' < m$ and $n' < n$.)

A simple method to restore the CFA pattern from a tampered color image $\mathbf{I}'_{(1)} = \left(\mathbf{I}'_{(1),\text{red}}, \mathbf{I}'_{(1),\text{green}}, \mathbf{I}'_{(1),\text{blue}}\right)$ is to straightly reapply the color filter interpolation [20],

$$\mathbf{I}'_{(\hat{0}),k} = \mathsf{interpolate}(\mathsf{select}(\mathbf{I}'_{(1),k}, k)) \quad \text{for} \quad k \in \{\text{red, green, blue}\}. \tag{24}$$

However, this procedure is far from optimal because it discards information in the unselected parts of each color channel, i.e., $\mathsf{select}(\mathbf{I}'_{(1),l}, k)$ with $k, l \in \{\text{red, green, blue}\}$ and $l \neq k$. To avoid this, the interpolation of each color channel can be formulated as a matrix product,

$$\vec{\mathbf{I}_k} = \mathbf{H}_k \vec{\mathbf{I}}. \tag{25}$$

Matrix $\mathbf{H}_k$ of dimension $nm \times nm$ holds the interpolation weights for color channel $k$. The notation $\vec{\mathbf{I}}$ denotes that matrices are vectorized to column vectors of length $nm$. Ignoring rounding errors for a moment, Eq. (25) should hold for authentic images whereas it is most likely violated after tampering. This can be expressed as a nonzero residual $\boldsymbol{\varepsilon}$,

$$\vec{\mathbf{I}'}_{(1),k} = \mathbf{H}_k \vec{\mathbf{I}'} + \boldsymbol{\varepsilon}. \tag{26}$$

The interpretation of this equation as a least-squares problem gives a method to synthesize CFA pattern with minimal distortion. An image $\mathbf{I}'$ that is compatible with $\mathbf{I}'_{(1)}$ and minimizes the $L_2$-norm $\|\boldsymbol{\varepsilon}\|$ can be found by solving

$$\vec{\mathbf{I}'_k} = \left(\mathbf{H}_k^{\mathsf{T}} \mathbf{H}_k\right)^{-1} \mathbf{H}_k^{\mathsf{T}} \vec{\mathbf{I}'}_{(1),k}. \tag{27}$$

The general solution to Eq. (27) is computationally impractical due to the inversion of an $nm \times nm$ matrix. The solution in [26] exploits structure and sparsity of $\mathbf{H}$, which allows to derive linear time algorithms and approximations. Once $\mathbf{I}'$ is obtained, it can be inserted in Eq. (23) to generate the color channels $\mathbf{I}'_{(\hat{0}),k}$ which

**(a)** original **(b)** 9 × 9 median filtered **(c)** CFA synthesis after median filtering

**Fig. 15** CFA peaks in the FFT-transformed linear predictor residue of a red color channel. The spectral images were enhanced with a maximum filter and scaled to the maximum contrast range for better printing quality (source: [26])

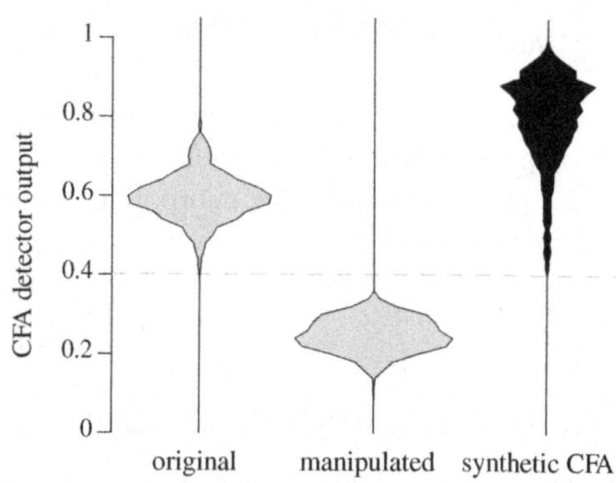


**Fig. 16** Violin plots for quantitative results of a CFA synthesis experiment. Distribution of CFA peak detector outputs from red channels of 1,000 images (data from [26])

contain a seemingly authentic CFA pattern and exhibit minimal distortion compared to $\mathbf{I}'_{(1),k}$.

Note that the minimum is found under the assumption of continuous signals. Discretization and rounding errors may lead to a slight divergence from the optimal solution of the discrete optimization problem. Finding algorithms to (approximately) solve this problem efficiently remains an open research question.

Figure 15 demonstrates this post-processing attack. The spectra are taken from linear predictor residue of the red color channel of one example image that has been taken as raw file from a digital camera. This is the method of choice for analyzing CFA pattern [39]. The four distinct peaks in Fig. 15 (a) appear exactly where expected for the red channel. Hence they indicate authenticity. After applying a median filter globally, the peaks disappear (Fig. 15 (b)). We have chosen the median filter as a simple global operation, but other manipulation steps result in a similar effect. Note that invalid CFA pattern can be identified even if the violations appear only locally. Figure 15 (c) finally shows the reappearing peaks after CFA synthesis. This spectrum is visually indistinguishable from the one belonging to the original image.

Quantitative results with a peak detector come to a slightly different conclusion. Figure 16 displays the detector response distribution for three sets of 1,000 images each. It is clearly visible that original and manipulated images can be separated perfectly with a decision threshold of $\tau = 0.4$. Moreover, the attack is successful in every single case. However, the shape of the distributions for original and synthesized CFA pattern differ substantially. The shift of probability mass toward higher values indicates that synthesized CFA pattern appear even more authentic than original CFA pattern. This over-fitting can lead to suspicion. It is most likely caused by the fact that CFA interpolation is not the last step in a chain of device-internal processing operations and the synthesis method is focused on restoring the traces of a single operation. We are not aware of research on remedies. A first step could be to blend parts of $\mathbf{I}'_{(\hat{0})}$ and $\mathbf{I}'_{(1)}$ to shape the detector response distribution.

The retained quality has been measured for the same set of images. The median gain in PSNR between the straight method of Eq. (24) and the distortion minimization method of [26] has been reported at 1.2 dB for the red channel and 0.9 dB for the green channel. The difference is because function select discards less information for the green than for the red and blue channels. Therefore, the quality gain for the blue channel is about as large as the gain for the red channel.

### 5.2.2 Substitution of Sensor Noise Pattern

A very powerful feature for forensic analyses are sensor defects in the image acquisition device (see Chap. 5 in this volume). Stable parts of the sensor noise, which emerges from variations in the hardware manufacturing process and sensor wear-out, leave traces that are unique for each individual sensor. This noise pattern is useful to link (parts of) images with acquisition devices, for examples by extracting a sensor fingerprint from the photo response non-uniformity (PRNU).

Here we describe an attack against a predecessor of the image source identification method presented in Chap. 5. Instead of the peak-to-correlation energy measure for multiplicative sensor fingerprints, a simpler correlation detector was proposed in the seminal publication [32],

$$C_* = \arg\max_{C_k \in \mathscr{C}} \operatorname{cor}(\mathbf{I}, \hat{\mathbf{K}}_{C_k}) \quad \text{with}$$
$$\operatorname{cor}(\mathbf{I}, \hat{\mathbf{K}}_{C_k}) = \sum_x \sum_y \mathsf{norm}\,(\mathbf{I} - F(\mathbf{I}))\,(x, y) \cdot \mathsf{norm}\left(\hat{\mathbf{K}}_{C_k}\right)(x, y). \tag{28}$$

Function $\mathsf{norm} : \mathbb{R}^{n \times m} \rightarrow \mathbb{R}^{n \times m}$ normalizes its argument by subtracting the mean and dividing by the variance. Function $F : \mathbb{R}^{n \times m} \rightarrow \mathbb{R}^{n \times m}$ is a low-pass filter to approximately separate image content from the noise residual. Matrix $\hat{\mathbf{K}}_{C_k}$ is an estimated sensor noise pattern obtained from simple averaging of noise residuals $\sum_i \left(\mathbf{I}_{(k),i} - F\left(\mathbf{I}_{(k),i}\right)\right)$ of a set of doubtlessly authentic reference images acquired with sensor $C_k$.

It is possible to mislead a detector based on Eq. (28) with the following procedure [17]. Suppose a given image $\mathbf{I}_{(k)}$ belonging to the class $\mathcal{C}_k$ shall be manipulated such that it is detected as belonging to the class $\mathcal{C}_l$, $l \neq k$, i.e., $\mathbf{I}_{(k)} \mapsto \mathbf{I}_{(\hat{l})}$. Under the assumption that the counterfeiter has access to the primary acquisition device, she can produce a sufficiently large number $L$ of

1. dark frames $\mathbf{I}^{\bullet}$ to estimate the dark current component of the sensor noise pattern $\hat{\mathbf{D}}_{(k)}$ by simple averaging, and
2. homogeneously illuminated frames $\mathbf{I}^{\circ}$ to estimate the flat-field frame $\hat{\mathbf{F}}_{(k)}$.

The flat-field frame $\mathbf{F}$ holds the PRNU component of the sensor pattern noise adjusted for the dark current component,

$$\hat{\mathbf{F}}_{(k)} = \frac{1}{L} \sum_{i=1}^{L} \left( \mathbf{I}^{\circ}_{(k),i} - \hat{\mathbf{D}}_{(k)} \right). \tag{29}$$

The tuple $\left(\hat{\mathbf{D}}, \hat{\mathbf{F}}\right)$ is a better representation of the sensor noise than the joint estimate $\hat{\mathbf{K}}$, which may contain parts of PRNU and dark current. Counterfeiting an images' origin now involves two steps, first the suppression of the original sensor noise pattern,

$$\mathbf{I}_{(\hat{\perp})} = \frac{\mathbf{I}_{(k)} - \hat{\mathbf{D}}_{(k)}}{\alpha \hat{\mathbf{F}}_{(k)}}, \tag{30}$$

where $\alpha$ is a scalar chosen to preserve the average luminance of the original image; second the insertion of a counterfeit sensor noise pattern,

$$\mathbf{I}_{(\hat{l})} = \alpha \mathbf{I}_{(\hat{\perp})} \hat{\mathbf{F}}_{(l)} + \hat{\mathbf{D}}_{(l)}. \tag{31}$$

If flat-field frame $\hat{\mathbf{F}}_{(l)}$ and dark frame $\hat{\mathbf{D}}_{(l)}$ are unavailable, for example because the counterfeiter has no access to the pretended acquisition device, an estimate $\hat{\mathbf{K}}_{(l)}$ from publicly available images can be used as substitute. However, this increases the risk of failure if the forensic investigator performs a triangle test with the same images (cf. [18] and Chap. 5 in this volume).

Figure 17 demonstrates how effectively the original noise pattern can be suppressed by the method of Eq. (30). The violin plot shows probability distributions of the correlation coefficient, Eq. (28), for 5×350 images acquired with four different devices. The reference pattern $\hat{\mathbf{K}}$ was estimated from images of the Canon S70 digital camera. As expected, only images taken with the Canon S70 produce significantly positive correlation coefficients. Observe that the probability mass of the black violin, representing Canon S70 images after suppression of the noise pattern, is almost as low as for the unrelated devices. In this example, the flat-field and dark frames were averaged from $L = 20$ images $\mathbf{I}^{\bullet}$ and $\mathbf{I}^{\circ}$.

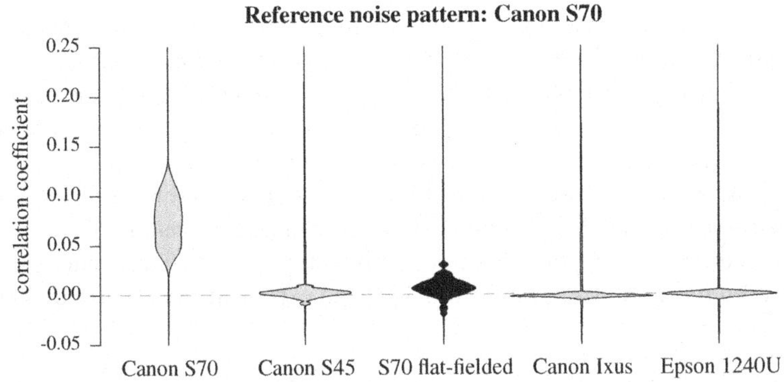


**Fig. 17** Violin plots showing quantitative results of effective suppression of sensor noise pattern by flat-fielding (data of 350 images from [17])

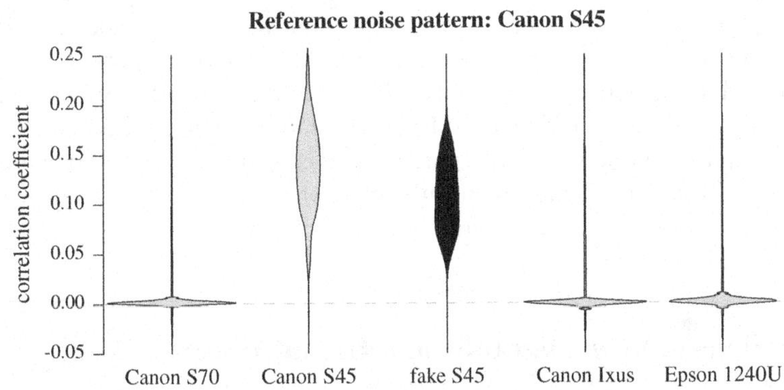


**Fig. 18** Violin plots showing quantitative results for false classifications after insertion of a counterfeit S45 sensor noise pattern into images acquired with the Canon S70 camera (data from [17])

Figure 18 continues the experiment and shows the results after a counterfeit noise pattern of the Canon S45 device has been inserted to the flat-fielded Canon S70 images using Eq. (31). Now both real and counterfeit Canon S45 images produce high correlation coefficients. Note that the shape of the distributions differ somewhat, though less than between different devices (compare Figs. 17 and 18). So in practice, a forensic investigator who has only a couple of images at her disposal, may have a hard time to extract useful information from the samples of different distributions.

The retained quality of this attack, if measured by PSNR between original $\mathbf{I}_{(k)}$ and counterfeited image $\mathbf{I}_{(\hat{l})}$, is below 30 dB on average. This seems notably worse than all other attacks surveyed in this chapter. A large part of this apparent degradation is due to the flat-fielding operation, which in fact is known as a technique to *improve* the visual quality of digital images. This improvement is later offset by the insertion of the counterfeit noise pattern, but both improvement and degradation are additive in the PSNR measure. So the overall quality loss is smaller than it appears.

This attack has been replicated several times, for example in [30], and in [45, 46] for cell-phone cameras. We are not aware of substantial refinements of this specific attack, for example adaptations to the peak-to-correlation energy detector.

The general potential weakness of PRNU-based image forensics to noise insertion attacks has already been mentioned in the seminal publication [32]. A defense strategy under the assumption that forensic investigator and counterfeiter share the images used for the attack is outlined in [33] and expanded and evaluated in [18].

### 5.2.3 Other Techniques

The above-described counter-forensic techniques to add traces of authentic images may need to be accompanied by additional adjustments. For example, the file structure and meta-data must be updated to be consistent with the pretended acquisition device. Due to the high number of mutually dependent specifications, this is not a trivial task. If the pretended acquisition device stores images as JPEG by default, then the JPEG quantization tables have to be adjusted to fit the manufacturer's internal table generation algorithm. A procedure to do this is to first suppress quantization artifacts (cf. Sect. 5.1.3) and subsequently recompress the images with quantization tables of a counterfeited acquisition device [44].

## 6 Relations to Steganography and Digital Watermarking

The previous sections introduced and discussed a formal framework for the description of counter-forensic attacks and surveyed the literature for concrete implementations. In this penultimate section, we broaden the perspective a bit further. Counter-forensics and its competition with digital image forensics should not be considered an isolated research field. It should rather be seen in close relation to established disciplines in the broader field of information hiding [25]. In this way, obvious parallels can help to foster and deepen the understanding of research questions in digital image forensics and counter-forensics that may have already been identified and discussed elsewhere.

Figure 19 illustrates how these disciplines relate to digital image forensics and counter-forensics. The figure serves as a blueprint for the following discussion. The process chain from image generation, via class change, to decision is depicted in columns from left to right. Important distinctions, such as between empirical acquisition and deterministic processing or the nature of constrains, are also reflected. Focal areas of the specific subfields of information hiding are arranged in rows so that corresponding hiding, detection, and deception techniques can be associated horizontally. Similarities of digital image forensics and counter-forensics with related

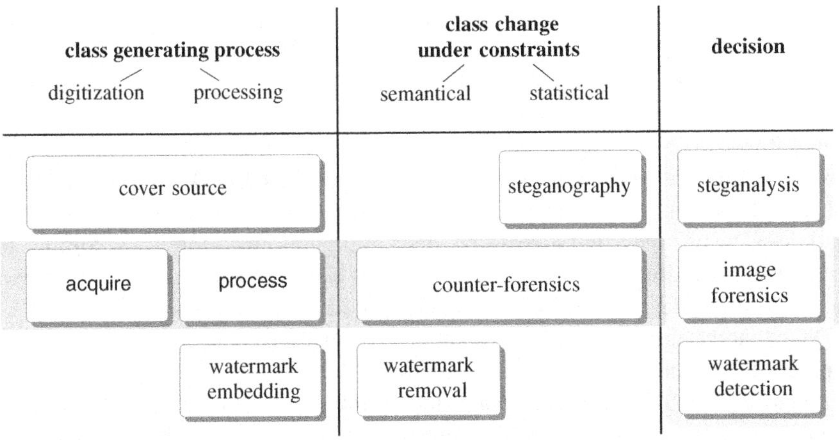


**Fig. 19** Relation of image forensics and counter-forensics to other fields in information hiding

tasks in steganographic communication and digital watermarking appear as vertical relations.

Counter-forensics shares common goals with *steganography*. By embedding a secret message, a cover image's class is changed from $\mathcal{C}_0$ to the class of stego images $\mathcal{C}_1$. Both steganography and counter-forensics try to hide the very fact of a class change, and their success can be measured by the Kullback–Leibler divergence between the two conditional probability distributions $\mathcal{P}_{\mathcal{C}_0}$ and $\mathcal{P}_{\mathcal{C}_1}$ (cf. Eq. (9)). This imposes statistical constraints on both. Steganography differs from counter-forensics in the amount and source of information to hide. Most steganographic algorithms are designed to embed a message by minimizing distortion, thereby preserving the cover's semantic meaning. Counter-forensics, by contrast, conceals the mere information that larger parts of the original image $\mathbf{I}_{(0)}$ have been modified, often with the aim to change its semantic meaning. The new semantic meaning of the counterfeit $\mathbf{I}'_{(1)}$ can be regarded as the 'message' to be transmitted. Unlike counter-forensics, steganography is not explicitly limited by perceptual constraints. Cover images are solely a means to communicate hidden messages. This leaves the steganographer with more degree of freedom to choose the cover, and thus stego image.

*Steganalysis*, as a counterpart to steganography, aims at unveiling the presence of a hidden message in a specific image without having access to the original cover. A general analogy between steganalysis and image forensics becomes evident if we consider the act of counterfeiting as information which is hidden inconspicuously in an image. This suggests that counter-forensics—similar to steganography, where capacity and security are considered as competing design goals—needs to tradeoff the amount of information to hide with achievable detectability. The stronger a manipulating operation interferes with the inherent image structure, the harder it is to feign an authentic image.

Another analogy exists between counter-forensics and *attacks against digital watermarking* schemes, where images of class $\mathcal{C}_1$ are generated by a *watermark embedding* process. In contrast to steganalysis, attacks against (robust) digital watermarks are designed to merely remove the embedded information, i.e., changing a watermarked image's class to $C_0$, while retaining the semantic meaning and to some degree the perceptual quality of the cover. In this sense, identifying traces in digital image forensics can be understood as inherent watermarks, which counter-forensics aim to suppress or change. Since attacks against digital watermarks do not specifically address statistical undetectability, the robustness issues of digital watermarking schemes may find a correspondent in digital image forensics. Similar to digital watermarking [22], forensic investigators also wish that the performance of their algorithms degrades gently as a function of image quality loss.

Note that the above parallels, on a technical level, correspond to the two different approaches to counter-forensics, as discussed in the Sect. 4.2. Integrated attacks are more closely related to steganography (hiding traces of a class change by design), whereas post-processing attacks have stronger similarities to attacks against digital watermarking (remove identifying traces).

Our brief excursion to related fields in information hiding highlights the strong overlap between counter-forensics, steganography and attacks against digital watermarking. While particularities of each field should not be disregarded, we believe that a lively and mutual interaction will prove beneficial to the whole field of information hiding. Not only can the rather young field of digital image forensics and counter-forensics learn from the more established branches, where research on adversaries and attacks is a common practice and widely accepted. Also steganography and steganalysis, which both have to cope with heterogenous image sources, can gain from findings in digital image forensics to conceive better, or at least adaptive, image models [3, 1]. Moreover, the literature now reports digital watermarking schemes that directly interact with the image generation process [34], which suggests to employ (counter-)forensic techniques as building blocks for the design of attacks against digital watermarks and the detection thereof. In summary, we see the relations in Fig. 19 as an outline for a comprehensive theory of information hiding. Formalizing unified theoretical foundations seems a promising goal to advance the field.

## 7 Countering Counter-Forensics: An Open Challenge

As image forensics matures and finds applications in practice, its results may be used to rule on momentous, and sometimes controversial, decisions in the real world. It is unavoidable that actors who might be impaired by certain outcomes will try to influence the decisions toward a more favorable outcome for themselves. Consequently, forensic methods will be under attack. To avoid that the field as a whole loses credibility, image forensics has to anticipate the existence of counter-forensics and come up with new countermeasures [18].

This chapter has shown that current forensic techniques are vulnerable to relatively simple targeted attacks. Forensics investigators have several options to react to counter-forensics. Attacks against a lack of robustness can be warded off by demanding higher image quality standards (recall the courtroom example of Sect. 3.3.3). Fixing the security weaknesses fundamentally requires better image models. While it is generally hard to find good image models, finding models that are 'good enough' to withstand simple counter-forensics seems much more feasible.

As a starting point, single-criterion detectors can be replaced with detectors that use image models with multiple (but not many) dimensions. This makes it considerably harder for the counterfeiter to move an image into the right detection region *in each dimension at the same time*. If the number of dimensions grows unmanageable, machine-learning techniques may be used as discrimination functions. It is remarkable—and somewhat reassuring from the forensic investigator's point of view—that so far no counter-forensic techniques are published against image forensics based on machine learning, such as camera model identification [16].

On a more general level, also the combination of indicators from several forensic techniques is a kind of dimensionality expansion. Already in their seminal paper, Popescu and Farid [38, p. 146] conjectured:

> *[...] as the suite of detection tools expands we believe that it will become increasingly harder to simultaneously foil each of the detection schemes.*

Nevertheless, it remains the responsibility of researchers in the field to substantiate this conjecture with facts, and measure how hard the counterfeiters' task will become. For this, image forensics has to be understood as a security technique, which is measured by its resistance to attacks.

Future proposals of new forensic techniques should consequently come with a security evaluation. Researchers should try to attack their schemes and argue or demonstrate how effective these attacks can be. In this way, the field will develop similar to the field of information hiding, where authors in the 1990s mainly focused on message integrity and imperceptibility, and rigorous security evaluations became the standard in the 2000s. New results on counter-forensics should in no way be understood as "helping the perpetrators", but they will become relevant contributions on their own, just like innovations in cryptanalysis serve as important benchmarks to evaluate new cryptographic systems.

**Acknowledgments** Thanks are due to Gang Cao, Thomas Gloe, and Miroslav Goljan for sharing the data of their original research papers.

## References

1. Barni M, Cancelli G, Esposito A (2010) Forensics aided steganalysis of heterogeneous images. In: IEEE international conference on acoustics, speech and signal processing, ICASSP 2010, 1690–1693 2010
2. Bishop CM (2006) Pattern recognition and machine learning. Springer, Berlin
3. Böhme R (2010) Advanced statistical steganalysis. Springer, Berlin
4. Böhme R, Freiling F, Gloe T, Kirchner M (2009) Multimedia forensics is not computer forensics. In: Geradts ZJ, Franke KY, Veenman CJ (eds) Computational forensics, 3rd interna-

tional workshop, IWCF 2009. Lecture notes in computer science, vol 5718. Springer, Berlin, pp 90–103

5. Cachin C (1998) An information-theoretic model for steganography. In: Aucsmith D (ed) Information hiding, second international workshop, IH 98. Springer, Berlin, pp 306–318
6. Cao G, Zhao Y, Ni R, Tian H (2010) Anti-forensics of contrast enhancement in digital images. In: Proceedings of the 12th ACM workshop on multimedia and security MM&Sec '10, pp 25–34. ACM Press, New York 2010
7. Cao G, Zhao Y, Ni R, Yu L, Tian H (2010) Forensic detection of median filtering in digital images. In: IEEE international conference on multimedia and EXPO, ICME 2010, pp 89–94. IEEE 2010
8. Chinese Academy of Sciences, Institute of Automation: Casia tampered image detection evaluation database (2009/2010). http://forensics.idealtest.org
9. Christlein V, Riess C, Angelopoulou E (2010) A study on features for the detection of copy-move forgeries. In: Freiling FC (ed) Sicherheit 2010, pp. 105–116. Gesellschaft für Informatik e.V., Bonn
10. Cox IJ, Miller ML, Bloom JA, Fridrich J, Kalker T (2008) Digital watermarking and steganography. Morgan Kaufmann, Burlington
11. Cromey DW (2010) Avoiding twisted pixels: ethical guidelines for the appropriate use and manipulation of scientific digital images. Sci Eng Ethics 16(4):639–667
12. Fan Z, Queiroz RL (2003) Identification of bitmap compression history: JPEG detection and quantizer estimation. IEEE Trans Image Process 12(2):230–235
13. Fridrich J (2009) Digital image forensics. IEEE Signal Process Mag 26(2):26–37
14. Fridrich J (2009) Steganography in digital media: principles, algorithms, and applications. Cambridge University Press, New York
15. Friedman G (1993) The trustworthy digital camera: restoring credibility to the photographic image. IEEE Trans Consum Electron 39(4):905–910
16. Gloe T, Borowka K, Winkler A (2009) Feature-based camera model identification works in practice: results of a comprehensive evaluation study. In: Katzenbeisser S, Sadeghi AR (eds) Information hiding, 11th international workshop, IH 2009. Lecture notes in computer science, vol 5806. Springer, Berlin, pp 262–276
17. Gloe T, Kirchner M, Winkler A, Böhme R (2007) Can we trust digital image forensics? In: Proceedings of the 15th international conference on multimedia, MULTIMEDIA '07, pp 78–86. ACM Press, New York, NY, 2007
18. Goljan M, Fridrich J, Chen M (2010) Sensor noise camera identification: Countering counter-forensics. In: Memon ND, Dittmann J, Alattar AM, Delp EJ (eds) Media forensics and security II, Proceedings of SPIE, vol 7541, p 75410S. SPIE, Bellingham, WA, 2010
19. Hsu YF, Chang SF (2006) Detecting image splicing using geometry invariants and camera characteristics consistency. In: IEEE international conference on multimedia and EXPO, ICME 2006, pp. 549–552. IEEE 2006
20. Huang Y (2005) Can digital image forgery detection be unevadable? A case study: color filter array interpolation statistical feature recovery. In: Li S, Pereira F, Shum HY, Tescher AG (eds) Proceedings of SPIE: visual communications and image processing, vol SPIE 5960, p 59602W, 2005
21. Johnson MK, Farid H (2007) Exposing digital forgeries in complex lighting environments. IEEE Trans Inf Forensics Secur 2(3):450–461
22. Kalker T (2001) Considerations on watermarking security. In: IEEE fourth workshop on multimedia signal processing, pp 201–206
23. Kerckhoffs A (1883) La cryptographie militaire. J des Sci Militaires IX 5–38:161–191
24. Kirchner M, Böhme R (2007) Tamper hiding: defeating image forensics. In: Furon T, Cayre F, Doërr G, Bas P (eds) Information hiding, 9th international workshop, IH 2007, Lecture notes in computer science, vol 4567. Springer, Berlin, pp 326–341
25. Kirchner M, Böhme R (2008) Hiding traces of resampling in digital images. IEEE Trans Inf Forensics Secur 3(4):582–592

26. Kirchner M, Böhme R (2009) Synthesis of color filter array pattern in digital images. In: Delp EJ, Dittmann J, Memon ND, Wong PW (eds) Media forensics and security, Proceedings of SPIE, vol 7254, p 72540K. SPIE, Bellingham, WA 2009
27. Kirchner M, Fridrich J (2010) On detection of median filtering in digital images. In: Memon ND, Dittmann J, Alattar AM, Delp EJ (eds) Media forensics and security II, Proceedings of SPIE-IS&T electronic imaging, vol SPIE 7541, p 754110. San Jose, CA, USA 2010
28. Lai S, Böhme R (2011) Countering counter-forensics: the case of JPEG compression. In: Filler T, Pevný T, Craver S, Ker A (eds) Information Hiding, 13th International Conference, IH 2011, vol 6958. Springer, Berlin, pp 285–298
29. Lam EY, Goodman JW (2000) A mathematical analysis of the DCT coefficient distributions for images. IEEE Trans Image Process 9(10):1661–1666
30. Li CT, Chang CY, Li Y (2010) On the repudiability of device identification and image integrity verification using sensor pattern noise. In: Weerasinghe D (ed) Information security and digital forensics, first international conference, ISDF 2009, Lecture notes of the institute for computer sciences, social informatics and telecommunications engineering, vol 41. Springer, Berlin, pp 19–25
31. Liew AWC, Yan H (2004) Blocking artifacts suppression in block-coded images using over-complete wavelet representation. IEEE Trans Circ Syst Video Technology 14(4):450–461
32. Lukáš J, Fridrich J, Goljan M (2005) Detecting digital image forgeries using sensor pattern noise. In: Said A, Apostolopoulos JG (eds) Proceedings of SPIE: image and video communications and processing 2005, vol 5685, pp 249–260
33. Lukáš J, Fridrich J, Goljan M (2006) Digital camera identification from sensor noise. IEEE Trans Inf Forensics Secur 1(2):205–214
34. Meerwald P, Uhl A (2009) Additive spread-spectrum watermark detection in demosaicked images. In: MM&Sec '09, Proceedings of the multimedia and security workshop 2009, pp 25–32
35. Ng TT, Chang SF (2004) A data set of authentic and spliced image blocks. Tech Rep ADVENT 203-2004-3, Department of Electrical Engineering, Columbia University, USA
36. Parrish D, Noonan B (2008) Image manipulation as research misconduct. Sci Eng Ethics 15(2):161–167
37. Pfitzmann A, Hansen M (2010) A terminology for talking about privacy by data minimization: Anonymity, unlinkability, undetectability, unobservability, pseudonymity, and identity management. http://dud.inf.tu-dresden.de/AnonTerminology.shtml 2010. (Version 0.34)
38. Popescu AC, Farid H (2004) Statistical tools for digital forensics. In: Fridrich J (ed) Information hiding, 6th international workshop, IH 2004. Lecture notes in computer science, vol 3200. Springer, Berlin, pp 128–147
39. Popescu AC, Farid H (2005) Exposing digital forgeries in color filter array interpolated images. IEEE Trans Signal Process 53(10):3948–3959
40. Sallee P (2005) Model-based methods for steganography and steganalysis. Int J Image Graph 5(1):167–189
41. Stamm MC, Liu KJR (2008) Blind forensics of contrast enhancement in digital images. In: IEEE international conference on image processing, ICIP 2008, pp 3112–3115
42. Stamm MC, Liu KJR (2010) Wavelet-based image compression anti-forensics. In: IEEE international conference on image processing, ICIP 2010
43. Stamm MC, Tjoa SK, Lin WS, Liu KJR (2010) Anti-forensics of JPEG compression. In: IEEE international conference on acoustics, speech and signal processing, ICASSP 2010, pp 1694–1697
44. Stamm MC, Tjoa SK, Lin WS, Liu KJR (2010) Undetectable image tampering through JPEG compression anti-forensics. In: IEEE international conference on image processing, ICIP 2010
45. Steinebach M, Liu H, Fan P, Katzenbeisser S Cell phone camera ballistics: attacks and countermeasures. In: Creutzburg R, Akopian D (eds) Multimedia on mobile devices 2010, Proceedings of SPIE, vol 7542, p 75420B. SPIE, Bellingham, WA (2010)

46. Steinebach M, Ouariachi ME, Liu H, Katzenbeisser S (2009) On the reliability of cell phone camera fingerprint recognition. In: Goel S (ed) Digital forensics and cyber crime, first international ICST conference, ICDF2C 2009. Lecture notes of the institute for computer sciences, social informatics and telecommunications engineering, vol 31. Springer, Berlin, pp 69–76
47. Thèvenaz P, Blu T, Unser M (2000) Image interpolation and resampling. In: Handbook of medical imaging, pp 393–420. Academic Press, Orlando, FL 2000
48. Ullerich C, Westfeld A (2008) Weaknesses of MB2. In: Shi YQ, Kim HJ, Katzenbeisser S (eds) Digital watermarking (Proceedings of IWDW 2007). Lecture notes in computer science, vol 5041. Springer-Verlag, Berlin, Heidelberg, pp 127–142
49. Valenzise G, Tagliasacchi M, Tubaro S (2011) The cost of JPEG compression anti-forensics. In: IEEE international conference on acoustics, speech and signal processing, ICASSP 2011, pp. 1884–1887. IEEE
50. Valenzise G, Nobile V, Tagliasacchi M, Tubaro S Countering JPEG anti-forensics. In: IEEE international conference on image processing, ICIP 2011. IEEE (in press)
51. Yuan HD (2011) Blind forensics of median filtering in digital images. IEEE Trans Inf Forensics Secur 6(4):1335–1345
52. Zhai G, Zhang W, Yang X, Lin W, Xu Y (2008) Efficient image deblocking based on postfiltering in shifted windows. IEEE Trans Circ Syst Video Tech 18(1):122–126

# Index



H. T. Sencar and N. Memon (eds.), *Digital Image Forensics*,
DOI: 10.1007/978-1-4614-0757-7,

Made in the USA
Monee, IL
13 September 2026

62976346R00210